Algebra and Trigonometry

SIXTH EDITION

KEEDY
BITTINGER
BEECHER

Instructor's Manual and Printed Test Bank

Algebra and Trigonometry

SIXTH EDITION

KEEDY
BITTINGER
BEECHER

Instructor's Manual and Printed Test Bank

Donna DeSpain

Addison-Wesley Publishing Company
Reading, Massachusetts • Menlo Park, California • New York
Don Mills, Ontario • Wokingham, England • Amsterdam • Bonn
Sydney • Singapore • Tokyo • Madrid • San Juan • Milan • Paris

Reproduced by Addison-Wesley from camera-ready copy supplied by the author.

ISBN: 0-201-52520-8

1 2 3 4 5 6 7 8 9 10-BK-95949392

TABLE OF CONTENTS

There are six alternate test forms for each chapter and the final examination. Alternate Test Forms A, B, C, and D are equivalent in length and difficulty. Synthesis questions occur at the end of Test Forms A, B, C, and D and are separated from the rest of the test by a solid line. Synthesis questions are meant to be more challenging like those problems found in the last part of each exercise set. The synthesis questions have been placed at the end to make it easy to omit them if the instructor wishes to do so.

Alternate Test Form E adds variety in style of questions and in the objectives tested. In most chapters Test Form E is more difficult than Test Forms A, B, C, D, and F.

With the exception of mathematical induction and proving identities, all questions on Test Form F are multiple choice. Effort was made to make the wrong answers as logically wrong as possible. In most cases answers were constructed to avoid students doing backward reasoning.

Special thanks are extended to Randy Forchee and Judy Salmon of Daniel Webster College for their tremendous job of checking the manuscript. Thanks also to Peggy Carter for her excellent typing and organizational skills.

Algebra and Trigonometry

SIXTH EDITION

KEEDY
BITTINGER
BEECHER

Instructor's Manual
and Printed Test Bank

CHAPTER 1 NAME ____________

TEST FORM A CLASS ______ SCORE ______ GRADE ______

Consider the numbers:

$$12,\ -\frac{4}{3},\ -\sqrt[4]{5},\ 0,\ 0.58,\ 8\frac{1}{2},\ \sqrt{50},\ -5.4,\ 7.0\overline{3},\ -11.$$

1. Which are whole numbers?

2. Which are irrational numbers?

3. Which are real numbers?

4. Which are rational numbers?

5. Which are natural numbers?

6. Which are integers?

Compute.

7. $-11 + |-11|$

8. $-3.9 + 7.6$

9. $(-15)(-3)$

10. $5 - (-14)$

11. $\frac{21}{-7}$

12. $\frac{32 \div 2^3 - 5^2 \cdot 3^2}{4(5 - 13) + 6 \cdot 5}$

Convert to decimal notation.

13. 5.401×10^{-3}

14. 5.8×10^4

15. 3.15 E 5

16. 6.17 E -3

ANSWERS

1. ____________
2. ____________
3. ____________
4. ____________
5. ____________
6. ____________
7. ____________
8. ____________
9. ____________
10. ____________
11. ____________
12. ____________
13. ____________
14. ____________
15. ____________
16. ____________

NAME ____________________

TEST FORM A

ANSWERS	
17. ______	Convert to scientific notation.
18. ______	17. 0.0000217 18. 593.21
19. ______	Compute. Write scientific notation for each answer.
20. ______	19. $(6.8 \times 10^4)(5.1 \times 10^{-15})$ 20. $\dfrac{4.76 \times 10^{-18}}{5.6 \times 10^{-12}}$
21. ______	Simplify.
22. ______	21. $(5x^3y^{-3})(-2x^4y^{-5})$ 22. $\dfrac{14ab^4c^{-2}}{21a^{-3}b^{-4}c^5}$
23. ______	23. $\sqrt[3]{-125}$ 24. $\sqrt[4]{81}$
24. ______	25. $\left(\sqrt{7} - \sqrt{3}\right)\left(\sqrt{7} + \sqrt{3}\right)$ 26. $\dfrac{\dfrac{1}{x^2} - \dfrac{1}{y^2}}{\dfrac{x^2 + 2xy + y^2}{xy}}$
25. ______	
26. ______	27. $(4a^3 + 3b)^2$
27. ______	28. $(5x^3y + 4x^2 - 2y^2 + 3) - (-2x^3y + 3xy - y^2 - 5)$
28. ______	29. $(2x - 1)^3$
29. ______	30. Write an expression containing a single radical:
30. ______	$\dfrac{\sqrt[3]{(a + b)^2} \cdot \sqrt{a + b}}{\sqrt[4]{(a + b)^3}}$.

ANSWERS

31. Convert to radical notation: $c^{-3/4}$.

31. ____________

Factor.

32. $45x^2 + 60x + 20$

33. $a^3 + 216$

32. ____________

33. ____________

34. $b^5 - 4b^3c^2$

35. $12m^4 - 26m^2 - 30$

34. ____________

36. $729 - 64a^6$

37. $5x + 10y - x^2 - 2xy$

35. ____________

38. $a^9 - 27a^3b^6$

36. ____________

39. Divide and simplify: $\dfrac{x^2 - x - 6}{x^2 - 6x + 9} \div \dfrac{(x + 2)^3}{x^2 - 9}$.

37. ____________

40. Add and simplify: $\dfrac{y}{y^2 + 4y - 21} + \dfrac{5}{y^2 - 9}$.

38. ____________

41. Rationalize the denominator: $\dfrac{8 + \sqrt{x}}{8 - \sqrt{x}}$.

39. ____________

40. ____________

42. Multiply: $(a^x + 3a^{-x})^3$.

41. ____________

43. Rationalize the numerator: $\dfrac{8 + \sqrt{x}}{8 - \sqrt{x}}$.

42. ____________

43. ____________

NAME ______________________

ANSWERS

44. ______

45. ______

46. ______

47. ______

48. ______

49. ______

44. Factor: $a^{1/2}b^{2/3} - a^{-1/2}b^{5/3}$.

45. The diagonal of a square has length $10\sqrt{2}$. Find the length of a side of the square.

46. Factor and simplify: $\dfrac{2x^2(3x - 1)^{1/2} - 3x^2(3x - 1)^{-1/2}}{[(3x - 1)^{1/3}]^3}$.

47. Change $50\ \dfrac{\text{ft}}{\text{sec}}$ to $\dfrac{\text{mi}}{\text{hr}}$.

48. Multiply: $(x^a - y^b)^3$.

49. Simplify: $\sqrt{2 + x} + \dfrac{1}{\sqrt{2 + x}}$.

CHAPTER 1 NAME ____________

TEST FORM B CLASS ______ SCORE ______ GRADE ______

ANSWERS

Consider the numbers:

$5.\overline{7}$, $-\sqrt[3]{9}$, $\frac{6}{7}$, -0.154, 13, $\sqrt{18}$, 0, $-4\frac{2}{3}$, -5, 4.1.

1. Which are whole numbers?

2. Which are irrational numbers?

3. Which are real numbers?

4. Which are rational numbers?

5. Which are natural numbers?

6. Which are integers?

Compute.

7. $-\frac{6}{5}\left(-\frac{2}{3}\right)$

8. 19 - 35

9. 5.2 + (-7.5)

10. $-3 + |-15|$

11. $-\frac{10}{3} \div \frac{4}{5}$

12. $\dfrac{75 \div 5^2 - 3^2 \cdot 4^2}{3(6 - 19) + 2 \cdot 7}$

Convert to decimal notation.

13. 6.13×10^5

14. 4.312×10^{-2}

15. 7.92 E 4

16. 9.02 E -5

1. ____
2. ____
3. ____
4. ____
5. ____
6. ____
7. ____
8. ____
9. ____
10. ____
11. ____
12. ____
13. ____
14. ____
15. ____
16. ____

NAME ______________________

TEST FORM B

ANSWERS	
17. ______	Convert to scientific notation.
18. ______	17. 0.0000032 18. 4374.52
19. ______	Compute. Write scientific notation for each answer.
20. ______	19. $(6.2 \times 10^{-3})(8.1 \times 10^{-6})$ 20. $\dfrac{1.26 \times 10^{3}}{2.8 \times 10^{-8}}$
21. ______	Simplify.
22. ______	21. $(3x)^4(2x)^3$ 22. $\dfrac{16x^2y^3z^{-4}}{24x^{-1}y^4z^{-3}}$
23. ______	23. $\sqrt[3]{-1000}$ 24. $\sqrt[5]{243}$
24. ______	25. $\left(\sqrt{6} + \sqrt{3}\right)\left(\sqrt{6} - \sqrt{3}\right)$ 26. $\dfrac{\frac{x^2 - y^2}{2xy}}{\frac{x + y}{y}}$
25. ______	
26. ______	27. $(5x^2 - 4y^3)^2$
27. ______	28. $(5x^2y + 2xy^2 - 5x - 2y) + (4x^3 - 3xy^2 - 9x - 6)$
28. ______	29. $(5y - 4)^3$
29. ______	
30. ______	30. Write an expression containing a single radical: $\dfrac{\sqrt[5]{c + d} \cdot \sqrt[4]{(c + d)^3}}{\sqrt[3]{(c + d)^2}}$.

NAME ______________________

TEST FORM B

ANSWERS

31. Convert to radical notation: $x^{4/5}$.

31. ______________

Factor.

32. $24x^2 - 120x + 150$

33. $b^3 - 512$

32. ______________

33. ______________

34. $28x^7 - 7x^3z^2$

35. $24a^4 + 30a^2 - 126$

34. ______________

35. ______________

36. $64 + 27a^6$

37. $8x - 4y - 2x^2 + xy$

36. ______________

37. ______________

38. $a^9 - 9a^3b^4$

38. ______________

39. Multiply and simplify: $\dfrac{3x^2 - 13x - 10}{2x^2 + 14x} \cdot \dfrac{x^2 - 49}{3x^2 - x - 2}$.

39. ______________

40. Subtract and simplify: $\dfrac{3x}{3x - 5y} - \dfrac{5y}{5y - 3x}$.

40. ______________

41. Rationalize the denominator: $\dfrac{4 - \sqrt{x}}{4 + \sqrt{x}}$.

41. ______________

42. Multiply: $(3a^x - 4b^{-x})^3$.

42. ______________

43. Rationalize the numerator: $\dfrac{4 - \sqrt{x}}{4 + \sqrt{x}}$.

43. ______________

NAME ______________________

ANSWERS

44. ______

45. ______

46. ______

47. ______

48. ______

49. ______

44. Factor: $a^{7/4}b^{-2/3} + a^{3/4}b^{1/3}$.

45. An airplane is flying at an altitude of 4200 ft. The slanted distance directly to the airport is 10,300 ft. How far horizontally is the airplane from the airport?

46. Factor and simplify: $\dfrac{x(x - 3)^{1/2} - (x + 3)(x - 3)^{-1/2}}{x - 3}$

47. Change $80\ \dfrac{\text{lb}}{\text{ft}^3}$ to $\dfrac{\text{ton}}{\text{yd}^3}$.

48. Multiply: $(2a + b + 3c)^3$.

49. Simplify: $\dfrac{(a - b)^3 + b^3}{b}$.

CHAPTER 1 NAME ____________

TEST FORM C CLASS ______ SCORE ________ GRADE ______

Consider the numbers:

$$-\sqrt{39},\ 415,\ 8\frac{1}{2},\ 0.\overline{21},\ -\sqrt[3]{4},\ -19,\ 0,\ \sqrt{13},\ -\frac{5}{8},\ -9.1.$$

1. Which are whole numbers?

2. Which are irrational numbers?

3. Which are real numbers?

4. Which are rational numbers?

5. Which are natural numbers?

6. Which are integers?

Compute.

7. $(-7.4) \times 8$

8. $\frac{-35}{-7}$

9. $15 + |-7|$

10. $-11.07 - 15.32$

11. $\frac{5}{8} + \left(-\frac{3}{5}\right)$

12. $\frac{5(19 - 3^2) - 4^2 \cdot 6}{7(5 - 13) + 3 \cdot 8}$

Convert to decimal notation.

13. 9.37×10^5

14. 6.12×10^{-4}

15. 4.09 E 7

16. 8.53 E -2

ANSWERS

1. ________

2. ________

3. ________

4. ________

5. ________

6. ________

7. ________

8. ________

9. ________

10. ________

11. ________

12. ________

13. ________

14. ________

15. ________

16. ________

CHAPTER 1 NAME ____________________

TEST FORM C

ANSWERS	
17. ________	Convert to scientific notation.
18. ________	17. 0.000537 18. 829.15
19. ________	Compute. Write scientific notation for each answer.
20. ________	19. $(5.9 \times 10^3)(9.2 \times 10^{-11})$ 20. $\dfrac{1.075 \times 10^{-15}}{4.3 \times 10^{-7}}$
21. ________	Simplify.
22. ________	21. $(-5x^3)^4$ 22. $\dfrac{8a^2b^3c^{-4}}{32ab^{-5}c}$
23. ________	23. $\sqrt[3]{-216}$ 24. $\sqrt[4]{256}$
24. ________	25. $\left(\sqrt{7} - 2\right)^2$ 26. $\dfrac{a - \frac{a}{b}}{a^2 - \frac{a^2}{b}}$
25. ________	
26. ________	27. $(2x - 3)(3x^2 - x + 4)$
27. ________	28. $(9x^4 + 2x^3y - 4xy^2 - 11) - (3x^3 + 5xy^2 - 2x + 4)$
28. ________	29. $(4a - 3)^3$
29. ________	
30. ________	30. Write an expression containing a single radical: $\dfrac{\sqrt{a - b} \cdot \sqrt[5]{(a - b)^4}}{\sqrt[4]{(a - b)^3}}$.

NAME ____________________

TEST FORM C

31. Convert to radical notation: $a^{-2/3}$.

Factor.

32. $75x^2 + 30x + 3$

33. $a^3 + 729$

34. $6x^2y^4 - 6x^2z^2$

35. $18m^4 + 51m^2 - 9$

36. $64 + 216b^6$

37. $4x - 20y + x^2 - 5xy$

38. $a^{11} + 8a^2b^9$

39. Divide and simplify: $\frac{x^2 + 11x + 28}{x^2 + 2x - 3} \div \frac{x^2 + 8x + 16}{x^2 + 4x - 5}$.

40. Add and simplify: $\frac{3x - 1}{5x^2 + 18x - 8} + \frac{6}{x^2 - x - 20}$.

41. Rationalize the denominator: $\frac{5 + \sqrt{x}}{5 - \sqrt{x}}$.

42. Multiply: $(2a^x - 4a^{-x})^3$.

43. Rationalize the numerator: $\frac{5 + \sqrt{x}}{5 - \sqrt{x}}$.

ANSWERS

31. ____________

32. ____________

33. ____________

34. ____________

35. ____________

36. ____________

37. ____________

38. ____________

39. ____________

40. ____________

41. ____________

42. ____________

43. ____________

NAME ____________________

TEST FORM C

ANSWERS

44. ____________

45. ____________

46. ____________

47. ____________

48. ____________

49. ____________

44. Factor: $a^{8/5}b^{-1/2} - a^{-2/5}b^{1/2}$.

45. A baseball diamond is actually a square 90 ft on a side. A catcher fields a bunt along the third-base line 15 ft from home plate. How far would the catcher have to throw the ball to first base?

46. Factor and simplify: $\dfrac{x^2(x^2 - 5)^{-1/2}x - (3x)(x^2 - 5)^{1/2}}{[(x^2 - 5)^{1/2}]^2}$.

47. Change $2400 \frac{g}{L}$ to $\frac{cg}{mL}$.

48. Multiply: $(a^{2x+y} \cdot a^{x-y})^4$.

49. Factor: $x^{16} - 625$.

CHAPTER 1 NAME ____________________

TEST FORM D CLASS ______ SCORE ________ GRADE ______

Consider the numbers:

$-5\frac{3}{4}$, 15.6, $\sqrt{80}$, -3.14, -19, $5.\overline{27}$, $-\sqrt[4]{52}$, 0, $\frac{11}{3}$.

1. Which are whole numbers?

2. Which are irrational numbers?

3. Which are real numbers?

4. Which are rational numbers?

5. Which are natural numbers?

6. Which are integers?

Compute.

7. $-5.6 + (-9.3)$

8. $-25.5 \div -0.3$

9. $\frac{17}{9} - \left(-\frac{2}{3}\right)$

10. $-815 + 409$

11. $|-17| - |-12|$

12. $\dfrac{14 \cdot 3^2 - 52 \div 2^2}{4(5 + 3^2) - 7 \cdot 3}$

Convert to decimal notation.

13. 2.175×10^{-2}

14. 9.1×10^3

15. 5.09 E 6

16. 1.32 E -4

ANSWERS

1. ____________
2. ____________
3. ____________
4. ____________
5. ____________
6. ____________
7. ____________
8. ____________
9. ____________
10. ____________
11. ____________
12. ____________
13. ____________
14. ____________
15. ____________
16. ____________

CHAPTER 1 NAME ____________________

TEST FORM D

ANSWERS	
17. ______	Convert to scientific notation.
18. ______	17. 0.00132 18. 574.9
19. ______	Compute. Write scientific notation for each answer.
20. ______	19. $(7.3 \times 10^{-6})(5.3 \times 10^{14})$ 20. $\dfrac{1.76 \times 10^{-4}}{3.2 \times 10^{-8}}$
21. ______	Simplify.
22. ______	21. $(-6x^3y^{-2})(2x^{-4}y)$ 22. $(-3c^{-2}d^3)^{-2}$
23. ______	23. $\sqrt[3]{-64}$ 24. $\sqrt[5]{32}$
24. ______	25. $\left(1 + \sqrt{5}\right)^2$ 26. $\dfrac{\dfrac{1}{x^2} - \dfrac{1}{y^2}}{\dfrac{2x^2 + 3xy + y^2}{xy}}$
25. ______	
26. ______	27. $(x^2 + ab)(x^2 - ab)$
27. ______	28. $(-4pq^2 + 7pq - 11p + 12q) + (11pq^2 - 14q^2 + 7q - 9)$
28. ______	29. $(3x - 2)^3$
29. ______	
30. ______	30. Write an expression containing a single radical: $$\frac{\sqrt[6]{(x + y)^5} \cdot \sqrt[4]{x + y}}{\sqrt{x + y}}.$$

CHAPTER 1 NAME ______________________

TEST FORM D

31. Convert to radical notation: $y^{3/8}$.

Factor.

32. $18x^2 - 48x + 32$

33. $c^3 - 1000$

34. $a^5 - 16a^3b^2$

35. $16p^4 + 26p^2 - 12$

36. $512 - 27a^9$

37. $6x + 24y - x^2 - 4xy$

38. $8a^6b^2 - b^8$

39. Multiply and simplify: $\frac{x^2 - x - 30}{x^2 - 1} \cdot \frac{x^2 - x - 2}{(x + 5)^3}$.

40. Subtract and simplify: $\frac{5y}{y^2 - 7y + 6} - \frac{3y}{y^2 - 4y - 12}$.

41. Rationalize the denominator: $\frac{6 - \sqrt{x}}{6 + \sqrt{x}}$.

42. Multiply: $(5x^t + x^{-t})^3$.

43. Rationalize the numerator: $\frac{6 - \sqrt{x}}{6 + \sqrt{x}}$.

ANSWERS

31. ______________

32. ______________

33. ______________

34. ______________

35. ______________

36. ______________

37. ______________

38. ______________

39. ______________

40. ______________

41. ______________

42. ______________

43. ______________

NAME ______________________

TEST FORM D

ANSWERS

44. ______

44. Factor: $a^{-1/3}b^{21/8} - a^{2/3}b^{5/8}$.

45. ______

45. How long must a wire be to reach from the top of a 16-m telephone pole to a point on the ground 9 m from the foot of the pole?

46. ______

46. Factor and simplify:

$$\frac{5x^2(2x + 1)^{-1/2} - x^3\left(\frac{2}{3}\right)(2x + 1)^{1/2}(3)}{[(2x + 1)^{1/4}]^4}.$$

47. ______

47. Change $40\ \frac{kg}{m}$ to $\frac{g}{cm}$.

48. ______

48. Convert to fractional notation: $4.\overline{24}$.

49. ______

49. Factor: $y^2 - \frac{15}{64} - \frac{1}{4}y$.

NAME ____________

TEST FORM E

CLASS ______ SCORE ______ GRADE ______

ANSWERS

1. See answer at left.
2. ______
3. ______
4. ______
5. ______
6. ______
7. ______
8. ______
9. ______
10. ______
11. ______
12. ______
13. ______
14. ______
15. ______
16. ______
17. ______
18. ______
19. ______
20. ______

1. In the following list of numbers, circle each irrational number.

   16.29, $4\frac{1}{3}$, 0, $\sqrt{50}$, 19.1, $\frac{13}{5}$, -47, $-\frac{2}{3}$, π, $\sqrt[3]{15}$, $\sqrt{169}$,

   0.043333 . . . (Numeral repeats),

   17.12112111211112 . . . (Numeral does not repeat)

2. Find $-(x^3 + 2x^2 - 9)$ when $x = -1$.

3. What property is illustrated by this sentence?

   $-17(ab) = (-17a)b$

Compute.

4. $|-7.9| + |14.2|$

5. $7 + (-18)$

6. $5(-7)(-4)$

7. $-19 - (-3)$

8. $\frac{-85}{-5}$

9. $\frac{2}{3} \div (-3)$

10. Convert to decimal notation: 9.7×10^{-3}.

11. Convert to scientific notation: 5,073,000.

Simplify.

12. $(-8x^2y^{-3})(-2xy^{-2})$

13. $\left(5\sqrt{a} - \sqrt{2}\right)\left(5\sqrt{a} + \sqrt{2}\right)$

14. $\sqrt[3]{343}$

15. $\frac{16r^{-9}s^8t^4}{20r^{-5}st^{-5}}$

16. $\frac{\frac{a}{a - b} - \frac{b}{a + b}}{\frac{ab}{a^2 - b^2}}$

17. $(2x^3 - y^2)^2$

18. $(2a - 7b)^3$

19. $\sqrt[5]{-1}$

20. $(2x^5 - 3x^2y - 5y^2 + 2) - (-x^5 - 5xy + y^2 - 9)$

NAME ______________________

TEST FORM E

ANSWERS	
21. ______	22. ______
23. ______	24. ______
25. ______	26. ______
27. ______	28. ______
29. ______	30. ______
31. ______	32. ______
33. ______	34. ______
35. ______	

21. Compute. Write scientific notation for the answer.

$$\frac{(2.9 \times 10^{-7})(1.29 \times 10^{4})}{4.3 \times 10^{8}}$$

Factor.

22. $48x^4 - 3$

23. $y^3 + 0.027$

24. $3x^2 - 25x - 18$

25. $8(3a - b)^2 + 2(3a - b) - 21$

26. Write an expression containing a single radical:

$$\frac{\sqrt[5]{(b + 3)^2} \cdot \sqrt[4]{b + 3}}{\sqrt[10]{(b + 3)^3}}.$$

27. Convert to radical notation: $z^{-9/11}$.

28. Convert to exponential notation and simplify:

$$\sqrt[8]{\frac{x^{24}y^{32}}{z^{40}}}.$$

29. Divide and simplify: $\dfrac{x^2 - 6x - 27}{x^2 + 5x - 14} \div \dfrac{x^2 + 3x}{x^2 - 49}$.

30. Subtract and simplify: $\dfrac{8}{a + b} - \dfrac{a - b}{a^2 - b^2}$.

31. Rationalize the denominator: $\dfrac{4\sqrt{a} - 5\sqrt{b}}{\sqrt{a} + \sqrt{b}}$.

32. Factor and simplify: $3x^{-1/4}y^{2/3} + 5x^{3/4}y^{-1/3}$.

33. A 6-m ladder is leaning against a building. The bottom of the ladder is 4 m from the building. How high is the top of the ladder?

34. Change $\dfrac{\$64}{\text{day}}$ to $\dfrac{¢}{\text{hr}}$.

35. Multiply: $(y^b + z^c)(y^b - z^c)$.

CHAPTER 1 NAME ____________________

TEST FORM F CLASS ______ SCORE ______ GRADE ______

Write the letter of your response on the answer blank.

ANSWERS

1. Add: $|-25| + (-9)$.

   a) -34 b) 16 c) 34 d) -16

2. Subtract: $-7 - (-13)$.

   a) -20 b) -6 c) 20 d) 6

3. Multiply: $6(-5)(-7)$.

   a) -210 b) -20 c) 210 d) 180

4. Divide: $\frac{-3.6}{-4}$.

   a) 9 b) -0.9 c) 0.9 d) -7.6

5. Find $-(3x^2 - 4x + 8)$ when $x = -2$.

   a) -28 b) 6 c) -4 d) 28

6. Convert to decimal notation: 8.217×10^{-4}.

   a) 82,170 b) 0.0008217 c) 8217 d) 8.217

7. Convert to scientific notation: 0.0000263.

   a) 0.263×10^{-3} b) 2.63×10^{-5}

   c) 2.63×10^{-4} d) 2.63×10^{4}

8. Simplify: $(3x^2y^{-3})(-5x^{-7}y)$.

   a) $\frac{15}{x^5y^2}$ b) $\frac{-15}{x^5y^4}$ c) $\frac{-2x^2}{y^4}$ d) $\frac{-15}{x^5y^2}$

9. Simplify: $\sqrt[5]{-243}$.

   a) -3 b) -9 c) 3 d) 7

10. Simplify: $\frac{24x^3y^4z^{-1}}{36xy^{-2}z^4}$.

   a) $\frac{2x^2y^6}{3z^5}$ b) $\frac{2xy^2}{3z^3}$ c) $\frac{2x^4y^6}{3z^2}$ d) $\frac{2x^2y^2}{3z^5}$

1. ________
2. ________
3. ________
4. ________
5. ________
6. ________
7. ________
8. ________
9. ________
10. ________

NAME ______________________

TEST FORM F

ANSWERS

11. ________

12. ________

13. ________

14. ________

15. ________

16. ________

17. ________

11. Multiply and simplify: $(4y^3 - x^2)^2$.

a) $8y^6 - 8x^2y^3 + x^4$ b) $16y^6 - x^4$

c) $16y^9 + x^4$ d) $16y^6 - 8x^2y^3 + x^4$

12. Multiply and simplify: $(3x - 5)^3$.

a) $27x^3 - 125$ b) $27x^3 - 135x^2 + 225x - 125$

c) $27x^2 - 30x + 125$ d) $27x^3 - 15x^2 + 15x - 125$

13. Multiply and simplify: $\left(\sqrt{3x} + y + \sqrt{5}\right)\left(\sqrt{3x} + y - \sqrt{5}\right)$.

a) $3x^2 + 6\sqrt{3y} - 5$ b) $9x^2 + 6xy + y^2 - 5$

c) $3x^2 + y^2 - 5$ d) $3x^2 + 2\sqrt{3xy} + y^2 - 5$

14. Simplify: $\dfrac{1 - \dfrac{3}{2x}}{x - \dfrac{9}{4x}}$.

a) $\dfrac{2}{2x + 3}$ b) $\dfrac{(3 - 2x)(4x - 1)}{9}$

c) $\dfrac{2(2x - 3)}{4x - 9}$ d) $\dfrac{4}{2x - 3}$

15. Subtract and simplify:
$(7x^2 - 5xy + 2y^2 - 4) - (3x^2 + 5xy - y^2 + 3)$.

a) $4x^2 - 10xy + 3y^2 - 7$ b) $4x^2 - y^2 - 1$

c) $4x^2 - 10xy + y^2 - 1$ d) $4x^2 - 10xy - y^2 - 7$

16. Factor: $5y^3 - 40$.

a) $5(y + 2)(y^2 - 2y - 4)$ b) $5(y + 2)(y^2 - 10y + 8)$

c) $5(y - 2)(y^2 + 2y + 4)$ d) $5(y - 2)(y^2 - 10y - 4)$

17. Factor: $5r^2 + 14r - 24$.

a) $(5r - 8)(r - 6)$ b) $(5r - 4)(r + 6)$

c) $(5r - 12)(r + 2)$ d) $(5r - 6)(r + 4)$

ANSWERS

18. Factor: $x^2 - 4xy + 4y^2 - 9z^2$.

a) $(x - 2y + 3z)(x - 2y + 3z)$

b) $(x - 2y - 3z)(x - 2y + 3z)$

c) $(x + 2y - 3z)(x + 2y + 3z)$

d) $x(x - 4y) + (2y - 3z)(2y + 3z)$

18. ____________

19. Write an expression containing a single radical.

$$\frac{\sqrt[5]{(x-2)^2}\,\sqrt{x-2}}{\sqrt[4]{(x-2)^3}}$$

a) $\sqrt[20]{(x-2)^7}$ b) $\sqrt[20]{(x-2)^3}$

c) $\sqrt[10]{(x-2)^3}$ d) $\sqrt[10]{(x-2)^7}$

19. ____________

20. Convert to a radical notation: $y^{5/8}$.

a) $\left(\sqrt{y^8}\right)^5$ b) $\left(\sqrt[5]{y}\right)^8$ c) $\sqrt[8]{5y}$ d) $\sqrt[8]{y^5}$

20. ____________

21. Factor and simplify: $7a^{2/3}b^{-1/2} - 4a^{-1/3}b^{1/2}$.

a) $\dfrac{7a - 4b}{a^{1/3}b^{1/2}}$ b) $\dfrac{7a^{1/3} - 4b^{1/2}}{a^{1/3}b^{1/2}}$

c) $\dfrac{7a - 4b}{a^{2/3}b^{1/2}}$ d) $\dfrac{7a - 4b}{a^{1/3}b^{2}}$

21. ____________

22. How long is a guy wire reaching from the top of a 14-m pole to a point 9 m from the base of the pole?

a) 10.72 m b) 16.64 m c) 196 m d) 277 m

22. ____________

23. Change 120 $\frac{\text{km}}{\text{hr}}$ to $\frac{\text{m}}{\text{min}}$.

a) 2000 b) 200 c) 2 d) $33\frac{1}{3}$

23. ____________

NAME ____________________

TEST FORM F

ANSWERS

24. ____________

25. ____________

26. ____________

27. ____________

28. ____________

29. ____________

24. Consider the numbers:

$-\sqrt{80},\ \pi,\ \frac{8}{5},\ \sqrt{121},\ 0,\ 0.\overline{14},\ 5\frac{2}{3},\ -7.7.$

List all the irrational numbers.

a) $-\sqrt{80},\ \sqrt{121}$ b) $0,\ 0.\overline{14}$

c) $-\sqrt{80},\ \pi$ d) $-\sqrt{80},\ 0.\overline{14},\ \pi$

25. Multiply and simplify: $\frac{x^2 - 4}{x + 3} \cdot \frac{x^2 - 2x - 15}{x^2 - 3x - 10}$.

a) $x - 2$ b) 2

c) $\frac{x^2 + x - 6}{x + 3}$ d) $x + 2$

26. Add and simplify: $\frac{9 + x}{x - 2} + \frac{5}{2 - x}$.

a) $\frac{14 + x}{x - 2}$ b) $\frac{5 - x}{x - 2}$ c) $\frac{14 + x}{2 - x}$ d) $\frac{x + 4}{x - 2}$

27. Divide and simplify: $\frac{x^2 + 3x - 28}{x^2 + 6x + 9} \div \frac{x - 4}{x + 3}$.

a) $\frac{(x + 7)(x - 1)}{(x + 3)(x + 2)}$ b) $\frac{x + 7}{x + 3}$

c) $\frac{x - 4}{x + 3}$ d) $\frac{x}{x + 3}$

28. Subtract and simplify: $\frac{1}{x^2 - 16} - \frac{7}{x^2 - 2x - 8}$.

a) $\frac{x - 5}{(x - 4)(x + 2)(x + 4)}$ b) $\frac{-2}{(x - 4)^2(x + 2)}$

c) $\frac{-2(3x + 13)}{(x - 4)(x + 2)(x + 4)}$ d) $\frac{-7x - 27}{(x - 4)(x + 2)(x + 4)}$

29. Rationalize the denominator: $\frac{5x}{7 - \sqrt{x}}$.

a) $\frac{5(7 - \sqrt{x})}{-1}$ b) $\frac{35x + 5}{7 - x}$

c) $\frac{35x + 5x\sqrt{x}}{49 - x}$ d) $\frac{35x + 5x\sqrt{x}}{49 + x}$

CHAPTER 2 NAME ____________________

TEST FORM A CLASS ________ SCORE __________ GRADE ______

Simplify.

1. i^{33}

2. $\sqrt{-7}\,\sqrt{-5}$

3. $(4 - i)(2 + 5i)$

4. $(1 + 7i) - (-2 - 3i)$

5. $\frac{4 + 5i}{2 - 3i}$

6. $(5 - i)(5 + i)$

7. Find the reciprocal of $2 - i$ and express it in the form $a + bi$.

8. Solve for x and y: $3x + 2 - 5i = 14 + (2 - y)i$.

Solve.

9. $(3y - 5)(y + 7)(y - 1) = 0$

10. $x^4 - 3x^2 - 4 = 0$

11. $3a^2 - 13a + 12 = 0$

12. $4t^2 - 2t - 7 = 0$

13. $\frac{20}{x + 3} = \frac{8}{x}$

14. $4x^2 - x + 5 = 0$

15. $\sqrt{y - 3} = \sqrt{y + 18} - 3$

16. $\frac{9}{x - 2} - \frac{4}{x - 1} = -1$

17. $(a - 4)(a - 3) - 30 = 0$

18. $18 - 5y < 28$

19. $x^3 + x^2 - 4x - 4 = 0$

20. $5x - 7(x - 6) = 2[x - 3(2x - 7)]$

ANSWERS

1. ______________
2. ______________
3. ______________
4. ______________
5. ______________
6. ______________
7. ______________
8. ______________
9. ______________
10. ______________
11. ______________
12. ______________
13. ______________
14. ______________
15. ______________
16. ______________
17. ______________
18. ______________
19. ______________
20. ______________

CHAPTER 2 NAME ______________________

TEST FORM A

ANSWERS

21. ______

22. ______

23. ______

24. ______

25. ______

26. ______

27. ______

28. ______

29. ______

30. ______

31. ______

32. ______

33. ______

21. The length of a rectangle is three times the width. The perimeter is 56 m. Find the dimensions.

22. The speed of a boat in still water is 10 mph. It travels 15 mi upstream and 15 mi downstream in a total time of 4 hr. What is the speed of the current?

23. Solve by completing the square. Show your work.

$2x^2 - 12x - 6 = 0$

24. Determine the nature of the solutions of $x^2 - 4x + 4 = 0$ by evaluating the discriminant.

25. Write a quadratic equation whose solutions are $\sqrt{2}$ and $-3\sqrt{2}$.

26. Solve $F = \frac{km_1m_2}{d^2}$ for m_1.

27. A polygon has 9 diagonals. How many sides does it have?

28. Find an equation of variation in which y varies jointly as x and z and inversely as the square of w, and y = 123 when x = 3, z = 2, and w = 4.

29. The volume of wood V in a tree trunk varies jointly as the height h and the square of the girth g. If the volume is 2160 ft^3 when the height is 80 ft and the girth is 6 ft, what is the height when the volume is 2400 ft^3 and the girth is 8 ft?

30. Determine whether these equations are equivalent:

$$4x - 2 = 17$$
$$3x + 1 = 16 - x.$$

31. Determine the meaningful replacements in the radical expression $\sqrt{15 - 3x}$.

Solve.

32. $x = \frac{2}{3 + x}$

33. $\sqrt{6x} - \sqrt{2x} = 4$

CHAPTER 2 *NAME* ______________________

TEST FORM B *CLASS* _______ *SCORE* _________ *GRADE* ______

Simplify.

1. i^{71}

2. $\sqrt{-15}\sqrt{-2}$

3. $(3 - 4i)(1 + i)$

4. $(9 + 3i) + (2 - i)$

5. $\dfrac{4 - 3i}{2 + i}$

6. $(7 + i)(7 - i)$

7. Find the reciprocal of $5 + 2i$ and express it in the form $a + bi$.

8. Solve for x and y: $-1 + (x - y)i = x + 3y + 7i$.

Solve.

9. $(y - 9)(y + 3)(3y - 1) = 0$

10. $(x^2 - x)^2 - 5(x^2 - x) + 6 = 0$

11. $2b^2 - 3b - 20 = 0$

12. $5t^2 - 2t - 4 = 0$

13. $\dfrac{8}{2x - 3} = \dfrac{16}{2 - 4x}$

14. $2x^2 - 5x + 9 = 0$

15. $\sqrt{31 - y} + \sqrt{y + 1} = 8$

16. $\dfrac{5x - 1}{3} - \dfrac{x + 2}{4} = 2$

17. $(a + 3)(a - 1) - 12 = 0$

18. $24 - 7x > 38$

19. $x^3 - 2x^2 - 9x + 18 = 0$

20. $x - 2(x - 10) = 5[x - 2(x - 2)]$

ANSWERS

1. ______________
2. ______________
3. ______________
4. ______________
5. ______________
6. ______________
7. ______________
8. ______________
9. ______________
10. ______________
11. ______________
12. ______________
13. ______________
14. ______________
15. ______________
16. ______________
17. ______________
18. ______________
19. ______________
20. ______________

CHAPTER 2 NAME ____________

TEST FORM B

ANSWERS

21. ____________

22. ____________

23. ____________

24. ____________

25. ____________

26. ____________

27. ____________

28. ____________

29. ____________

30. ____________

31. ____________

32. ____________

33. ____________

21. Suppose that $1500 is invested at 8%, compounded annually. What amount will be in the account at the end of 5 years?

22. A can do a certain job in 5 hr, B can do the same job in 6 hr, and C can do the same job in 10 hr. How long would the job take with all three working together?

23. Solve by completing the square. Show your work.

$5x^2 - 20x - 10 = 0$

24. Determine the nature of the solutions of $2x^2 - x - 4 = 0$ by evaluating the discriminant.

25. Write a quadratic equation whose solutions are -2 and 2.

26. Solve $C = 2\pi r$ for π.

27. The hypotenuse of a triangle is 12 ft longer than one leg. The other leg of the triangle measures 36 ft. Find the length of the hypotenuse.

28. Find an equation of variation in which y varies jointly as x and z and inversely as the square of w, and $y = 20$ when $x = \frac{1}{2}$, $z = 4$, and $w = 5$.

29. The length ℓ of rectangles of fixed area is inversely proportional to the width w. Suppose that the length is 80 cm when the width is 5 cm. Find the length when the width is 8 cm.

30. Determine whether these equations are equivalent:

$$7x - 3 = 4$$
$$9x + 1 = 2x + 8.$$

31. Determine the meaningful replacements in the radical expression $\sqrt{x - 4}$.

32. A car is driven 280 miles. If it had gone 14 mph faster, it could have made the trip in 1 hr less time. What was the original speed?

33. Solve for x: $kx^2 + (3k - 2)x - 6 = 0$.

CHAPTER 2 NAME ____________________

TEST FORM C CLASS ________ SCORE __________ GRADE ______

Simplify.

1. i^{19} 2. $-\sqrt{-5}\sqrt{-3}$ 3. $(5 - 2i)(4 + i)$

4. $(10 + 2i) - (-4 - 5i)$ 5. $\frac{6 - i}{4 + 3i}$

6. $(3 - i)(3 + i)$

7. Find the reciprocal of $6 - 5i$ and express it in the form $a + bi$.

8. Solve for x and y: $-3 + (x + y)i = 3x - 9y + 7i$.

Solve.

9. $(5y - 2)(y + 4)(y - 5) = 0$ 10. $x - 2\sqrt{x} - 8 = 0$

11. $5a^2 - 13a - 6 = 0$ 12. $3x^2 - 4x - 2 = 0$

13. $\frac{10}{3x - 4} = \frac{5}{x + 3}$ 14. $2x^2 + x + 4 = 0$

15. $\sqrt{y + 15} - \sqrt{14 - y} = 3$ 16. $\frac{2x + 4}{7} + \frac{x - 1}{4} = 3$

17. $(a + 2)(a - 5) - 8 = 0$ 18. $29 - 3x < 17$

19. $x^3 + 3x^2 - x - 3 = 0$

20. $3x - 5(x + 6) = 3[x - 5(2 - x)]$

ANSWERS

1. ____________
2. ____________
3. ____________
4. ____________
5. ____________
6. ____________
7. ____________
8. ____________
9. ____________
10. ____________
11. ____________
12. ____________
13. ____________
14. ____________
15. ____________
16. ____________
17. ____________
18. ____________
19. ____________
20. ____________

NAME ______________________

TEST FORM C

ANSWERS

21. ______

22. ______

23. ______

24. ______

25. ______

26. ______

27. ______

28. ______

29. ______

30. ______

31. ______

32. ______

33. ______

21. In triangle ABC, angle A is twice as large as angle B. Angle C measures 20° less than angle B. Find the measures of the angles.

22. A boat travels 215 km downstream in the same time that it takes to travel 150 km upstream. The speed of the current in the stream is 6.5 km/h. Find the speed of the boat in still water.

23. Solve by completing the square. Show your work.

$2x^2 - 8x - 4 = 0$

24. Determine the nature of the solutions of $3x^2 - x + 4 = 0$ by evaluating the discriminant.

25. Write a quadratic equation whose solutions are 4 and $-\frac{3}{2}$.

26. Solve $P = 2\ell + 2w$ for ℓ.

27. The area of a triangle is 42 cm^2. The base is 5 cm longer than the height. Find the height.

28. Find an equation of variation in which y varies jointly as x and z, and $y = 36$ when $x = 2$ and $z = 4$.

29. The distance s that an object falls from some point above the ground varies directly as the square of the time t that it falls. If the object falls 36 ft in 1.5 sec, how long will it take the object to fall 144 ft?

30. Determine whether these equation are equivalent:

$$3x + 7 = 2$$
$$x + 5 = 2x.$$

31. Determine the meaningful replacements in the radical expression $\sqrt{5x - 1}$.

Solve.

32. $3y^2 - (2y - 3)(y + 1) = 5$

33. $(x - 6)^{2/3} = 3$

NAME ______________________

TEST FORM D

CLASS ________ SCORE ________ GRADE ______

Simplify.

1. i^{45}

2. $-\sqrt{-10}\,\sqrt{-3}$

3. $(6 - i)(2 + 3i)$

4. $(-5 - 2i) + (7 - 3i)$

5. $\frac{5 - 2i}{3 - 4i}$

6. $(4 - i)(4 + i)$

7. Find the reciprocal of $7 + 3i$ and express it in the form $a + bi$.

8. Solve for x and y: $5x + 3 + 2i = 8 + (3 + y)i$.

Solve.

9. $(2y - 3)(y - 4)(y - 1) = 0$

10. $y^{2/3} - 2y^{1/3} - 3 = 0$

11. $7b^2 + 26b - 8 = 0$

12. $2x^2 - 5x - 4 = 0$

13. $\frac{15}{x + 3} = \frac{6}{x}$

14. $3x^2 - 2x + 3 = 0$

15. $\sqrt{y - 2} + 1 = \sqrt{y + 9}$

16. $\frac{6}{2x - 1} + \frac{3}{x + 4} = 1$

17. $(a + 4)(a - 2) - 16 = 0$

18. $5 - 9x > -13$

19. $x^3 + x^2 - 16x - 16 = 0$

20. $5x - 7(x - 4) = 2[x - 7(x - 2)]$

ANSWERS

1. ____________
2. ____________
3. ____________
4. ____________
5. ____________
6. ____________
7. ____________
8. ____________
9. ____________
10. ____________
11. ____________
12. ____________
13. ____________
14. ____________
15. ____________
16. ____________
17. ____________
18. ____________
19. ____________
20. ____________

NAME ____________________

TEST FORM D

ANSWERS

21. ________

22. ________

23. ________

24. ________

25. ________

26. ________

27. ________

28. ________

29. ________

30. ________

31. ________

32. ________

33. ________

21. An investment is made at 6% compounded annually. It grows to $1404.50 at the end of one year. How much was originally invested?

22. Bill can mow a lawn twice as fast as Sam. Working together it takes them 2 hours to mow the lawn. How long would it take each of them, working alone, to do the job?

23. Solve by completing the square. Show your work.

$3x^2 - 12x - 9 = 0$

24. Determine the nature of the solutions of $4x^2 - 5x - 3 = 0$ by evaluating the discriminant.

25. Write a quadratic equation whose solutions are 3i and -3i.

26. Solve $\frac{P_1V_1}{T_1} = \frac{P_2V_2}{T_2}$ for V_1.

27. The hypotenuse of a triangle is 39 ft. One leg is 6 ft longer than twice the other. What are the lengths of the legs?

28. Find an equation of variation in which y varies jointly as x and z and inversely as the product of w and p, and $y = \frac{5}{8}$ when $x = 2$, $z = 4$, $w = 5$, and $p = 9$.

29. The time t required to drive a fixed distance varies inversely as the speed r. It takes 6 hr at 55 mph to drive a fixed distance. How long would it take to drive the fixed distance at 40 mph?

30. Determine whether these equations are equivalent:

$$5x - 9 = -2$$

$$3x + 5 = 12 - 2x.$$

31. Determine the meaningful replacements in the radical expression $\sqrt{3x - 4}$.

32. For interest compounded annually, what is the interest rate when $7550 grows to $9580 in 3 years?

33. Solve: $\sqrt{x - 5} - \sqrt[4]{x - 5} = 6$.

CHAPTER 2 NAME ______________________

TEST FORM E CLASS ________ SCORE __________ GRADE ______

Matching. In Questions 1 - 5, simplify. In Questions 6 - 15, solve. Place the letter of the answer in the answer blank. Some letters may be used more than once, and some may not be used.

1. $(\sqrt{3} - 4i)(\sqrt{3} + 4i)$

2. $\frac{i}{5 - i}$

3. $\frac{3 + 3i}{(3 - 3i)^2}$

4. $(9 - 27i) - (2 - 19i)$

5. $i^{20} - i^{43} + i^{17} - i^{105}$

6. $5t^2 - 3t + 2 = 0$

7. $\sqrt{x + 25} = \sqrt{40 - x} + 3$

8. $x^4 - 13x^2 + 36 = 0$

9. $\frac{5x}{x + 3} + \frac{15}{x} = \frac{45}{x^2 + 3x}$

10. $6z^2 - 10z - 5 = 0$

11. $-7x + 5 > -9$

12. $\frac{7}{x + 1} = \frac{3}{x + 3}$

13. $(a + 4)(a + 1) - 18 = 0$

14. $(y^2 - 5) + 2y = 5y - 1$

15. $4[2(x - 3) - 4(x + 2)] = 4x - 2(x - 2)$

ANSWERS

1. ________
2. ________
3. ________
4. ________
5. ________
6. ________
7. ________
8. ________
9. ________
10. ________
11. ________
12. ________
13. ________
14. ________
15. ________

ANSWERS

a) $-\frac{1}{6} + \frac{1}{6}i$ b) $x < -2$ c) $-7, 2$

d) $11 + 46i$ e) $7 - 8i$ f) $-\frac{1}{4} - \frac{1}{4}i$

g) $\frac{2 \pm 2i\sqrt{3}}{3}$ h) 19 i) $1 + i$

j) 2, 4, -4 k) 2, -2, 4, -4 ℓ) $-\frac{2}{5}, 1$

m) 0, -3 n) $\frac{3 \pm i\sqrt{31}}{10}$ o) $-4\frac{1}{2}$

p) $-\frac{1}{26} + \frac{5}{26}i$ q) -1, 4 r) 24

s) $x < 2$ t) $\frac{5 \pm \sqrt{55}}{6}$ u) 24, -9

v) -6 w) -2, 2, -3, 3 x) -13

y) 0 z) ∅

NAME ______________________

TEST FORM E

ANSWERS
16. ______
17. ______
18. ______
19. ______
20. ______
21. ______
22. ______
23. ______
24. ______
25. ______
26. ______
27. ______
28. ______
29. ______
30. ______

16. Solve for x and y: $6 - (3 - y)i = x - 1 + 7i$.

17. Ship A can fill an oil storage tank 3 times as fast as ship B. When working together they can fill the oil storage tank in 6 hours. How long would it take ship A to fill the oil storage tank alone?

18. A boat can move at a speed of 20 km/h in still water. It will move 84 km downstream in a river in the same time it takes to move 36 km upstream. What is the speed of the river?

19. An investment is made at 11% compounded annually. It grows to $3885 at the end of one year. How much was originally invested?

20. Solve by completing the square. Show your work.

    $4x^2 - 8x + 1 = 0$

21. Determine the nature of the solutions of

    $5z^2 - 7z + 2 = 0$ by evaluating the discriminant.

22. Without solving, find the sum and the product of the solutions of $x^2 - 7x + 3 = 0$.

23. Write a quadratic equations whose solutions are $1 - i$ and $1 + i$.

24. Solve $a = \sqrt{2b - c}$ for b.

25. Solve $\frac{1}{a} = \frac{1}{b} + \frac{1}{c}$ for c.

26. Find the length of the side of a square inscribed in a circle of diameter 12.

27. Find an equation of variation where y varies directly as the cube of x and inversely as the sum of s and w, and $y = \frac{2}{5}$ when $x = 2$, $s = 1$, and $w = 3$.

28. In a circuit with constant voltage the current varies inversely as the resistance of the circuit. In a specially designed circuit the current is 7 amps when the resistance is 4 ohms. How many amps are there in the circuit when the resistance is 6 ohms?

29. Determine the meaningful replacements in the radical expression $\sqrt{5 + 12x}$.

30. Determine whether these equations are equivalent:

    $\frac{y^2 - 9}{y + 3} = y - 3$ $\qquad$ $y - 3 = y - 3$.

CHAPTER 2 NAME ______________________

TEST FORM F CLASS ________ SCORE ________ GRADE ______

Write the letter of your response on the answer blank.

ANSWERS

1. Simplify: $1^{27} - i^{15}$.

   a) $-i$ b) -1 c) $2i$ d) 0

2. Simplify: $(4 - i)(-2 + 3i)$.

   a) $-6 + 3i$ b) $-5 + 14i$ c) 9 d) $-11 + 14i$

3. Simplify: $(2 - 4i) - (-2 - i)$.

   a) $-5i$ b) $4 - 5i$ c) $6i$ d) $4 - 3i$

4. Simplify: $\dfrac{3 + i}{2 - 4i}$.

   a) $\frac{1}{10} + \frac{7}{10}i$ b) $\frac{1}{6} + \frac{7}{6}i$ c) $\frac{1}{10} + \frac{1}{2}i$ d) $\frac{1}{3} - \frac{2}{3}i$

5. Solve for x and y: $3x - 4 - 6i = 12 + (5 + y)i$.

   a) $x = \frac{8}{3}$, $y = -1$ b) $x = 2$, $y = -11$

   c) $x = 3$, $y = -1$ d) $x = \frac{16}{3}$, $y = -11$

6. Solve: $10x^2 + 3 = 17x$. Find the larger solution.

   a) $\frac{1}{5}$ b) $\frac{2}{3}$ c) $\frac{3}{2}$ d) 3

7. Solve: $\dfrac{4}{x - 3} - \dfrac{9}{x + 1} = 0$.

   a) $\frac{5}{8}, 3$, or $\frac{22}{5}$ b) $5, -\frac{31}{5}$, or $-\frac{5}{8}$

   c) $-1, \frac{31}{5}$, or 7 d) There are no solutions.

1. ____________

2. ____________

3. ____________

4. ____________

5. ____________

6. ____________

7. ____________

NAME ______________________

TEST FORM F

ANSWERS	
8. ______	8. Solve: $y(3y + 1)(y - 7) = 0$. a) $-1,0,5$ b) $-\frac{1}{3},7$ c) $0,7$ d) $-\frac{1}{3},0,7$
9. ______	9. Solve: $4x^2 + 1 = -4x$. a) There is no solution. b) There is just one solution, and it is negative. c) There are two solutions, one positive and one negative. d) There are two solutions, both positive.
10. ______	10. Solve: $2x - 5(x - 1) = 2[x - 3(x + 2)]$. a) -1 b) 7 c) -17 d) $\frac{17}{7}$
11. ______	11. Solve: $-3x + 7 < -14$. a) $x > 7$ b) $x < -7$ c) $x > -7$ d) $x > \frac{7}{3}$
12. ______	12. Solve and describe the solutions: $5(y - 4) = (y - 4)(y - 1)$. a) There are two solutions, both positive. b) There are two solutions, one positive and one negative. c) There is just one solution. d) There is no solution.
13. ______	13. Solve: $4x^2 - 6x + 3 = 0$. a) $\frac{3 \pm 2i\sqrt{3}}{4}$ b) $\frac{-3 \pm i\sqrt{5}}{4}$ c) $\frac{-2 \pm i\sqrt{3}}{4}$ d) $\frac{3 \pm i\sqrt{3}}{4}$
14. ______	14. Solve: $\sqrt{3x - 2} - \sqrt{x - 5} = 3$. a) The answer is one odd integer. b) The answer is one negative integer. c) The answer is two odd integers. d) The answer is one even integer and one odd integer.

ANSWERS

15. Grain flows through spout A twice as fast as through spout B. When grain flows through both spouts, a grain bin is filled in 6 hours. How many hours would it take to fill the grain bin if grain flows through spout B alone?

a) 4 b) 18 c) 8 d) 16

15. ____________

16. The speed of car A is 15 mph faster than the speed of car B. Car A travels 300 mi in the same time it takes car B to travel 225 mi. Find the speed of car A.

a) 60 mph b) 45 mph c) 50 mph d) 65 mph

16. ____________

17. Determine the nature of the solutions of $3x^2 - 7x + 10 = 0$ by evaluating the discriminant.

a) Two real solutions b) One real solution

c) Two non-real solutions d) One integer solution

17. ____________

18. Find a quadratic equation having the solutions $2 - 5i$, $2 + 5i$.

a) $x^2 - 4x + 29 = 0$ b) $x^2 + 4 = 0$

c) $x^2 + 4x + 29 = 0$ d) $x^2 + 25 = 0$

18. ____________

19. Solve for x: $C = \sqrt{2 - \frac{x^2}{y^2}}$.

a) $x = \frac{\sqrt{C^2 - y^2}}{2}$ b) $x = y\sqrt{2 - C^2}$

c) $x = \sqrt{y^2 + C^2y^2}$ d) $x = y\sqrt{C^2 - 2}$

19. ____________

ANSWERS

20. ______

21. ______

22. ______

23. ______

24. ______

20. Solve: $\frac{7}{x} - \frac{3}{x^2} = 2$. Find the sum of the roots.

a) $2\frac{1}{2}$ b) $1\frac{1}{2}$ c) $-3\frac{1}{2}$ d) $3\frac{1}{2}$

21. A picture frame measures 10 cm by 18 cm; 48 cm^2 of picture shows. Which of the following equations can be used to find the width of the frame?

a) $2x + 28 = 48$ b) $(18 + 2x)(10 + 2x) = 48$

c) $x^2 - 14x + 33 = 0$ d) $10x \cdot 18x = 48$

22. Find an equation of variation in which y varies inversely as the cube of x, and y = 5 when x = 2.

a) $y = 10x^3$ b) $y = \frac{40}{x^3}$ c) $y = \frac{1.6}{x^3}$ d) $y = 0.625x^3$

23. The force of wind on a sail varies jointly as the area of the sail and the square of the wind velocity. On a square foot of a sail the force is 4 lb when the wind velocity is 20 mph. Find the force of a 16-mph wind on a sail of area 50 sq. ft.

a) 128 lb b) 400 lb c) 320 lb d) 160 lb

24. Solve: $x^3 - x^2 = 1 - x$.

a) There is just one solution.

b) There are two solutions, both negative.

c) There are three solutions.

d) There are two solutions, one negative and one positive.

CHAPTER 3 NAME ____________

TEST FORM A CLASS ______ SCORE ______ GRADE ______

Consider the relation {(2,8), (-1,-5), (8,7), (5,1)} for Questions 1 - 3.

1. Determine whether the relation is a function.
2. Find the range.
3. Find the domain.

Graph.

4. $f(x) = (x + 1)^2$
5. $y = |x - 2|$
6. $f(x) = 3\ \text{INT}(x)$

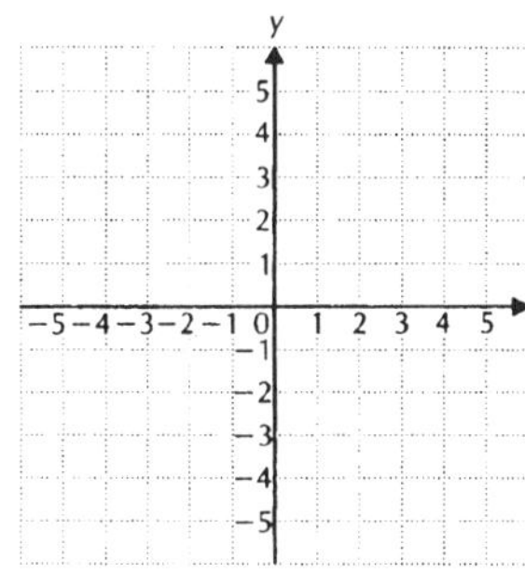

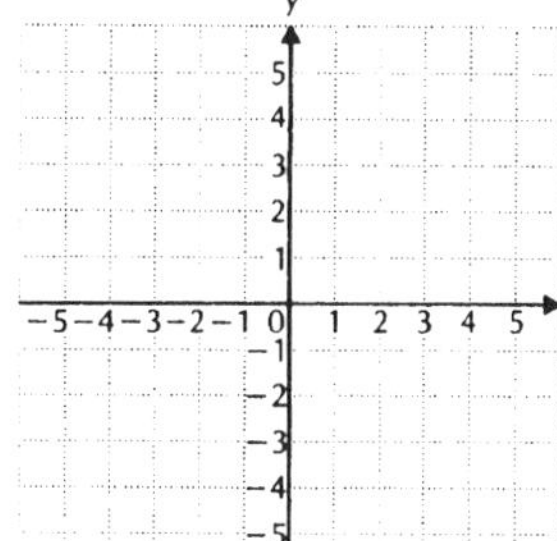

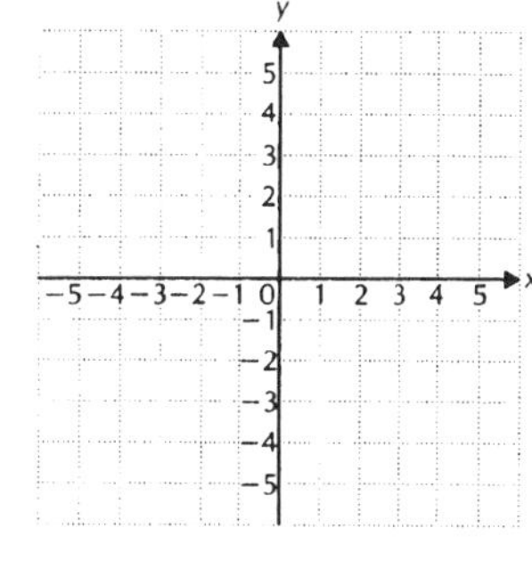

Consider the following relations for Questions 7 and 8.

a) $3x - \frac{2}{y} = 0$ b) $x^3 + 2y = y^2$ c) $x = -2$

d) $x^2 + y^2 = 1$ e) $y = x^2 - 4$ f) $4y = |2x|$

7. Which are symmetric with respect to the origin?
8. Which are symmetric with respect to the x-axis?
9. Find the domain of the function f given by $\sqrt{x - 2}$.

Consider the functions f and g given by $f(x) = 2x^2$ and $g(x) = 3x - 5$ for Questions 10 and 11.

10. Find $(f + g)(x)$, $(f - g)(x)$, $fg(x)$, $(f/g)(x)$, $f \circ g(x)$, and $g \circ f(x)$.
11. Find the domain of f, g, f + g, f - g, fg, f/g, $f \circ g$, and $g \circ f$.
12. Which of the following are graphs of functions?

a)

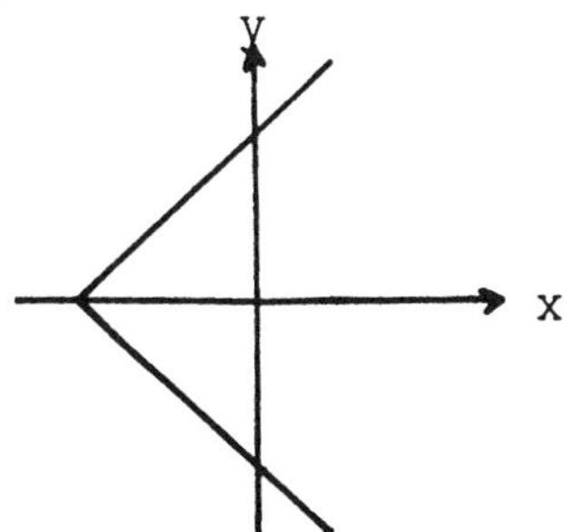

b)

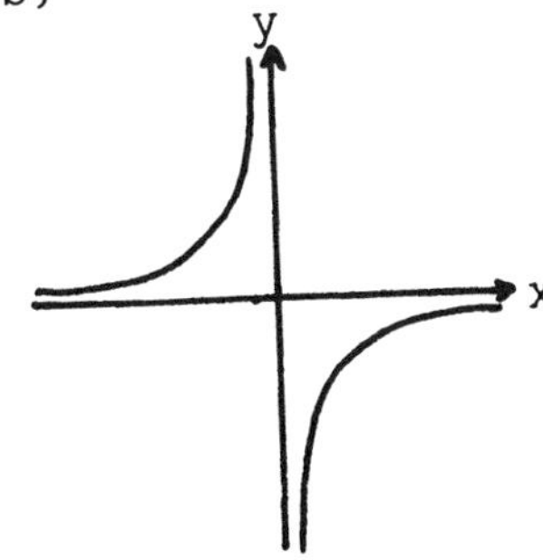

ANSWERS

1. ____________
2. ____________
3. ____________
4. See graph.
5. See graph.
6. See graph.
7. ____________
8. ____________
9. ____________
10. ____________
11. ____________
12. ____________

NAME ______________________

TEST FORM A

ANSWERS	
13. ________	Use $g(x) = x^2 - 3x + 4$ for Questions 13 - 16. Find:
14. ________	13. $g(-2)$ 14. $g(0)$
15. ________	15. $g(a - 2)$ 16. $\frac{g(a + h) - g(a)}{h}$
16. ________	17. Given $R(x) = 5x^2 - 17x - 42$ and $C(x) = 3.9x^2 + 15x - 12$, find $P(x)$.
17. ________	18. Find the slope and the y-intercept of the line $15 = 7x - 2y$.
18. ________	19. Find an equation of the line through (-1,7) with $m = 2$.
19. ________	20. Find an equation of the line containing (4,-1) and (6,2).
20. ________	21. Find the distance between (3,-4) and (7,-2).
21. ________	22. Find the midpoint of the segment with endpoints (3,-2) and (6,1).
22. ________	23. Determine whether these lines are parallel, perpendicular, or neither. $5x - y = 11$; $x = \frac{y}{5} + 3$
23. ________	
24. See graph.	24. Graph $y = \frac{2}{3}x - 1$ using the slope and the y-intercept.
25. ________	25. Find an equation of the line containing the given point and perpendicular to the given line. $(3,-1)$; $y = \frac{1}{4}x - 5$
26. ________	26. Find an equation of the circle with center (3,-4) and radius 2.
27. ________	27. Find the center and the radius of the circle $x^2 + y^2 - 8x + 2y + 1 = 0$.

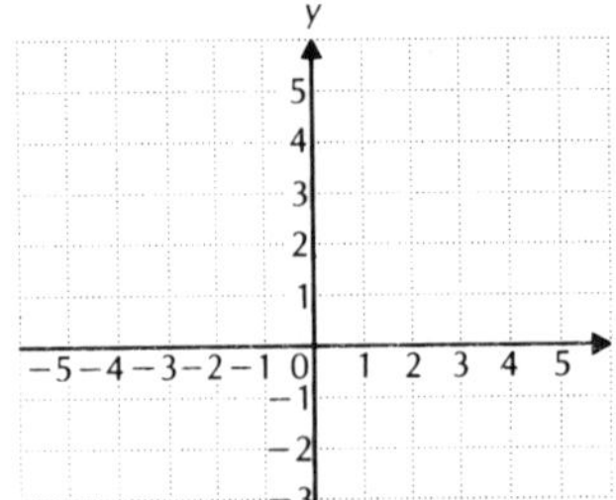

NAME ______________________

TEST FORM A

28. Here is a graph of $y = f(x)$. Sketch the graph of each of the following.

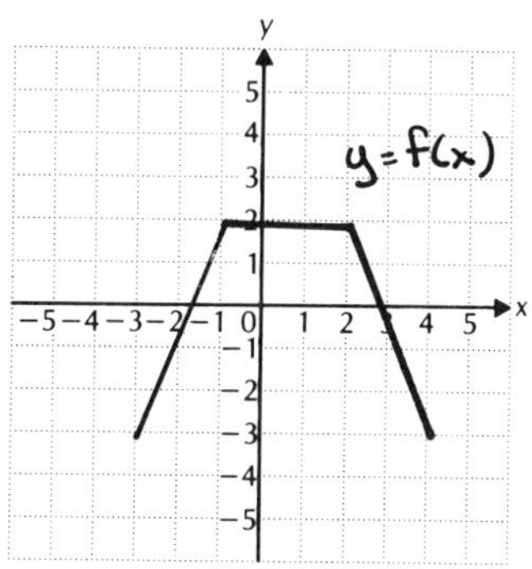

a) $y = f(x + 1)$ b) $y = f(2x)$ c) $y = 3 + f(x)$

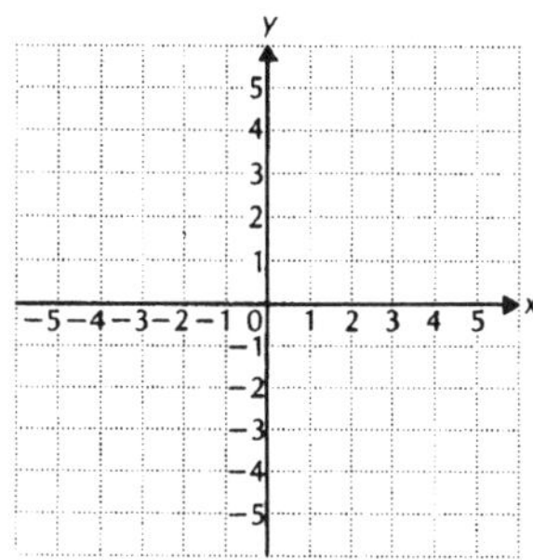

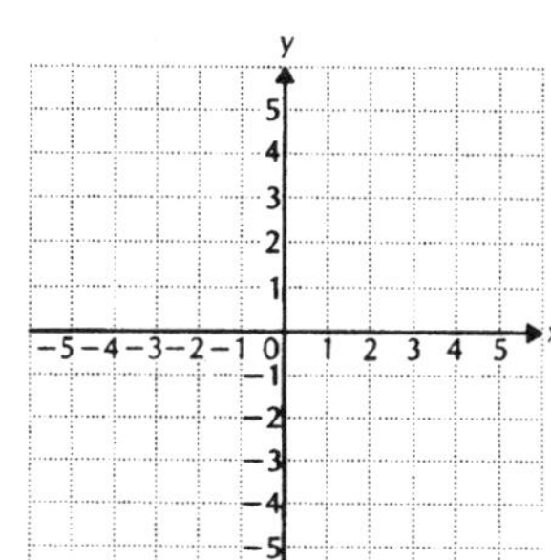

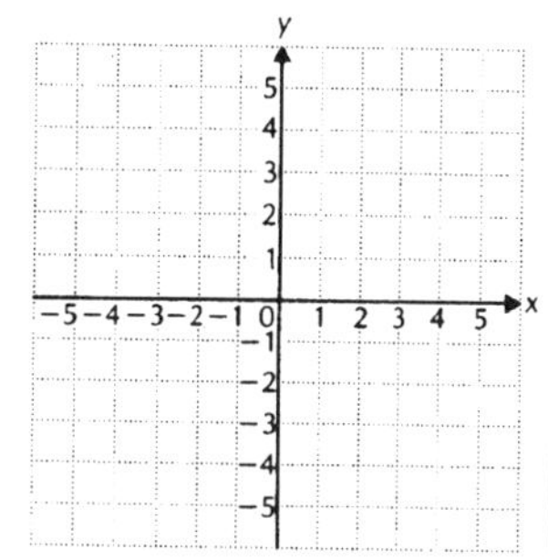

Use the following for Questions 29 - 31.

a)

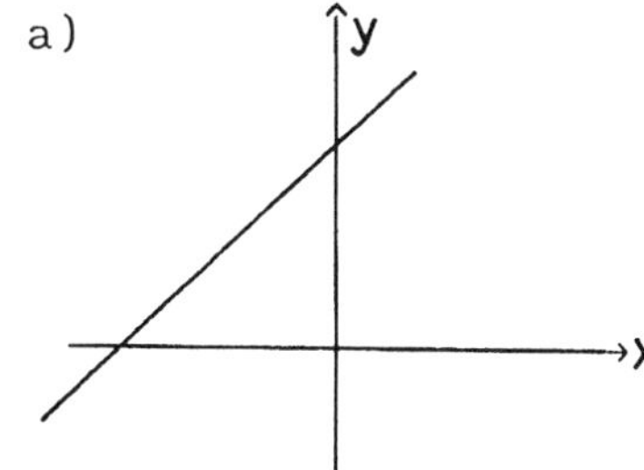

b)

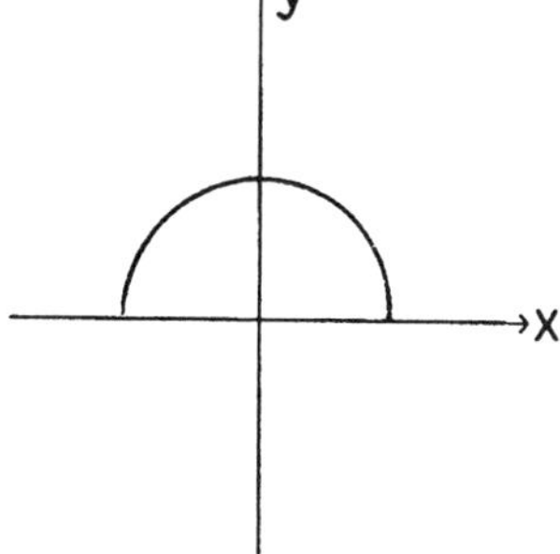

c) $f(x) = 3x^2 + 5x$

d) $f(x) = x^3 - 4x + 1$

e) $f(x) = \sqrt{x}$

f) $f(x) = x^9 - x^3$

29. Which are even?

30. Which are odd?

31. Which are neither even nor odd?

ANSWERS

28. a) See graph.

b) See graph.

c) See graph.

29. ____________

30. ____________

31. ____________

NAME ______________________

TEST FORM A

ANSWERS	
32. ______	Write interval notation for the set. 32. $\{x \mid -1 < x < 3\}$ 33. $\{x \mid x \geq -7\}$
33. ______	Use the following for Questions 34 and 35.
34. ______	
35. ______	34. Which are decreasing? 35. Which are increasing?
36. See graph.	36. Graph. $f(x) = \begin{cases} x^2, & \text{for } x < -1, \\ x, & \text{for } -1 \leq x < 3, \\ x + 2, & \text{for } x \geq 3 \end{cases}$
37. ______	37. A rectangle of dimensions x by y is inscribed in a circle of radius 6 ft. Express the area A as a function of x.
38. a) ______ b) ______	38. Find f(x) and g(x) such that $h(x) = f \circ g(x)$. a) $h(x) = 5\lvert 3x - 2\rvert$ b) $h(x) = 18 - 2(2x - 3)^2$
39. ______	39. Find the domain: $f(x) = (x - 4x^{-1})^{-1}$.

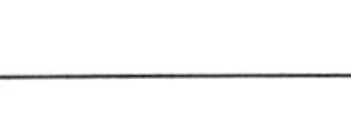

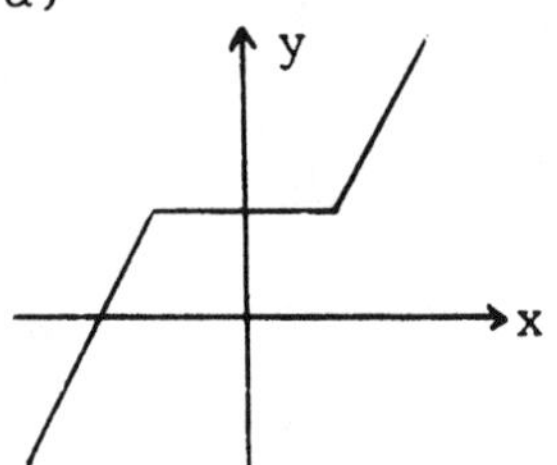

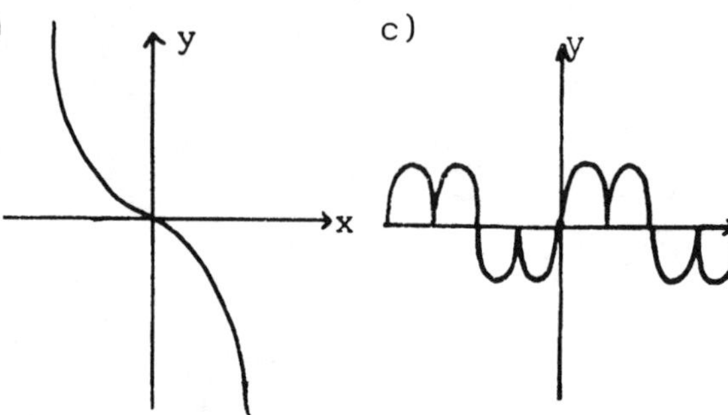

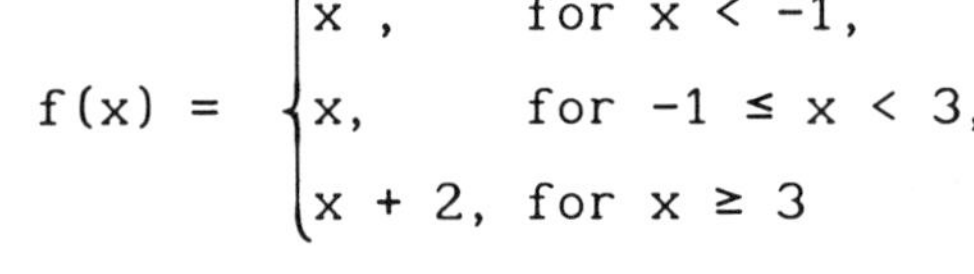

CHAPTER 3 NAME ________________

TEST FORM B CLASS ______ SCORE ________ GRADE ______

Consider the relation {(7,1), (2,5), (-1,4), (7,-3)} for Questions 1 - 3.

1. Determine whether the relation is a function.
2. Find the range.
3. Find the domain.

Graph.

4. $f(x) = 2 - x^2$
5. $x = |y - 3|$
6. $f(x) = \text{INT}(x + 2)$

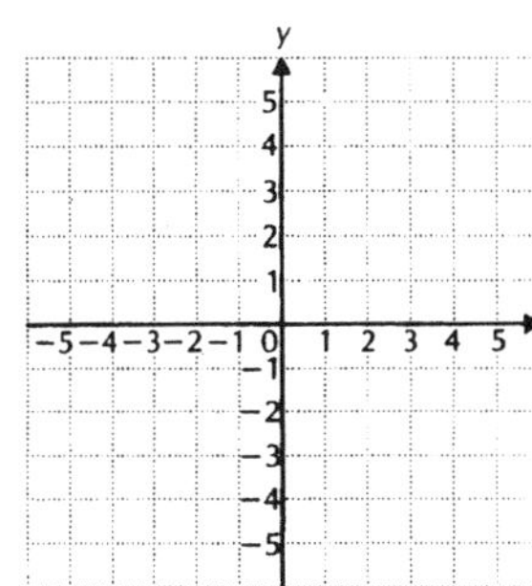

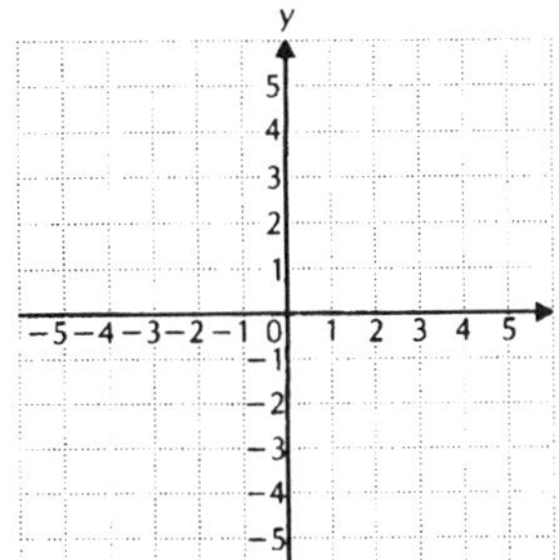

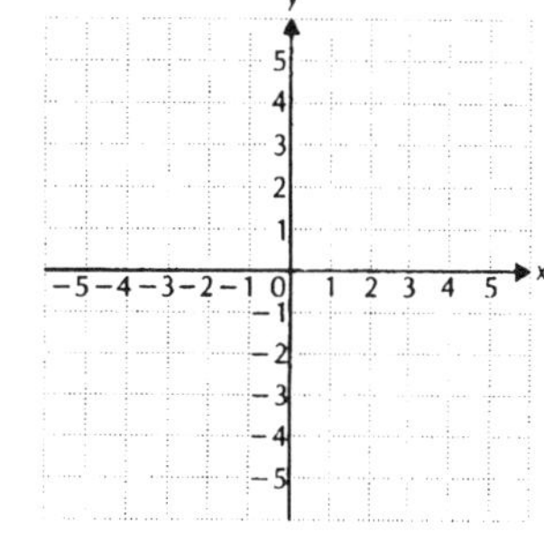

Consider the following relations for Questions 7 and 8.

a) $y = x^4 - 5$ b) $x^3 - x = y^2$ c) $x^2 - 3 = y$

d) $3x - \frac{5}{y} = 0$ e) $2x = |y|$ f) $x = -1$

7. Which are symmetric with respect to the origin?
8. Which are symmetric with respect to the x-axis?
9. Find the domain of the function f given by $\frac{1}{8 - x^3}$.

Consider the functions f and g given by $f(x) = 5x + 2$ and $g(x) = x^2 - 1$ for Questions 10 and 11.

10. Find $(f + g)(x)$, $(f - g)(x)$, $fg(x)$, $(f/g)(x)$, $f \circ g(x)$, and $g \circ f(x)$.
11. Find the domain of f, g, f + g, f - g, fg, f/g, $f \circ g$, and $g \circ f$.
12. Which of the following are graphs of functions?

a)

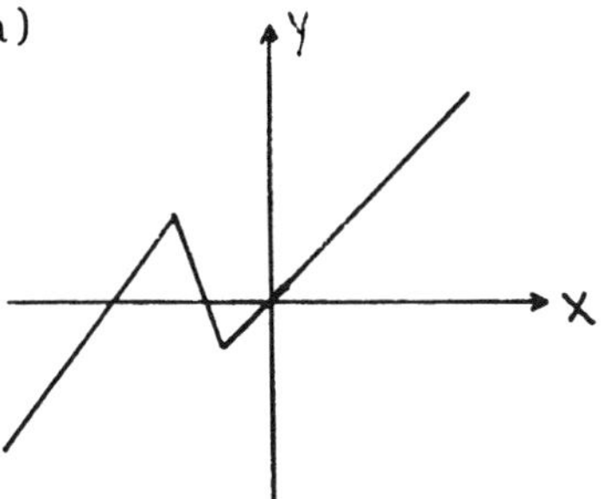

b)

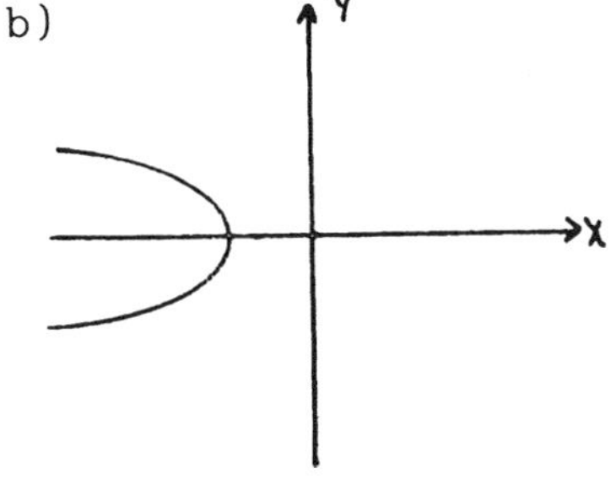

ANSWERS

1. ________________
2. ________________
3. ________________
4. See graph.
5. See graph.
6. See graph.
7. ________________
8. ________________
9. ________________
10. ________________
11. ________________
12. ________________

NAME ________________________

TEST FORM B

Use $g(x) = 3x^2 + 8x$ for Questions 13 - 16. Find:

13. $g(-1)$

14. $g(0)$

15. $g(a + 1)$

16. $\frac{g(a + h) - g(a)}{h}$

17. Given $R(x) = 4x^2 + 45x + 200$ and $C(x) = 1.7x^2 + 8x - 90$, find $P(x)$.

18. Find the slope and the y-intercept of the line $20 - 3x = 2y$.

19. Find an equation of the line through $(-1,4)$ with $m = -2$.

20. Find an equation of the line containing $(-2,-3)$ and $(4,-2)$.

21. Find the distance between $(8,-3)$ and $(4,-2)$.

22. Find the midpoint of the segment with endpoints $(3,-7)$ and $(6,1)$.

23. Determine whether these lines are parallel, perpendicular, or neither.

$$3y - 2x = 7,$$

$$2y = 8 - 3x$$

24. Graph $y = \frac{3}{4}x + 2$ using the slope and the y-intercept.

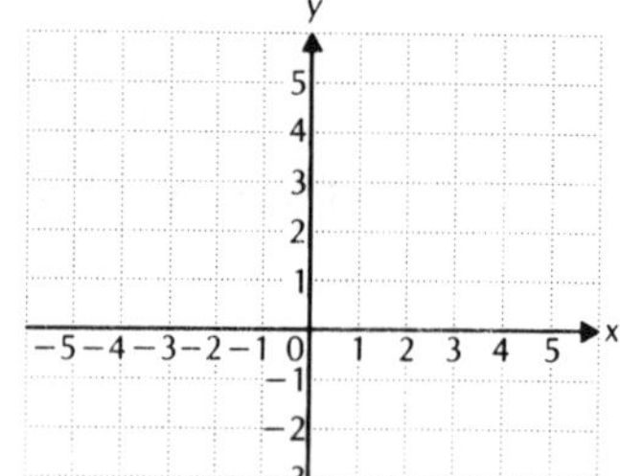

25. Find an equation of the line containing the given point and parallel to the given line.

$$(-4,2);\; y + 3 = 2x$$

26. Find an equation of the circle with center $(-1,-3)$ and radius 5.

27. Find the center and the radius of the circle

$$x^2 + y^2 - 4x + 10y + 4 = 0.$$

ANSWERS

13. ____________

14. ____________

15. ____________

16. ____________

17. ____________

18. ____________

19. ____________

20. ____________

21. ____________

22. ____________

23. ____________

24. See graph.

25. ____________

26. ____________

27. ____________

NAME ______________________

TEST FORM B

28. Here is a graph of $y = f(x)$. Sketch the graph of each of the following.

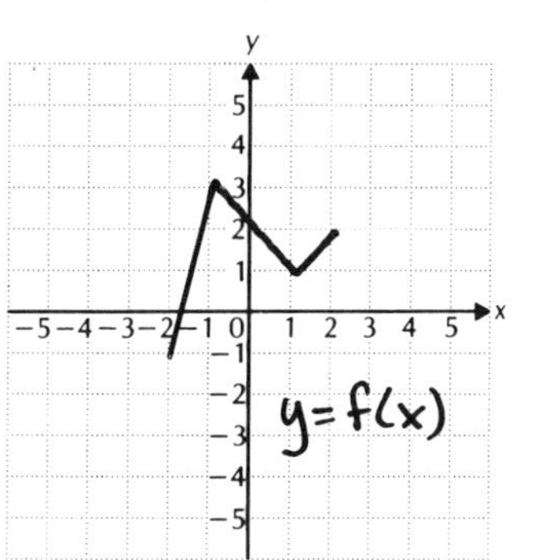

a) $y = f(x) - 1$ b) $y = \frac{1}{2}f(x)$ c) $y = 3 + f(x)$

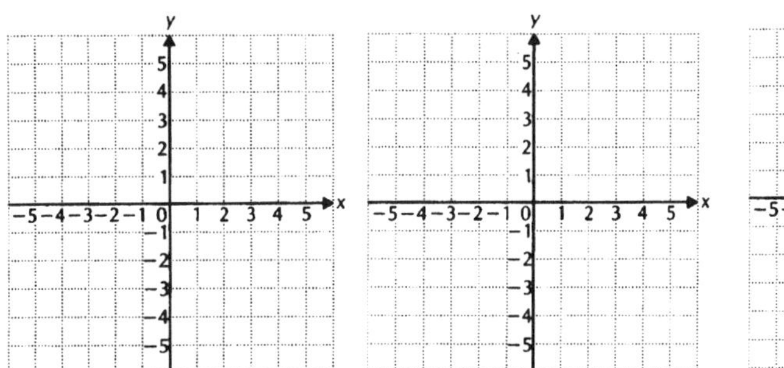

Use the following for Questions 29 - 31.

a)

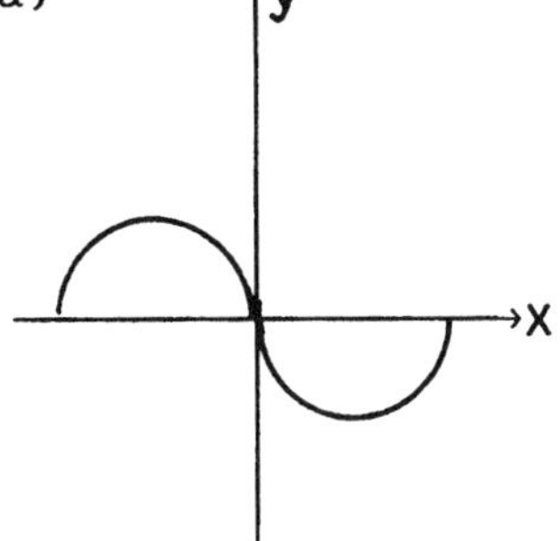

b)

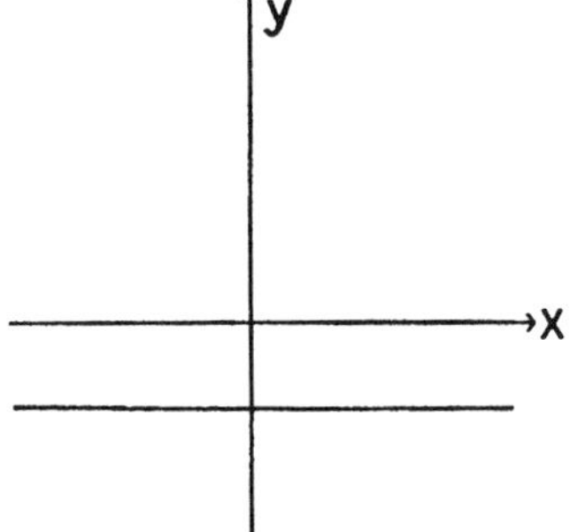

c) $f(x) = \frac{1}{x}$ d) $f(x) = |2x|$

e) $f(x) = 4x^2 + 3x^4 - 2$ f) $f(x) = x^3 - 2x + 5$

29. Which are even? 30. Which are odd?

31. Which are neither even nor odd?

ANSWERS

28. a) See graph.

b) See graph.

c) See graph.

29. ______________

30. ______________

31. ______________

NAME ________________

TEST FORM B

ANSWERS	
32. ________	Write interval notation for the set.
	32. $\{x \mid x \le -1\}$ 33. $\{x \mid -9 < x \le 4\}$

Use the following for Questions 34 and 35.

33. ________

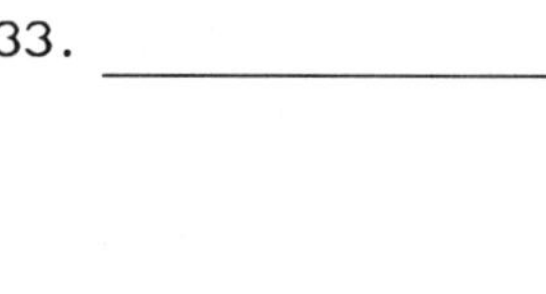

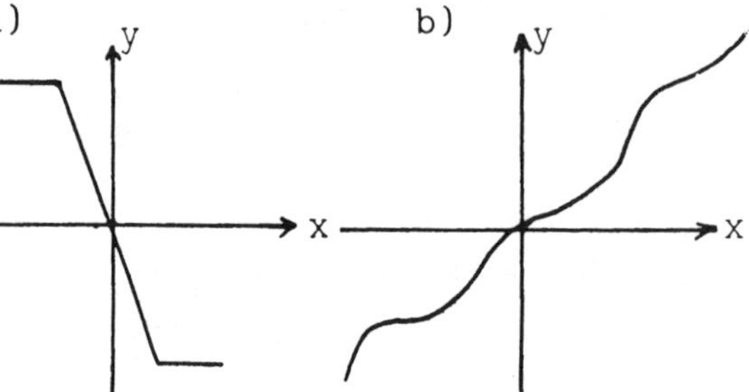

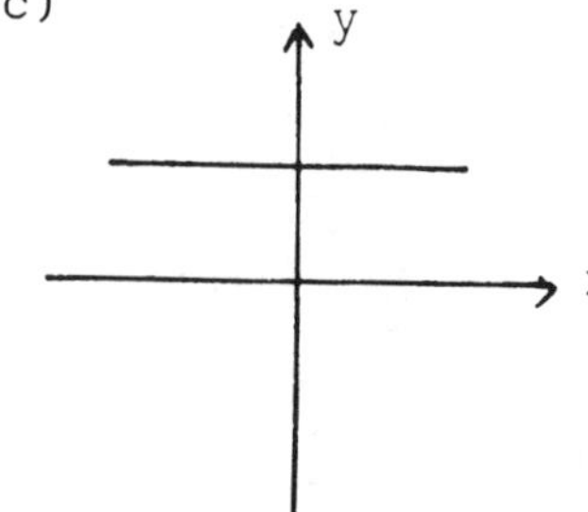

34. ________

35. ________

34. Which are increasing?

35. Which are neither increasing nor decreasing?

36. See graph.

36. Graph.

$$f(x) = \begin{cases} \frac{x}{2}, & \text{for } x \le 0, \\ x^2 - 1, & \text{for } 0 < x \le 2, \\ x + 3, & \text{for } x > 2 \end{cases}$$

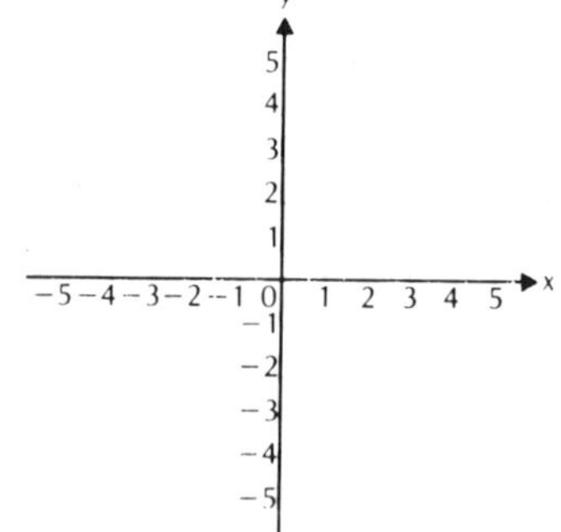

37. ________

37. The base b of a triangle is 4 less than twice the height h. Express the area A of the triangle as a function of the height h.

38. a) ________

b) ________

38. Find f(x) and g(x) such that $h(x) = f \circ g(x)$.

a) $h(x) = \dfrac{1}{(x + 3)^2}$

b) $h(x) = 5(x - 3)^3 + 2$

39. ________

39. Find k so that the line containing (-2,k) and (4,2) is parallel to the line containing (7,1) and (2,-4).

CHAPTER 3 NAME ______________

TEST FORM C CLASS ________ SCORE ________ GRADE ______

Consider the relation {(3,5), (-1,-6), (5,3), (0,-1)} for Questions 1 - 3.

1. Determine whether the relation is a function.
2. Find the range.
3. Find the domain.

Graph.

4. $f(x) = x^2 - 5$
5. $y = |x + 2|$
6. $f(x) = 2\ \mathrm{INT}(x - 1)$

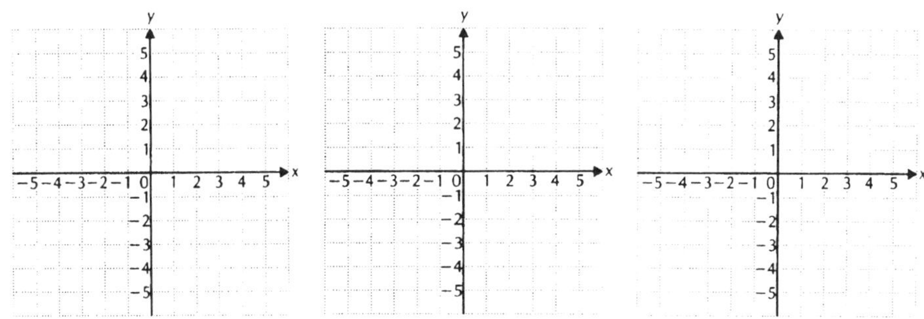

Consider the following relations for Questions 7 and 8.

a) $x^3 + 2 = y$ b) $\frac{2y}{3} - x = 4$ c) $x = |y|$

d) $y = 4$ e) $x = y^2 + 3$ f) $y^2 + x^2 = 5$

7. Which are symmetric with respect to the y-axis?
8. Which are symmetric with respect to the origin?
9. Find the domain of the function f given by $3 - \frac{5}{x}$.

Consider the functions f and g given by $f(x) = 9 - 4x$ and $g(x) = x^2 + 3$ for Questions 10 and 11.

10. Find $(f + g)(x)$, $(f - g)(x)$, $fg(x)$, $(f/g)(x)$, $f\circ g(x)$, and $g\circ f(x)$.
11. Find the domain of f, g, f + g, f - g, fg, f/g, $f\circ g$, and $g\circ f$.
12. Which of the following are graphs of functions?

a)

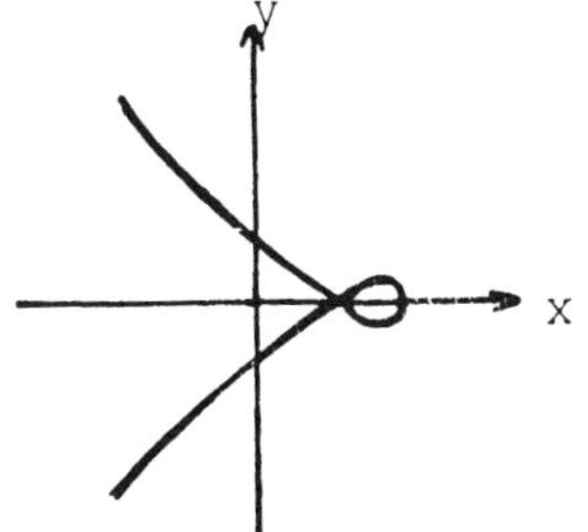

b)

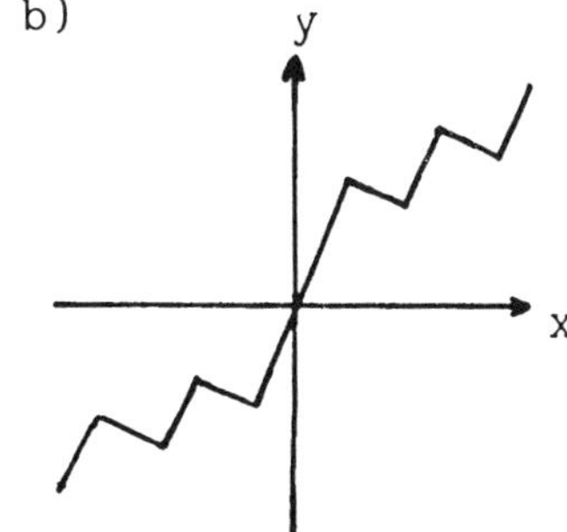

ANSWERS

1. ______________
2. ______________
3. ______________
4. See graph.
5. See graph.
6. See graph.
7. ______________
8. ______________
9. ______________
10. ______________

____

____

____

____

____

11. ______________

____

____

____

____

____

____

____

12. ______________

NAME ______________________

TEST FORM C

ANSWERS	
	Use $g(x) = x^2 + 5x - 1$ for Questions 13 - 16. Find:
13. ______	13. $g(-3)$ 14. $g(0)$
	15. $g(b + 2)$ 16. $\dfrac{g(a + h) - g(a)}{h}$
14. ______	17. Given $R(x) = 2x^2 + 50x - 10$ and $C(x) = 1.5x^2 - 14x + 15$, find $P(x)$.
15. ______	18. Find the slope and the y-intercept of the line $2x - 5y = 9$.
16. ______	19. Find an equation of the line through $(0,-7)$ with $m = 2$.
	20. Find an equation of the line containing $(4,-2)$ and $(-1,5)$.
17. ______	21. Find the distance between $(3,4)$ and $(-4,-2)$.
18. ______	22. Find the midpoint of the segment with endpoints $(-4,-2)$ and $(5,-3)$.
19. ______	23. Determine whether these lines are parallel, perpendicular, or neither. $x - 3y = 7$, $3x - y = 2$.
20. ______	
21. ______	24. Graph $y = -\frac{2}{3}x + 4$ using the slope and the y-intercept.
22. ______	
23. ______	
24. See graph.	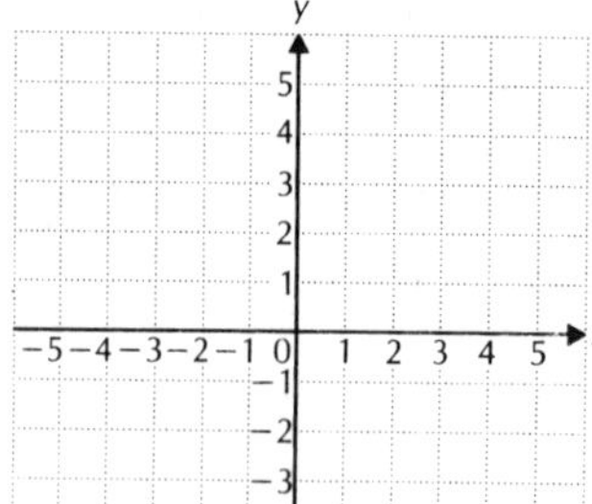
25. ______	25. Find an equation of the line containing the given point and perpendicular to the given line. $(0,-5)$; $2y = x - 1$
26. ______	26. Find an equation of the circle with center $(-1,5)$ and radius 3.
27. ______	27. Find the center and the radius of the circle $x^2 + y^2 - 6x + 4y + 8 = 0$.

NAME ______________________

TEST FORM C

28. Here is a graph of $y = f(x)$. Sketch the graph of each of the following.

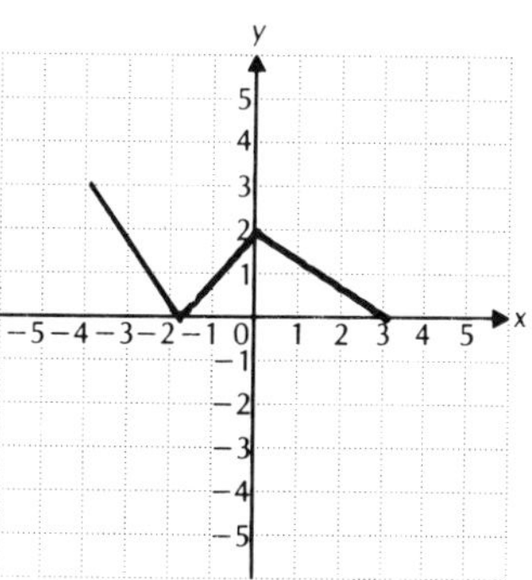

a) $y = f(x - 1)$ 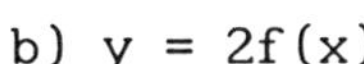 b) $y = 2f(x)$ c) $y = 1 - f(x)$

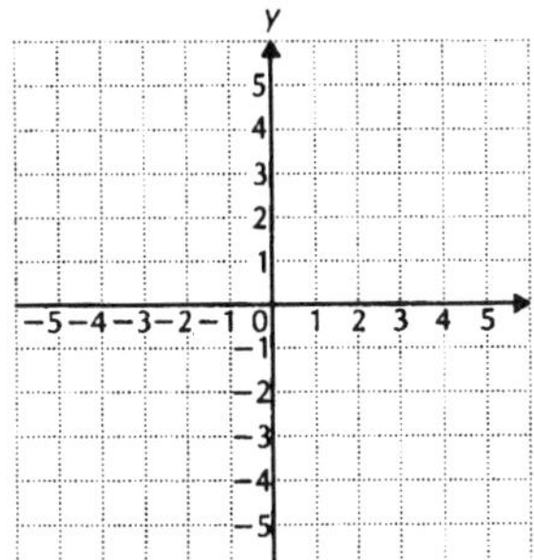

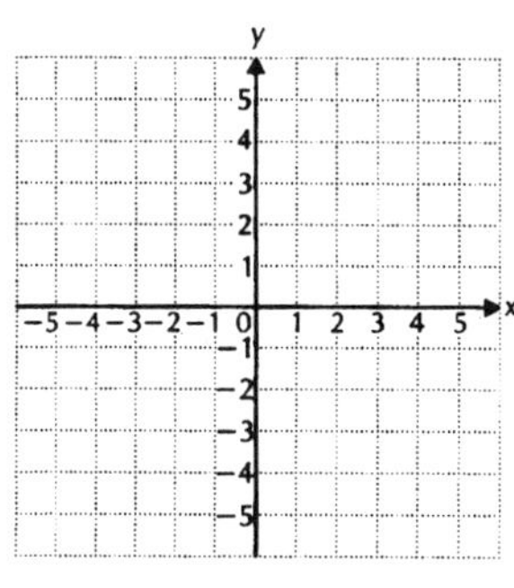

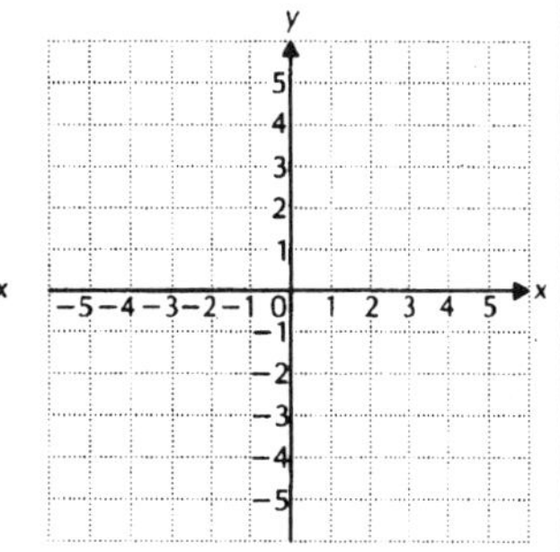

Use the following for Questions 29 - 31.

a)

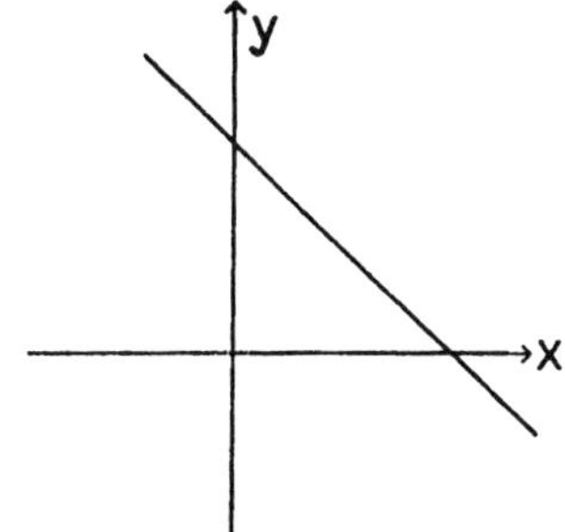

b)

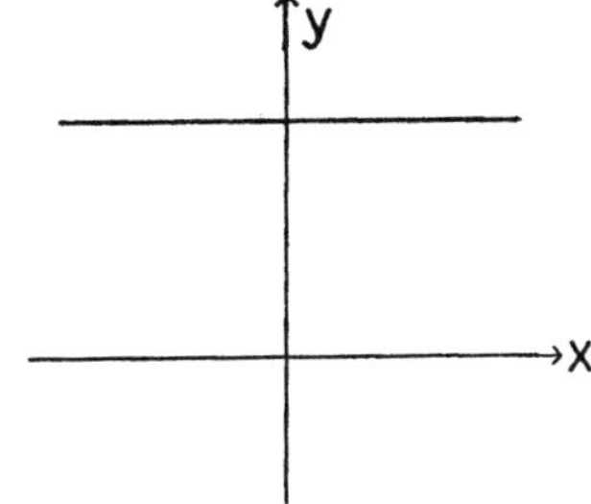

c) $f(x) = -2x^3 + 5x$

d) $f(x) = 3|x| - x$

e) $f(x) = x^9$

f) $f(x) = 5x^4 + 3x^2 - 6$

29. Which are even?

30. Which are odd?

31. Which are neither even nor odd?

ANSWERS

28. a) See graph.

b) See graph.

c) See graph.

29. ______________

30. ______________

31. ______________

ANSWERS

32. ____________

33. ____________

34. ____________

35. ____________

36. See graph.

37. ____________

38. a) ____________

b) ____________

39. ____________

Write interval notation for the set.

32. $\{x \mid -2 < x \le 5\}$

33. $\{x \mid x < -4\}$

Use the following for Questions 34 and 35.

a)

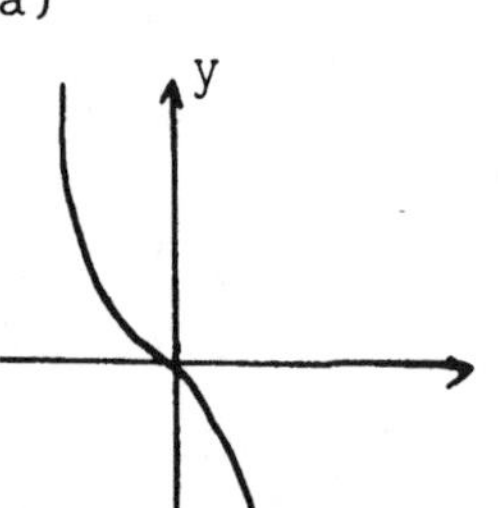

b)

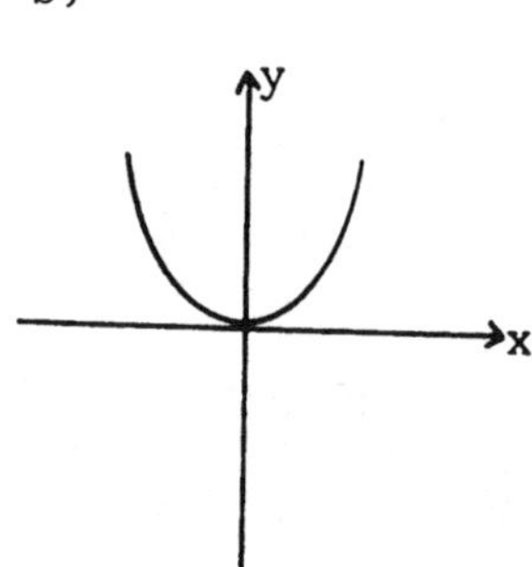

c)

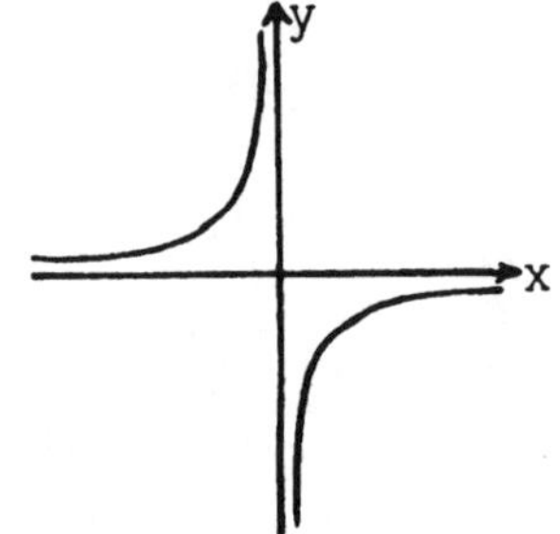

34. Which are decreasing?

35. Which are neither increasing nor decreasing?

36. Graph.

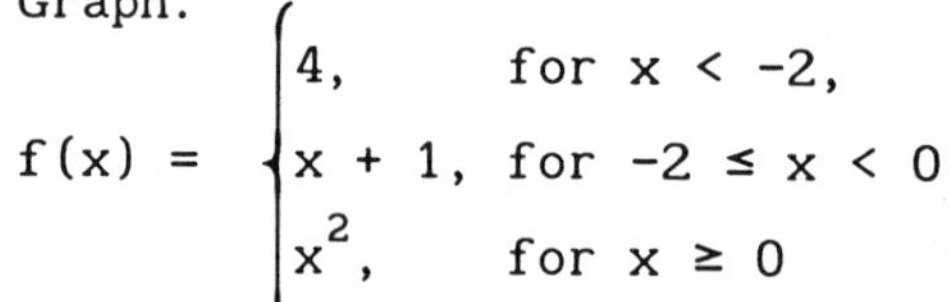

$$f(x) = \begin{cases} 4, & \text{for } x < -2, \\ x + 1, & \text{for } -2 \le x < 0, \\ x^2, & \text{for } x \ge 0 \end{cases}$$

37. A rectangle of dimensions x by y has a perimeter of 20 ft. Express the area A as a function of x.

38. Find f(x) and g(x) such that $h(x) = f \circ g(x)$.

a) $h(x) = \sqrt[3]{3x - 4}$

b) $h(x) = 2(x - 4)^2 + (x - 4) + 7$.

39. Find the point on the x-axis that is equidistant from the points (2,7) and (10,1).

CHAPTER 3 NAME ______________________

TEST FORM D CLASS ______ SCORE ________ GRADE ______

Consider the relation {(3,-1), (5,3), (-1,4), (3,5)} for Questions 1 - 3.

1. Determine whether the relation is a function.
2. Find the range.
3. Find the domain.

Graph.

4. $f(x) = 4 - x^2$ 5. $x = |y + 1|$ 6. $f(x) = \text{INT}\ (x - 3)$

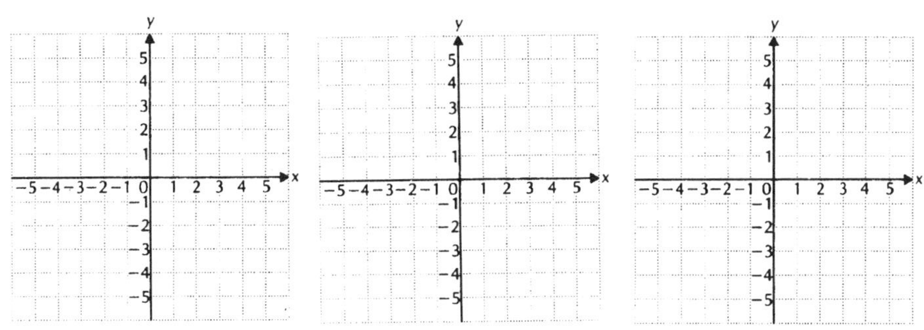

Consider the following relations for Questions 7 and 8.

a) $x^5 - 5 = y^2$ b) $\frac{3x}{2} - y = 5$ c) $x = 9$

d) $y = |x|$ e) $x = y^3 - 4$ f) $x^2 - y^3 = 2$

7. Which are symmetric with respect to the y-axis?
8. Which are symmetric with respect to the origin?
9. Find the domain of the function f given by $\sqrt{2x + 3}$.

Consider the functions f and g given by $f(x) = 3x^2 - 1$ and $g(x) = 2x + 4$ for Questions 10 and 11.

10. Find $(f + g)(x)$, $(f - g)(x)$, $fg(x)$, $(f/g)(x)$, $f \circ g(x)$, and $g \circ f(x)$.
11. Find the domain of f, g, f + g, f - g, fg, f/g, $f \circ g$, and $g \circ f$.
12. Which of the following are graphs of functions?

a)

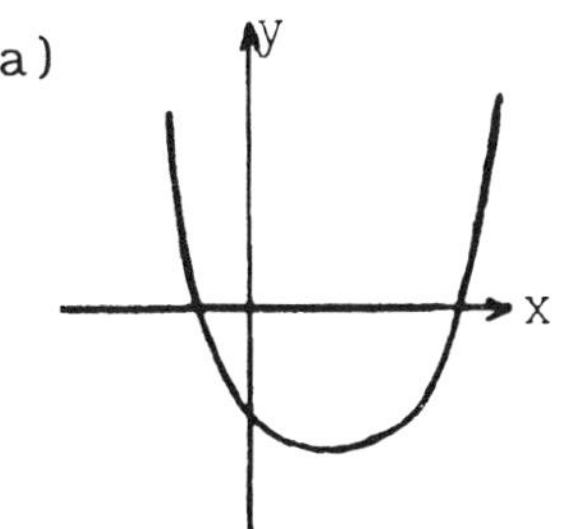

b)

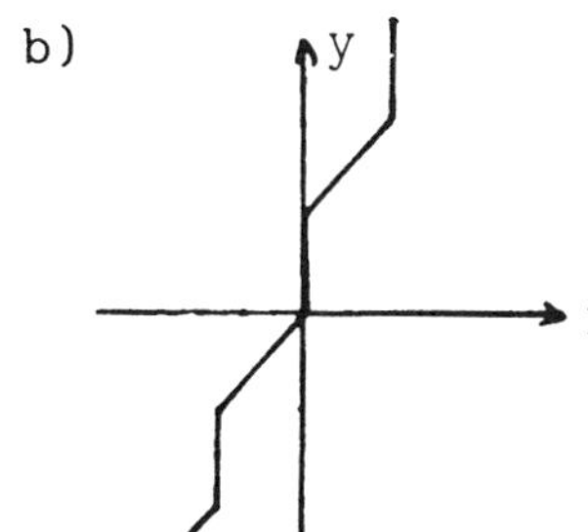

ANSWERS

1. ______
2. ______
3. ______
4. See graph.
5. See graph.
6. See graph.
7. ______
8. ______
9. ______
10. ______
______
______
______
______
______
11. ______
______
______
______
______
______
______
______
12. ______

CHAPTER 3 NAME ______________________

TEST FORM D

ANSWERS	
13. ______	Use $g(x) = 2x^2 - 5x$ for Questions 13 - 16. Find:
14. ______	13. $g(-4)$ 14. $g(0)$
15. ______	15. $g(a - 1)$ 16. $\frac{g(a + h) - g(a)}{h}$
16. ______	17. Given $R(x) = 2x^2 + 13x - 17$ $C(x) = 0.9x^2 - 5x + 12$, find $P(x)$.
17. ______	18. Find the slope and the y-intercept of the line $15 - 2x = 5y$.
18. ______	19. Find an equation of the line through (1,-2) with $m = -3$.
19. ______	20. Find an equation of the line containing (3,5) and (-4,1).
20. ______	21. Find the distance between (7,-2) and (-1,3).
21. ______	22. Find the midpoint of the segment with endpoints (2,-4) and (6,3).
22. ______	23. Determine whether these lines are parallel, perpendicular, or neither. $2x - y = 7$, $5x = -10y + 3$.
23. ______	24. Graph $y = \frac{1}{2}x - 5$ using the slope and the y-intercept.
24. See graph.	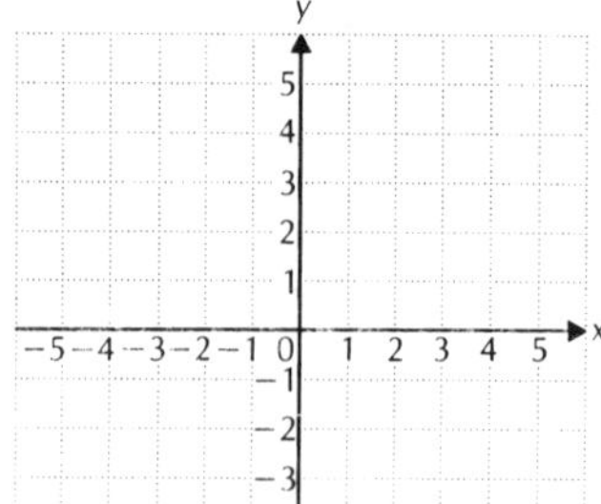
25. ______	25. Find an equation of the line containing the given point and parallel to the given line. $(1,-3)$; $2x = 4y - 3$
26. ______	26. Find an equation of the circle with center (2,-6) and radius 4.
27. ______	27. Find the center and the radius of the circle $x^2 + y^2 - 2x + 12y + 1 = 0$.

NAME ______________________

TEST FORM D

28. Here is a graph of $y = f(x)$. Sketch the graph of each of the following.

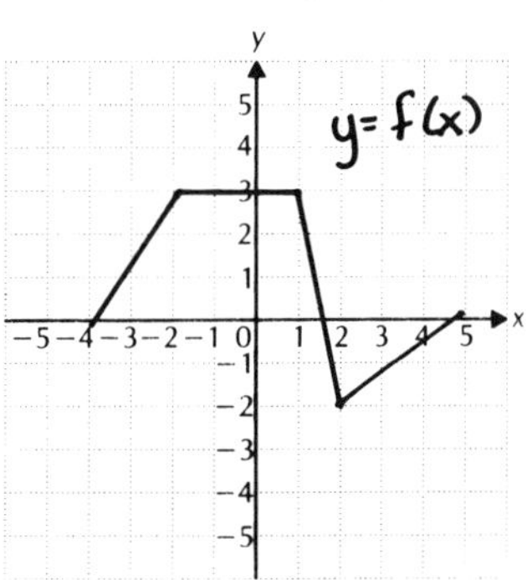

a) $y = \frac{1}{2}f(x)$ b) $y = f(x) + 1$ c) $y = f(x - 2)$

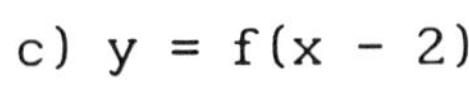

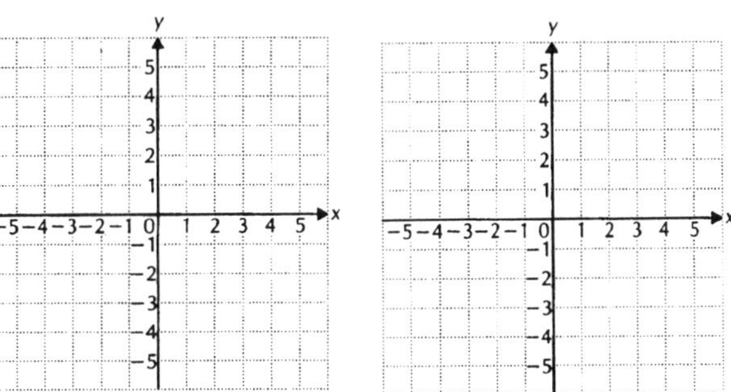

Use the following for Questions 29 - 31.

a)

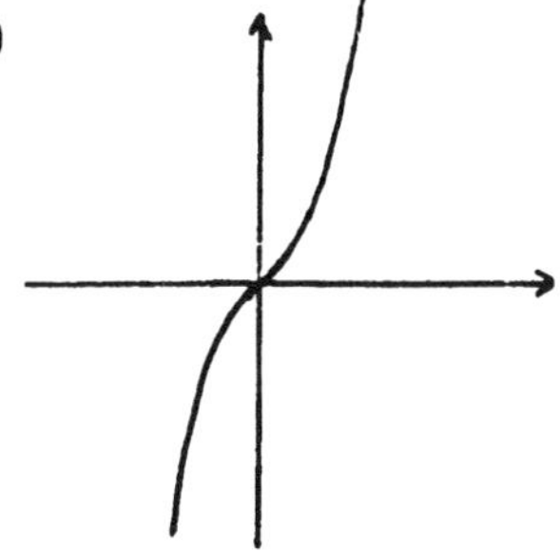

b)

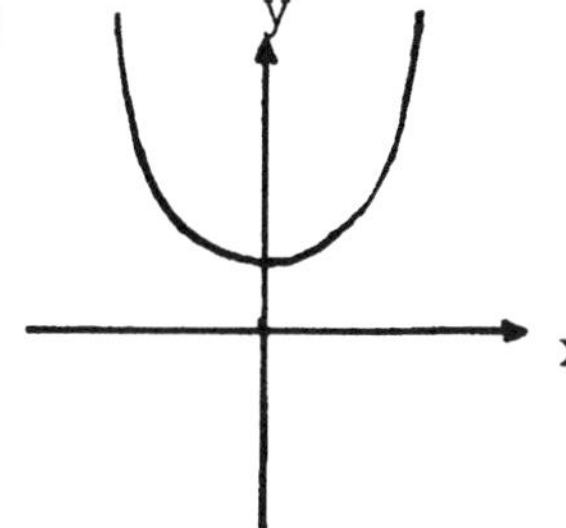

c) $f(x) = \frac{3}{x}$

d) $f(x) = x^3 - 2x^2 + 1$

e) $f(x) = \sqrt{x + 2}$

f) $f(x) = 2x^4 - 3x^2$

29. Which are even?

30. Which are odd?

31. Which are neither even nor odd?

ANSWERS

28. a) See graph.

b) See graph.

c) See graph.

29. ______

30. ______

31. ______

NAME ______________________

TEST FORM D

ANSWERS

32. ______________________

33. ______________________

34. ______________________

35. ______________________

36. See graph.

37. ______________________

38. a) ______________________

b) ______________________

39. ______________________

Write interval notation for the set.

32. $\{x | -5 < x \leq 2\}$

33. $\{x | x \leq 6\}$

Use the following for Questions 34 and 35.

a)

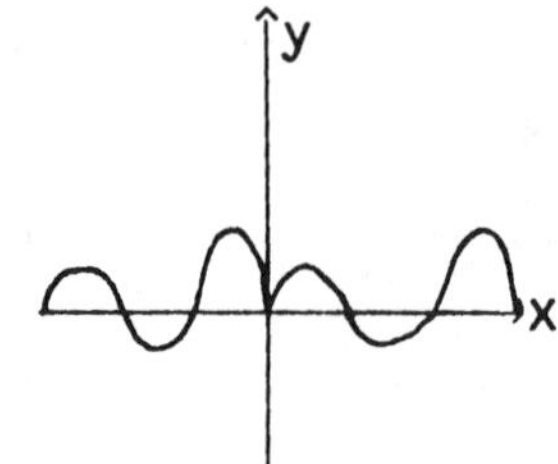

b)

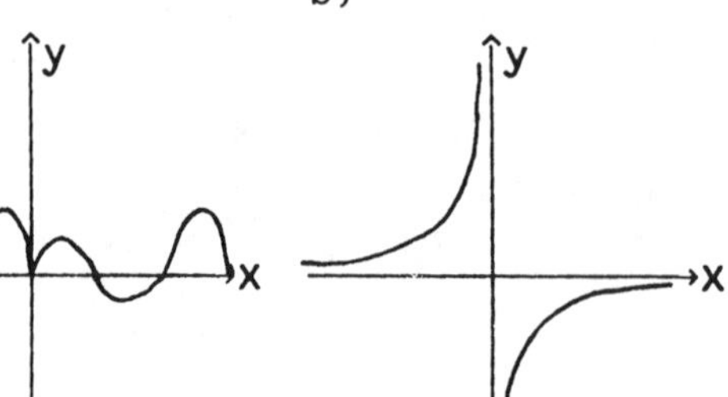

c)

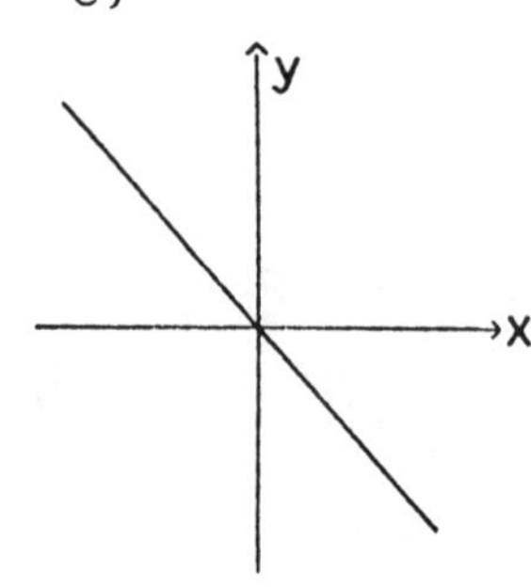

34. Which are decreasing?

35. Which are increasing?

36. Graph.

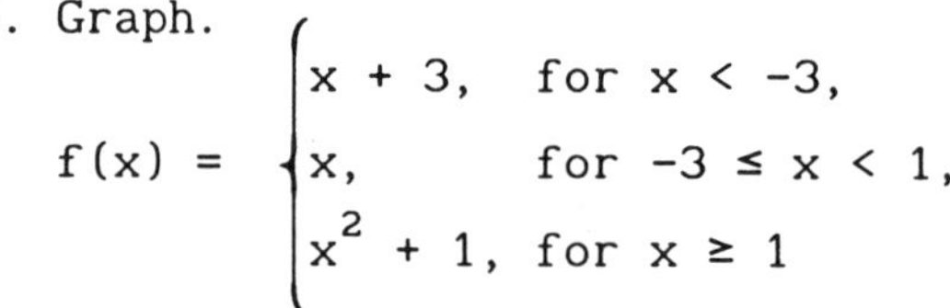

$$f(x) = \begin{cases} x + 3, & \text{for } x < -3, \\ x, & \text{for } -3 \leq x < 1, \\ x^2 + 1, & \text{for } x \geq 1 \end{cases}$$

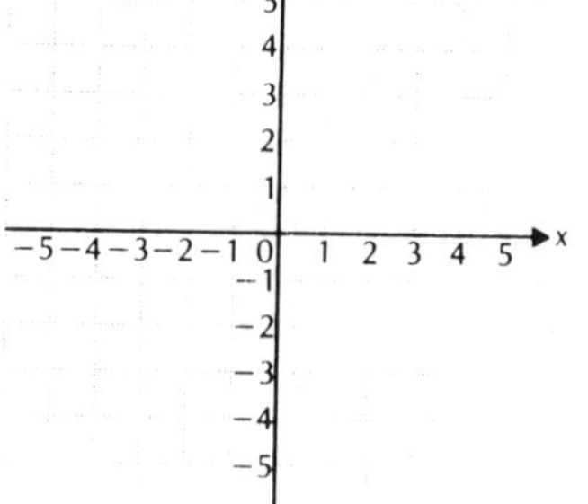

37. A rectangle of dimensions x by y is inscribed in a circle of radius 10 ft. Express the area A as a function of x.

38. Find f(x) and g(x) such that $h(x) = f \circ g(x)$.

a) $h(x) = \dfrac{1}{\sqrt{2x - 1}}$

b) $h(x) = 6(x + 2)^2 + 5$

39. Find k so that the line containing (-1,k) and (3,1) is perpendicular to the line containing (6,2) and (-1,4).

CHAPTER 3 NAME ______________________

TEST FORM E CLASS ______ SCORE ______ GRADE ______

Consider the relation {(0,-1), (5,2), (-5,-1), (3,5), (-2,4)} for Questions 1 - 3.

1. Determine whether the relation is a function.

2. Find the domain.

3. Find the range.

Graph.

4. $x = |y + 1|$ 5. $y = x^2 + 2$ 6. $f(x) = \text{INT}(x - 1)$

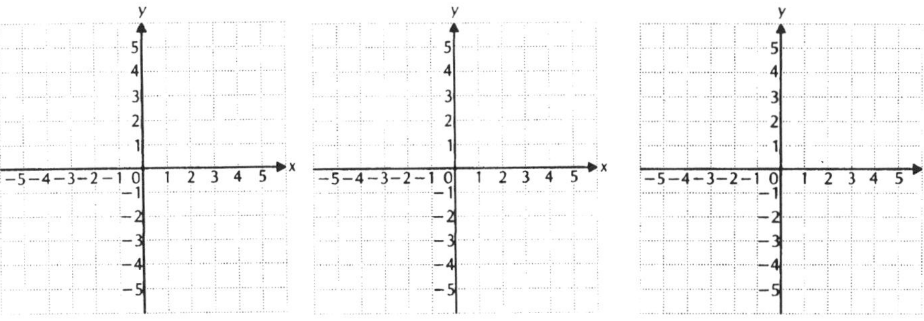

Consider the following relations for Questions 7 - 9.

a) $y = x^3$ b) $y = 3x^2 - 2x + 5$

c) $y = 7$ d) $x = \frac{2}{y}$

e) $4x - y = 6$ f) $4x^2 + 4y^2 = 8$

7. Which are symmetric with respect to the x-axis?

8. Which are symmetric with respect to the origin?

9. Which are symmetric with respect to the y-axis?

10. Find the domain: $f(x) = \dfrac{x - 4}{x^2 - 3x + 4}$.

Consider the functions f and g given by $f(x) = 2x - 3$ and $g(x) = 4x^2 - 1$ for Questions 11 and 12.

11. Find $(f + g)(x)$, $(f - g)(x)$, $fg(x)$, $(f/g)(x)$, $f \circ g(x)$, and $g \circ f(x)$.

12. Find the domain of f, g, f + g, f - g, fg, f/g, $f \circ g$, and $g \circ f$.

ANSWERS

1. ______
2. ______
3. ______
4. See graph.
5. See graph.
6. See graph.
7. ______
8. ______
9. ______
10. ______
11. ______
12. ______

CHAPTER 3 *NAME* ______________________

TEST FORM E

ANSWERS

13. ______________

14. ______________

15. ______________

16. ______________

17. ______________

18. ______________

13. Which of the following are graphs of functions?

a)

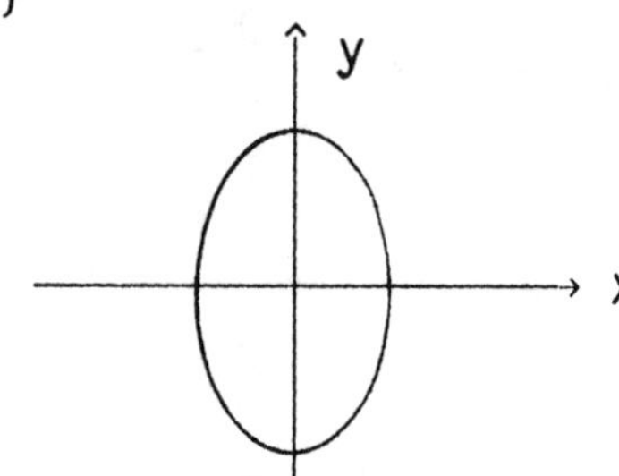

b)

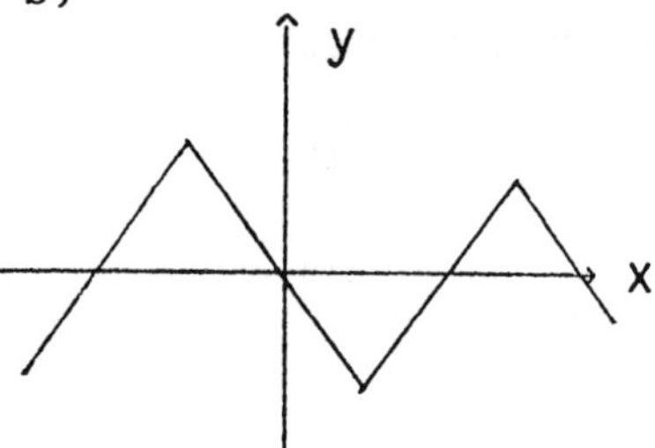

c)

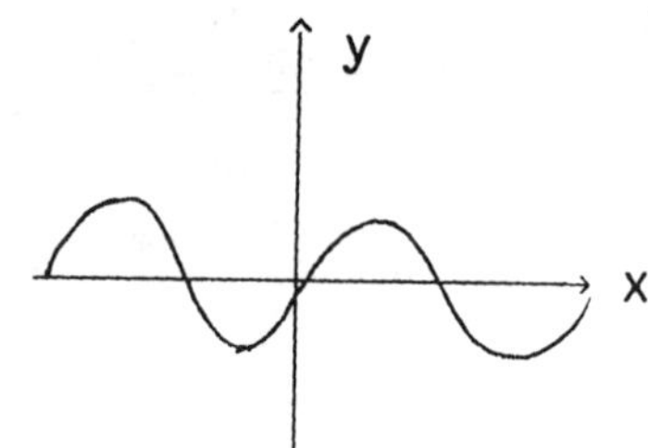

d)

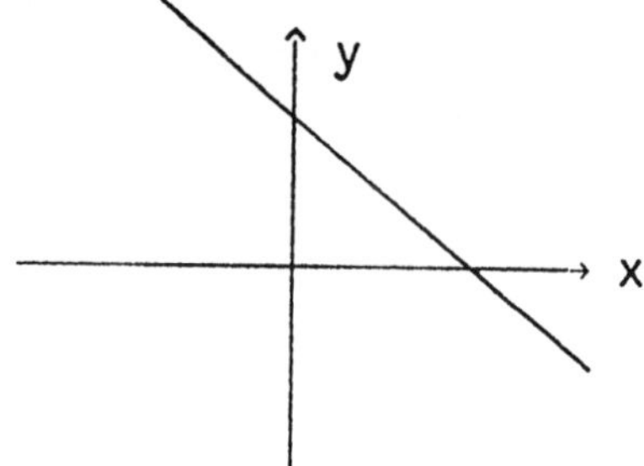

Use $g(x) = 3x^2 - x + 5$ for Questions 14 - 17. Find:

14. $g(4)$

15. $g(-3)$

16. $g(a - 2)$

17. $\dfrac{g(a + h) - g(a)}{h}$

18. Given $R(x) = 30x - 0.2x^2$ and $C(x) = 3x - 2$, find $P(x)$.

NAME ______________________

TEST FORM E

19. Here is a graph of $y = f(x)$. Sketch the graph of each of the following.

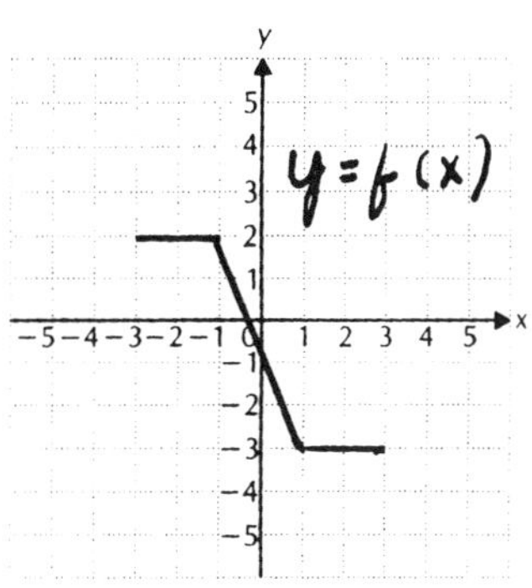

a) $y = f(x + 1)$

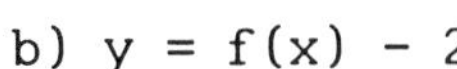

b) $y = f(x) - 2$

c) $y = 2f(x)$

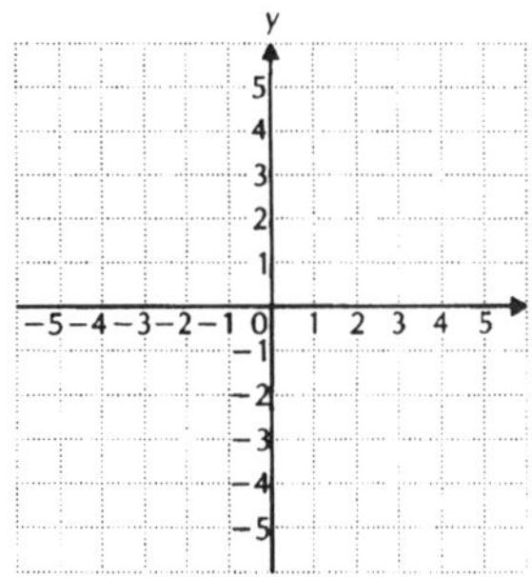

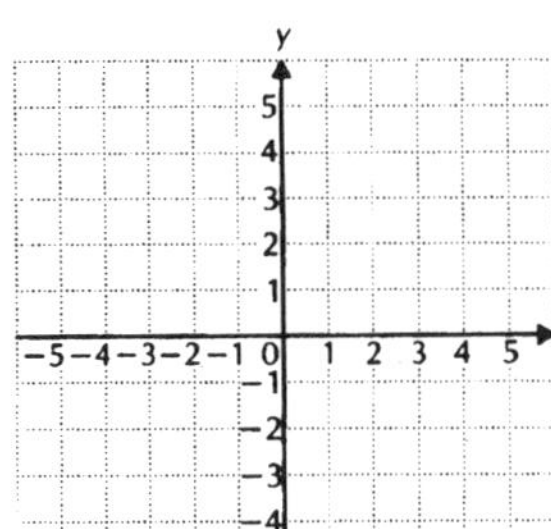

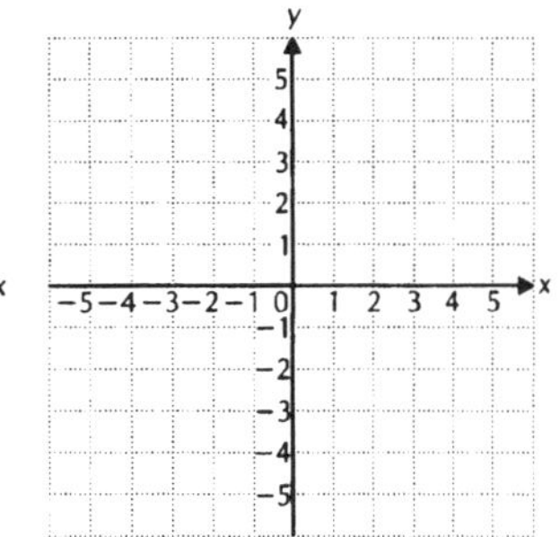

Use the following for Questions 20 - 22.

a)

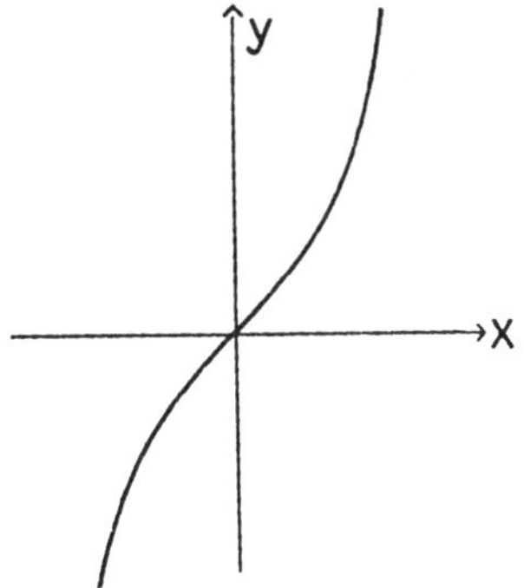

b)

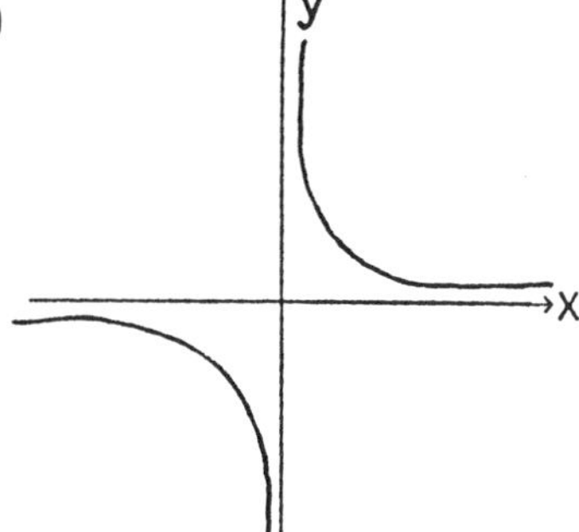

c) $f(x) = x^2 + 7$

d) $f(x) = x^3 - 2x - 8$

e) $f(x) = |3x|$

f) $f(x) = 3x^5 - 2x$

20. Which are even?

21. Which are odd?

22. Which are neither even nor odd?

Write interval notation for the set.

23. $\{x | -2 \le x < 5\}$

24. $\{x | x > -3\}$

ANSWERS

19. a) See graph.

b) See graph.

c) See graph.

20. ______________

21. ______________

22. ______________

23. ______________

24. ______________

NAME ______________________

TEST FORM E

ANSWERS

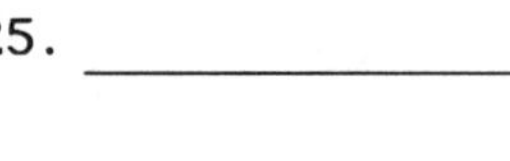

25. ______________

26. ______________

27. ______________

28. ______________

29. ______________

30. See graph.

31. ______________

32. ______________

33. ______________

34. ______________

Use the following for Questions 25 and 26.

a) 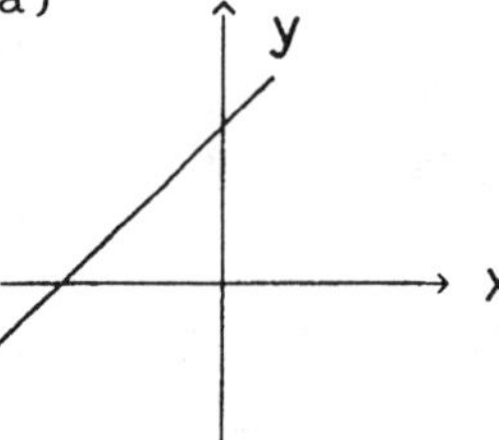

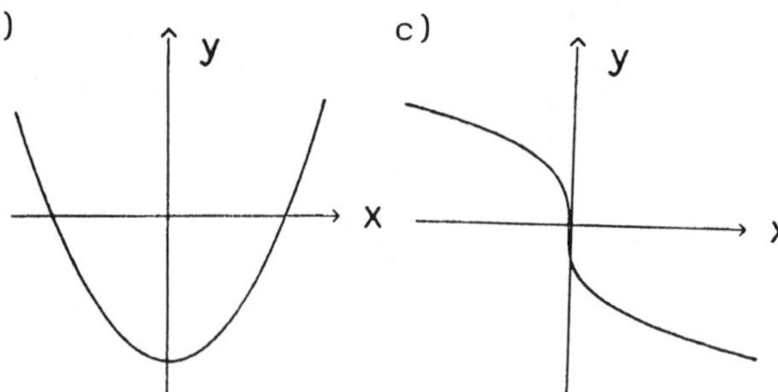

25. Which are increasing?

26. Which are neither increasing nor decreasing?

27. Find the slope and y-intercept of the line $5 = \frac{y}{2} + 3x$.

28. Find an equation of the line containing (0,-7) and (2,3).

29. Find the distance between (7,-1) and (6,2).

30. Graph.

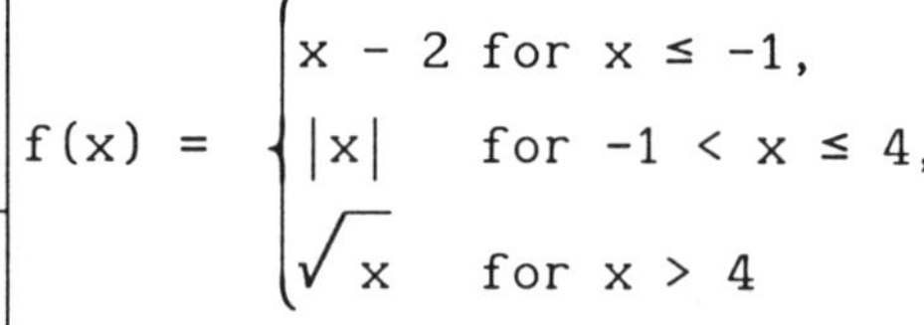

$$f(x) = \begin{cases} x - 2 & \text{for } x \le -1, \\ |x| & \text{for } -1 < x \le 4, \\ \sqrt{x} & \text{for } x > 4 \end{cases}$$

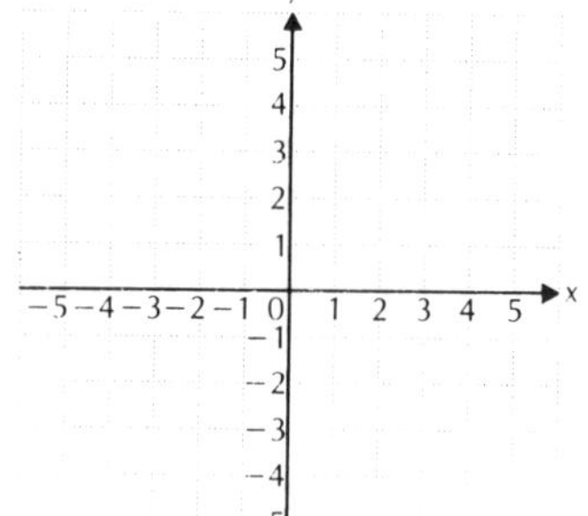

31. A rectangle of dimensions s by t is inscribed in a circle of radius 8 cm. Express the area A of the rectangle as a function of t.

32. Find an equation of the line containing the given point and perpendicular to the given line:

$$(-1,4);\; x = \frac{2}{3}y - 1.$$

33. Find f(x) and g(x) such that $h(x) = f \circ g(x)$.

$$h(x) = \frac{5\sqrt{3x - 8}}{2}$$

34. Find an equation of the circle with center (-1,5) and radius 12.

CHAPTER 3 NAME ____________________

TEST FORM F CLASS ______ SCORE ________ GRADE ______

Write the letter of your response on the answer blank.

ANSWERS

1. List the range of the relation whose ordered pairs are (2,4), (-2,4), (4,3), (0,-4), and (-2,3).

a) {-4,0,3,4} b) {-4,3,4}

c) {-4,-4,0,3,4} d) {2,2,3,4,4}

1. ____________

Consider the following relations for Questions 2 - 4.

a) $y = x^5 - x^3$ b) $x^4 + y^4 = 2$ c) $\frac{y}{2} = \frac{3}{x}$

d) $x = 7$ e) $y = |x|$ f) $y = x^2 + 5$

2. Which are symmetric with respect to the origin?

a) a, c, f b) b, d c) b, c d) a, b, c

2. ____________

3. Which are symmetric with respect to the x-axis?

a) b, e, f b) b, d c) a, b, c d) b, c

3. ____________

4. Which are symmetric with respect to the y-axis?

a) c,e b) a, c, d c) b, e, f d) b, f

4. ____________

5. Classify this function as even, odd, or neither even nor odd; $f(x) = 4x^7 - 3x^3$.

a) Even b) Odd c) Neither

5. ____________

6. Find the domain: $f(x) = \dfrac{3x + 1}{x^3 - 4x}$.

a) $\{x|x \neq 0\}$ b) $\{x|x \neq -2,2\}$

c) $\{x|x \neq 0, x \neq -2, x \neq 2\}$ d) $\{x|x \neq 2\}$

6. ____________

NAME ____________

TEST FORM F

ANSWERS

7. ______

8. ______

9. ______

10. ______

11. ______

7. Given $f(x) = 3x - 2$ and $g(x) = x^2 + x - 1$, find $f \circ g(x)$.

a) $3x^2 + 3x - 5$ b) $9x^2 - 11x + 3$

c) $3x^2 + 3x - 3$ d) $9x^2 - 9x + 1$

8. Which of the following are graphs of functions?

a) b) c) d)

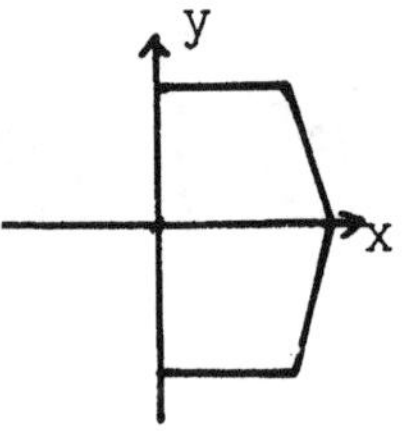

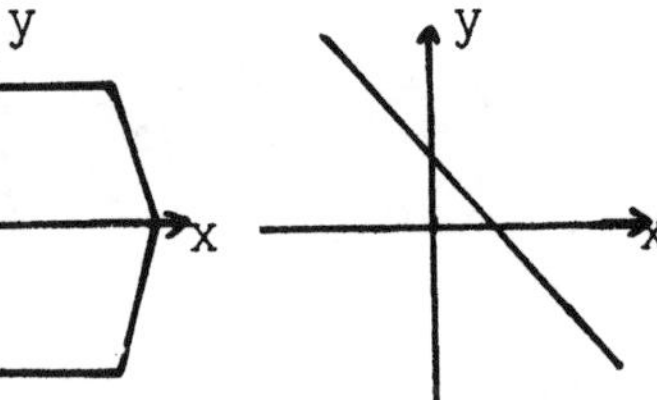

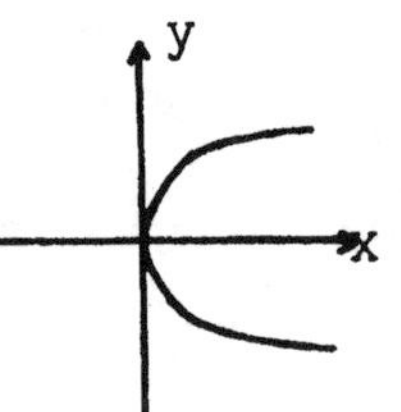

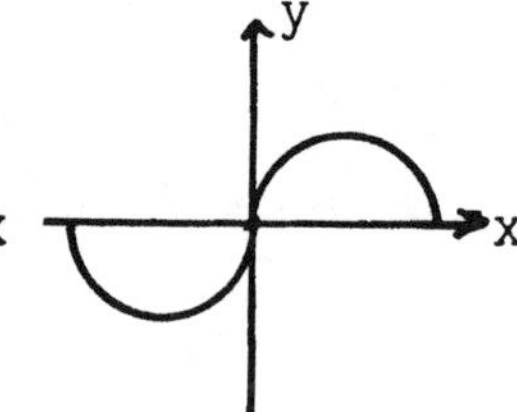

a) b, d b) a, d c) b d) b, c

9. Find $\frac{f(a + h) - f(a)}{h}$ when $f(x) = \frac{1}{x}$.

a) $\frac{1}{a(a + h)}$ b) $\frac{1}{h^2}$ c) $\frac{-h^2}{a(a + h)}$ d) $\frac{-1}{a(a + h)}$

10. Find the slope of the line $3 - 4y = -6x$.

a) -6 b) -4 c) $\frac{3}{2}$ d) $-\frac{2}{3}$

11. Find the distance between $(1,-4)$ and the midpoint of the segment with endpoints $(2,3)$ and $(-4,5)$.

a) 4 b) $2\sqrt{10}$ c) $2\sqrt{13}$ d) $2\sqrt{17}$

NAME ______________________

TEST FORM F

ANSWERS

12. Graph: $x = |y - 2|$.

a)

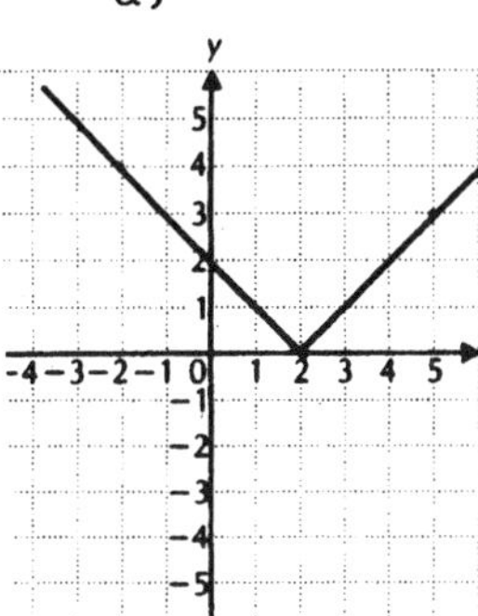

b)

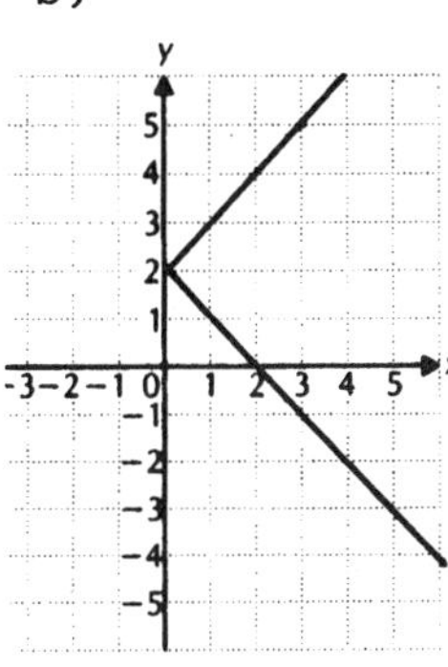

c)

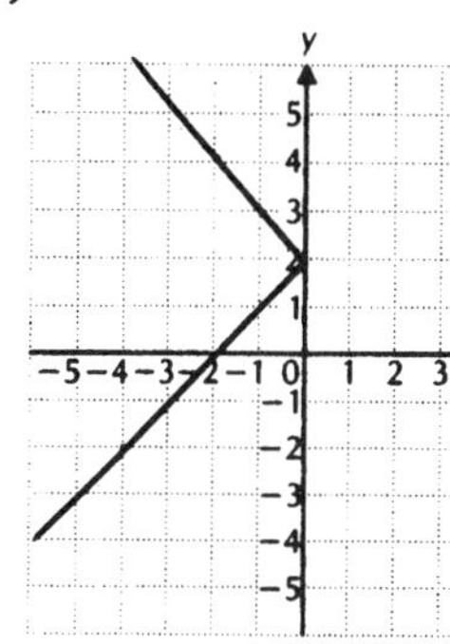

12. ______________

13. Graph: $y = 1 - x^2$.

a)

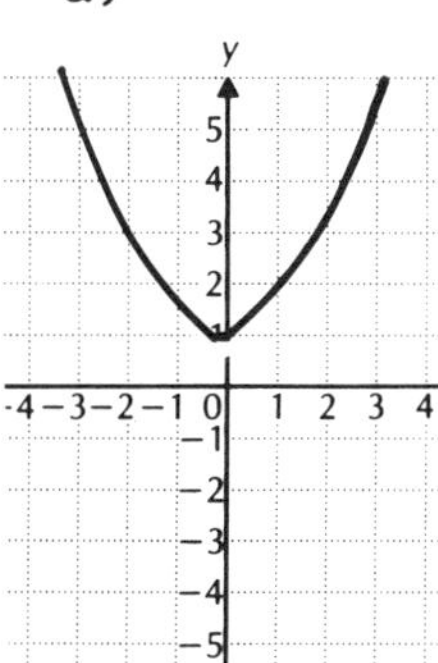

b)

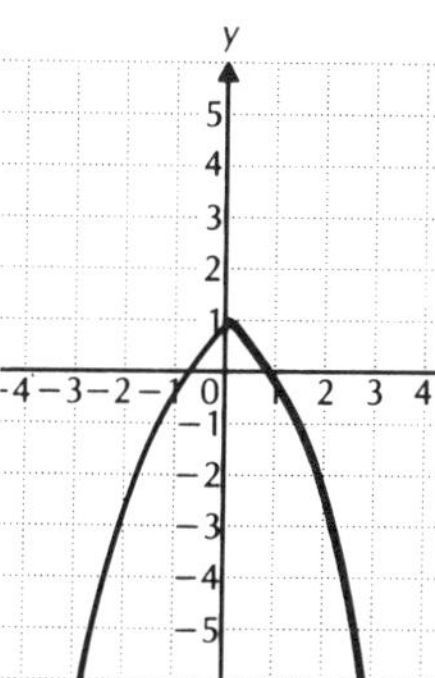

c)

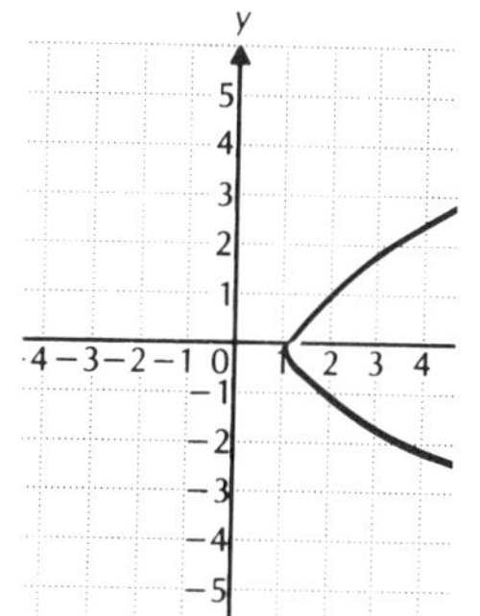

13. ______________

14. Find an equation of the line containing the given point and parallel to the given line.

$(-2,1);\ 2y - 3x = 5$

a) $y = \frac{3}{2}x - \frac{3}{2}$

b) $y = \frac{2}{3}x - 2$

c) $y = \frac{3}{2}x + 4$

d) $y = \frac{3}{2}x - \frac{7}{2}$

14. ______________

15. Write interval notation for $\left\{x \middle| 0 \le x < \frac{5}{2}\right\}$.

a) $\left[0,\frac{5}{2}\right]$ b) $\left(0,\frac{5}{2}\right]$ c) $\left(0,\frac{5}{2}\right)$ d) $\left[0,\frac{5}{2}\right)$

15. ______________

16. Classify this function as increasing, decreasing, or neither increasing nor decreasing.

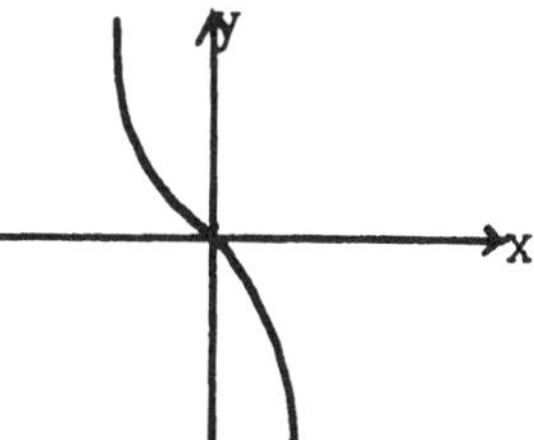

a) Increasing

b) Decreasing

c) Neither

16. ______________

NAME ______________________________

TEST FORM F

ANSWERS

17. ______________

18. ______________

17. Here is a graph of $y = f(x)$.
Sketch the graph of $y = f(x) - 3$.

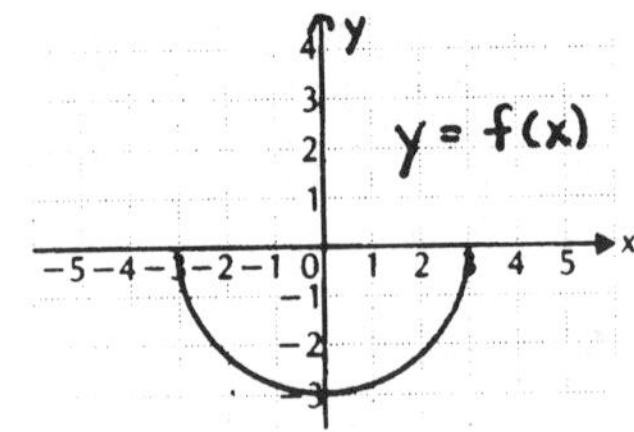

a)

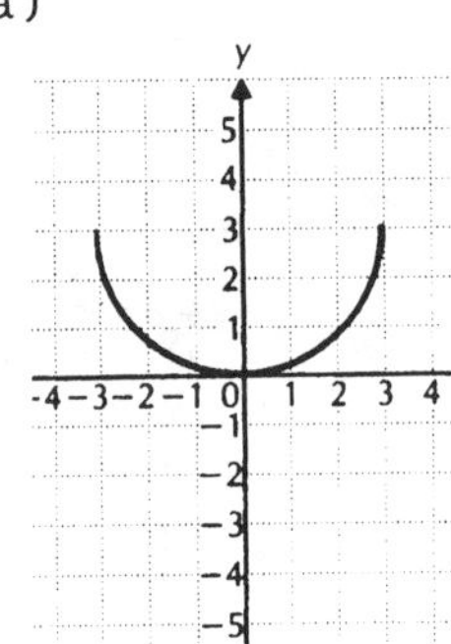

b)

c)

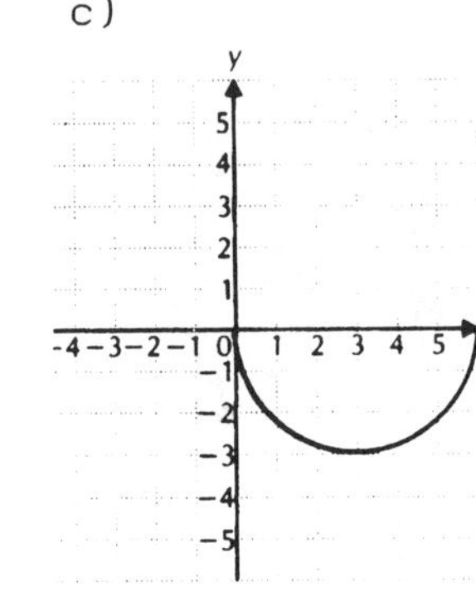

18. Here is a graph of $y = f(x)$.
Sketch the graph of $y = -\frac{1}{2}f(x + 1)$.

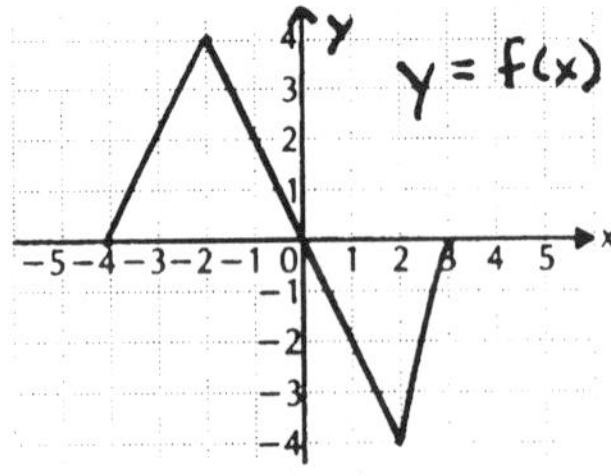

a)

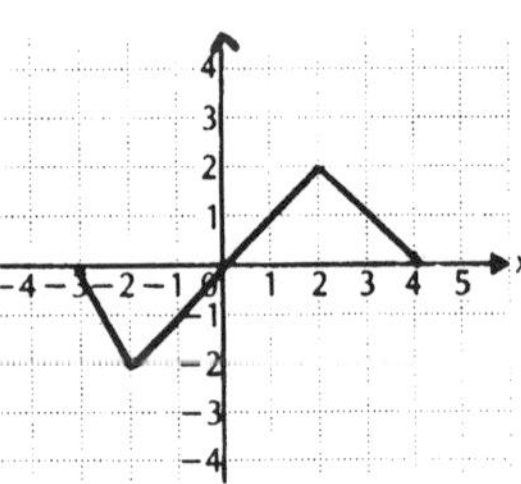

b)

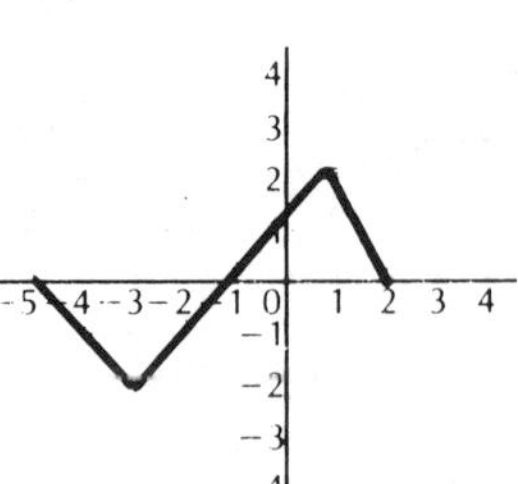

c)

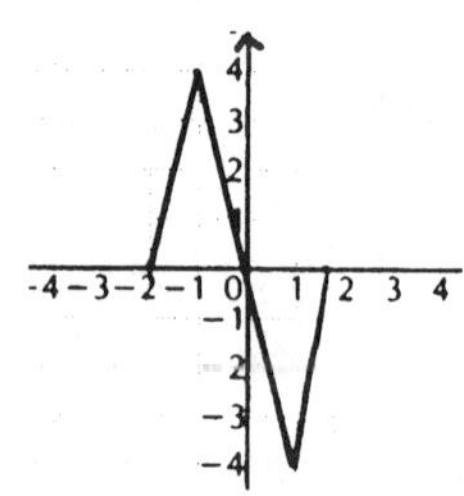

CHAPTER 4 NAME ______________________

TEST FORM A CLASS ________ SCORE ________ GRADE ______

For the functions in Questions 1 and 2:

a) use completing the square to put each equation into the form $f(x) = a(x - h)^2 + k$;

b) find the vertex; and

c) determine whether there is a maximum or minimum function value and find that value.

1. $f(x) = 2x^2 - 8x + 1$

2. $f(x) = -3x^2 + 2x - 1$

3. Graph $f(x) = -2x^2 + 4x + 3$.

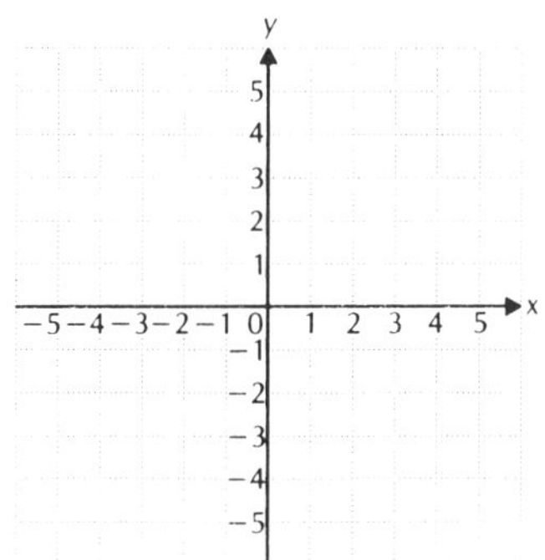

4. Find the x-intercepts of $f(x) = 4x^2 - 3x - 2$.

5. Find $\{2,7,9,11\} \cup \{1,2,5,9\}$.

6. Graph $\{x|x \leq 5\} \cap \{x|x > 3\}$. 

Solve.

7. $x + 2 > -6$ and $-3x + 1 \geq 4$

8. $|2x - 5| \geq 7$

9. $|x - 7| < 3$

10. $|3x - 2| = 8$

11. $x^2 - 10x + 9 > 0$

12. $3x^2 + x - 2 < 0$

13. $\frac{x - 3}{x + 4} < 2$

ANSWERS

1. a) ______
   b) ______
   c) ______

2. a) ______
   b) ______
   c) ______

3. See graph.

4. ______

5. ______

6. See graph.

7. ______

8. ______

9. ______

10. ______

11. ______

12. ______

13. ______

NAME ____________________

TEST FORM A

ANSWERS	
14. ______	14. Find two numbers whose sum is -16 and whose product is a maximum.
15. See graph.	Sketch a graph of the polynomial function. 15. $f(x) = x^4 - 4x^2$ 16. $f(x) = x^4 - 4x^2 + 2$
16. See graph.	
17. ______	17. Find the remainder when $2x^4 - 3x^2 + 9x - 1$ is divided by $x - 4$.
18. ______	18. Determine whether $x - i$ is a factor of $2x^3 + 2x$.
19. ______	19. Use synthetic division to find the quotient and the remainder. Show all your work. $(4x^3 - 3x^2 + 2x - 2) \div (x - 1)$
20. ______	20. Use synthetic division to find P(-2). $P(x) = 3x^4 - 6x^3 + 2x^2 - 4$

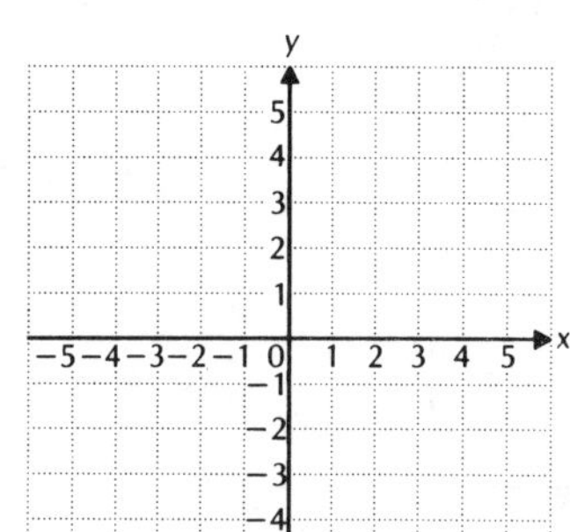

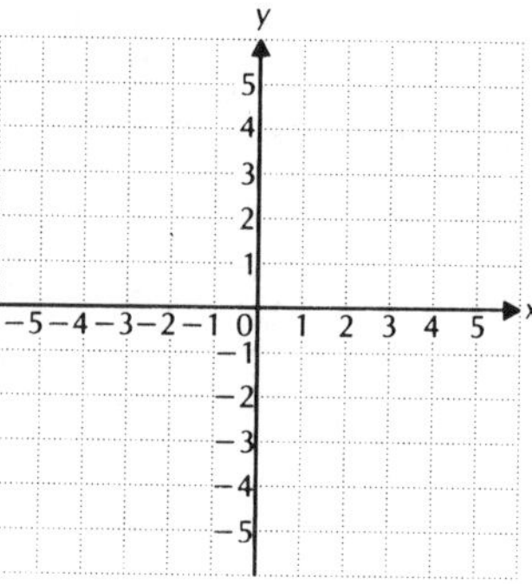

 NAME ____________________

ANSWERS

21. Factor the polynomial $P(x)$. Then solve the equation $P(x) = 0$.

$$P(x) = x^3 - 2x^2 - 11x + 12$$

21. ____________

22. Find a polynomial of degree 4 with roots 3, -3, 1 + i, and 1 - i.

22. ____________

23. Find a polynomial of lowest degree having roots 2 and -3, and having 1 as a root of multiplicity 2 and 5 as a root of multiplicity 3.

23. ____________

24. Find a polynomial of lowest degree with rational coefficients that has 4 - i and 3 as two of its roots.

24. ____________

25. List all possible rational roots of $4x^5 - 2x^3 - 3x^2 + 6$.

25. ____________

26. What does Descartes' rule of signs tell you about the number of positive real roots and the number of negative real roots of the polynomial?

$$8x^6 + 3x^4 - 9x^3 + x - 7$$

26. ____________

CHAPTER 4 NAME ______________________

TEST FORM A

ANSWERS

27. ______________

28. ______________

29. See graph.

30. ______________

31. ______________

32. ______________

27. Find the smallest positive integer that is guaranteed by Theorem 15 to be an upper bound to the roots of

$$4x^5 - 3x^3 + 2x - 8.$$

Then find the largest negative integer that is guaranteed to be a lower bound.

28. Approximate the irrational roots of $P(x) = x^4 - 5x^2 - 1$.

29. Graph: $f(x) = \dfrac{3x + 1}{x}$.

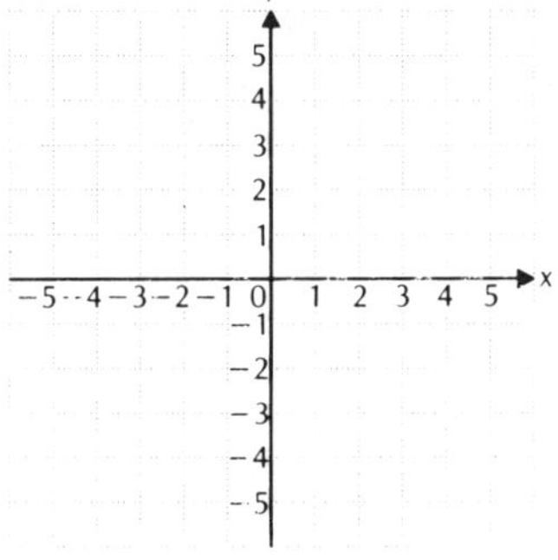

30. Find k such that $f(x) = kx^2 + 5x - 2$ has a minimum value at $x = -3$.

31. Solve: $|x + 4| \le |x - 1|$.

32. Solve: $x^3 - 5x^2 - 2x + 24 > 0$.

CHAPTER 4   *NAME* ______________________

TEST FORM B   *CLASS* ______ *SCORE* ______ *GRADE* ______

For the functions in Questions 1 and 2:

a) use completing the square to put each equation into the form $f(x) = a(x - h)^2 + k$;

b) find the vertex; and

c) determine whether there is a maximum or minimum function value and find that value.

1. $f(x) = -3x^2 + 12x - 4$ 2. $f(x) = 5x^2 + 2x + 1$

3. Graph $f(x) = 3x^2 + 6x + 5$.

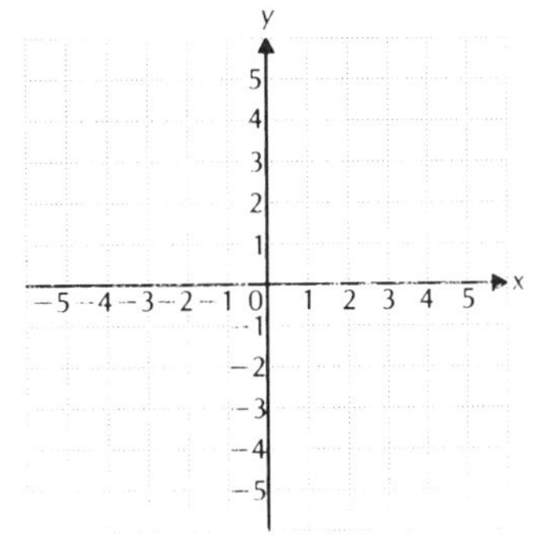

4. Find the x-intercepts of $f(x) = 2x^2 + 5x - 9$.

5. Find $\{2,5,7,11,15\} \cap \{5,10,15,20\}$.

6. Graph $\{x|x \le 2\} \cup \{x|x > 4\}$. 

Solve.

7. $x - 1 > -5$ and $-2x - 5 \ge 7$ 8. $|-3x + 1| > 4$

9. $|2x - 1| \le 5$ 10. $|5x - 1| = 21$

11. $x^2 + 2x - 8 > 0$ 12. $10x^2 + 3x - 1 < 0$

13. $\dfrac{x - 1}{3x + 2} > 1$

ANSWERS

1. a) ______
   b) ______
   c) ______

2. a) ______
   b) ______
   c) ______

3. See graph.

4. ______

5. ______

6. See graph.

7. ______

8. ______

9. ______

10. ______

11. ______

12. ______

13. ______

CHAPTER 4 NAME ____________________

TEST FORM B

ANSWERS

14. ____________

15. See graph.

16. See graph.

17. ____________

18. ____________

19. ____________

20. ____________

14. Of all the numbers whose difference is 18, find the two that have the minimum product.

Sketch a graph of the polynomial function.

15. $f(x) = x^4 - 3x^2$

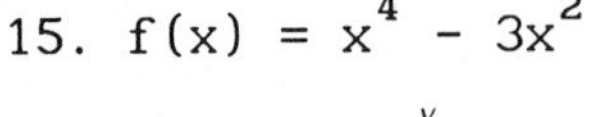

16. $f(x) = x^4 - 3x^2 + 4$

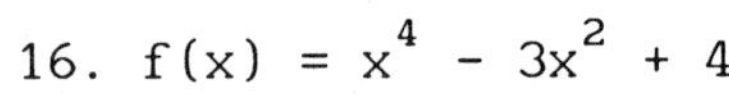

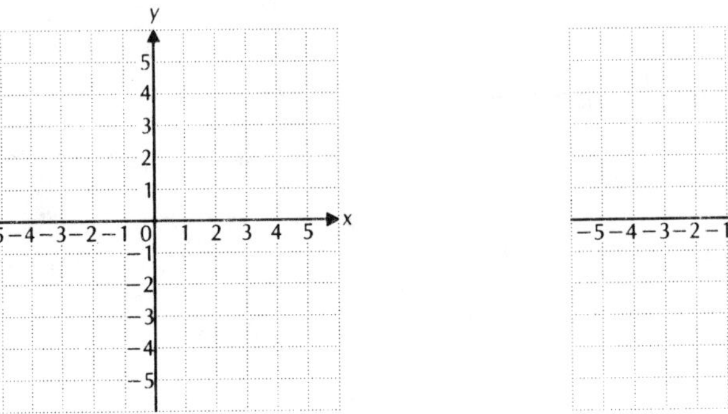

17. Find the remainder when $7x^5 - 2x^3 + x^2 - 2$ is divided by $x - 1$.

18. Determine whether $x - 2$ is a factor of $x^4 - 3x^2 - 4$.

19. Use synthetic division to find the quotient and the remainder. Show all your work.

$$(3x^4 - 2x^2 + 3x - 1) \div (x + 2)$$

20. Use synthetic division to find P(4):

$$P(x) = 2x^4 - 3x^3 + 2x - 5.$$

 NAME ____________

TEST FORM B

	ANSWERS
21. Factor the polynomial $P(x)$. Then solve the equation $P(x) = 0$. $$P(x) = x^3 - 2x^2 - 13x - 10$$	21. ____________
22. Find a polynomial of degree 3 with roots 5, $\sqrt{3}$, and $-\sqrt{3}$.	22. ____________
23. Find a polynomial of lowest degree having roots 0 and 5, and having 3 as a root of multiplicity 2 and −1 as a root of multiplicity 4.	23. ____________
24. Find a polynomial of lowest degree with rational coefficients that has 7 and $3 + \sqrt{2}$ as two of its roots.	24. ____________
25. List all possible rational roots of $2x^6 - 5x^3 + 8x^2 - 8$.	25. ____________
26. What does Descartes' rule of signs tell you about the number of positive real roots and the number of negative real roots of the polynomial? $$-5x^6 + 3x^4 - x^3 + 2x + 5$$	26. ____________

NAME ________________________

TEST FORM B

ANSWERS	
27. ________	27. Find the smallest positive integer that is guaranteed by Theorem 15 to be an upper bound to the roots of $$5x^4 - 3x^2 + 7x - 2.$$ Then find the largest negative integer that is guaranteed to be a lower bound.
28. ________	28. Approximate the irrational roots of $P(x) = x^4 - x^2 - 7$.
29. See graph.	29. Graph: $f(x) = \dfrac{x^2 - x - 1}{x - 2}$. 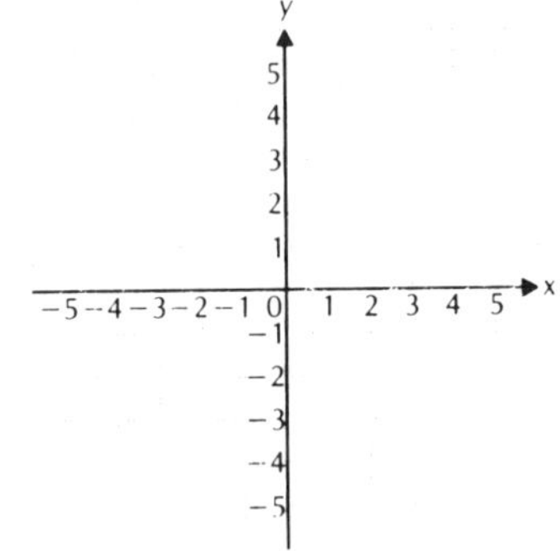
30. ________	30. Find c such that $f(x) = -3x^2 + 6x + c$ has a maximum value of 39.
31. ________	31. Solve: $x^4 - 5x^2 \leq 0$.
32. ________	32. When $x^2 - 5x + 2k$ is divided by $x + 3$, the remainder is -2. Find the value of k.

CHAPTER 4 NAME ______________________

TEST FORM C CLASS ______ SCORE ______ GRADE ______

For the functions in Questions 1 and 2:

a) use completing the square to put each equation into the form $f(x) = a(x - h)^2 + k$;

b) find the vertex; and

c) determine whether there is a maximum or minimum function value and find that value.

1. $f(x) = 6x^2 + 12x - 5$

2. $f(x) = -2x^2 + x + 3$

3. Graph $f(x) = -x^2 - 6x - 5$.

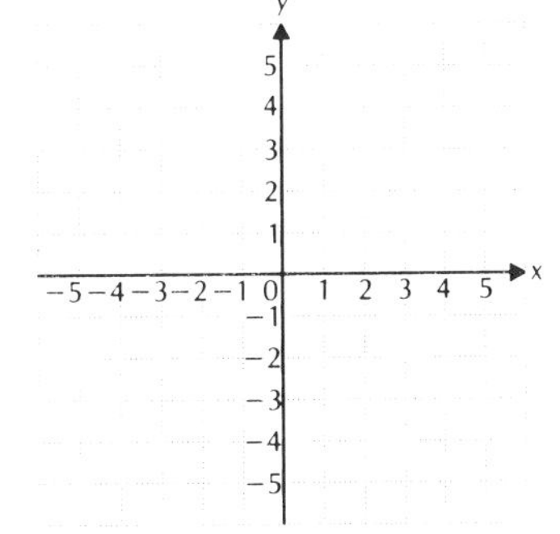

4. Find the x-intercepts of $f(x) = 3x^2 - 7x - 1$.

5. Find $\{1,4,7,8,16,20\} \cup \{1,7,9,11,18\}$.

6. Graph $\{x|x < 5\} \cap \{x|x \geq 0\}$.

Solve.

7. $2x + 3 \geq 1$ and $5x - 1 < 4$

8. $|5x - 3| \geq 7$

9. $|x + 12| < 4$

10. $|8x - 3| = 27$

11. $x^2 + 4x - 12 > 0$

12. $12x^2 - 5x - 2 < 0$

13. $\frac{x - 5}{x + 1} > 3$

ANSWERS

1. a) ______
   b) ______
   c) ______
2. a) ______
   b) ______
   c) ______
3. See graph.
4. ______
5. ______
6. See graph.
7. ______
8. ______
9. ______
10. ______
11. ______
12. ______
13. ______

NAME ______________________

TEST FORM C

ANSWERS

14. ______________________

15. See graph.

16. See graph.

17. ______________________

18. ______________________

19. ______________________

20. ______________________

14. The sum of the base and height of a triangle is 28 cm. Find the dimensions for which the area is a maximum.

Sketch a graph of the polynomial function.

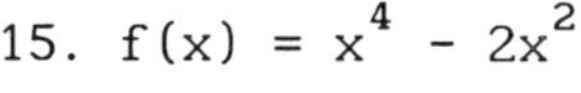

15. $f(x) = x^4 - 2x^2$

16. $f(x) = x^4 - 2x^2 - 3$

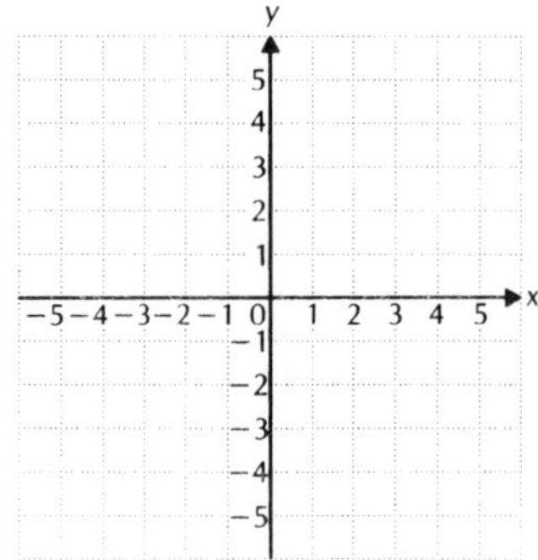

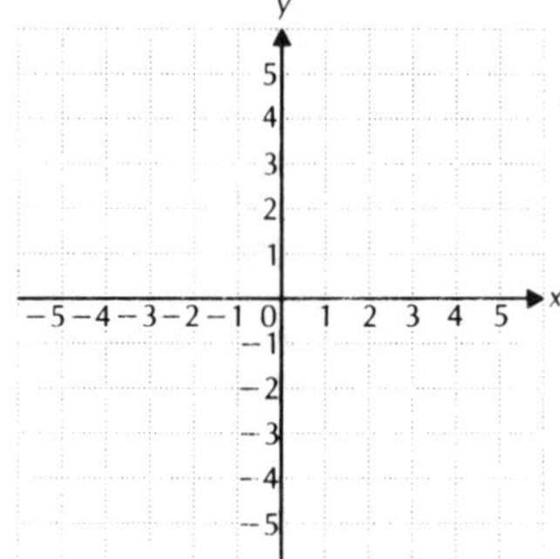

17. Find the remainder when $2x^4 - 4x^3 - 2x^2 + 1$ is divided by $x + 2$.

18. Determine whether $x - i$ is a factor of $x^4 + x$.

19. Use synthetic division to find the quotient and the remainder. Show all your work.

$$(5x^3 - 7x^2 + 3x - 4) \div (x - 4)$$

20. Use synthetic division to find $P(-3)$:

$$P(x) = 5x^4 - 2x^2 - 9x + 7.$$

NAME ______________________

TEST FORM C

ANSWERS

21. Factor the polynomial P(x). Then solve the equation P(x) = 0.

$$P(x) = x^3 + x^2 - 34x + 56$$

21. ______________

22. Find a polynomial of degree 4 with roots 4, -4, 3i, and -3i.

22. ______________

23. Find a polynomial of lowest degree having roots -4 and -3, and having 5 as a root of multiplicity 2 and -2 as a root of multiplicity 3.

23. ______________

24. Find a polynomial of lowest degree with rational coefficients that has -2 and -3 + 5i as two of its roots.

24. ______________

25. List all possible rational roots of $6x^5 - 3x^3 + 2x^2 - 4$.

25. ______________

26. What does Descartes' rule of signs tell you about the number of positive real roots and the number of negative real roots of the polynomial?

$$6x^5 - 2x^4 + 3x^2 + 2x + 7$$

26. ______________

NAME ______________________

TEST FORM C

ANSWERS

27. ____________

28. ____________

29. See graph.

30. ____________

31. ____________

32. ____________

27. Find the smallest positive integer that is guaranteed by Theorem 15 to be an upper bound to the roots of

$$3x^4 - 2x^3 + 7x - 1.$$

Then find the largest negative integer that is guaranteed to be a lower bound.

28. Approximate the irrational roots of $P(x) = x^4 - x - 3$.

29. Graph: $f(x) = \dfrac{x - 1}{x^2 - 3x - 10}$.

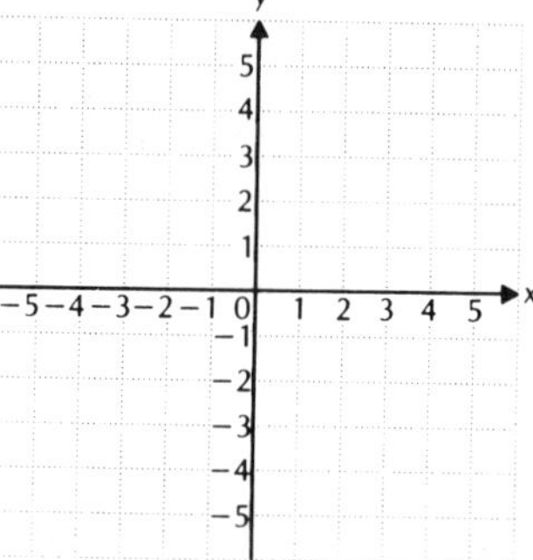

30. Solve: $(x - 2)^2 > x(x - 5)$.

31. The base of a triangle is 6 cm greater than the height. Find the possible heights h such that the area of the triangle will be greater than 15 cm.

32. Find k so that $x - 3$ is a factor of $x^3 + kx^2 - 2x + 5k$.

CHAPTER 4 NAME ________________

TEST FORM D CLASS ______ SCORE ________ GRADE ______

For the functions in Questions 1 and 2:

a) use completing the square to put each equation into the form $f(x) = a(x - h)^2 + k$;

b) find the vertex; and

c) determine whether there is a maximum or minimum function value and find that value.

1. $f(x) = -2x^2 - 8x + 3$

2. $f(x) = 4x^2 + 5x - 2$

3. Graph $f(x) = 2x^2 - 4x + 5$.

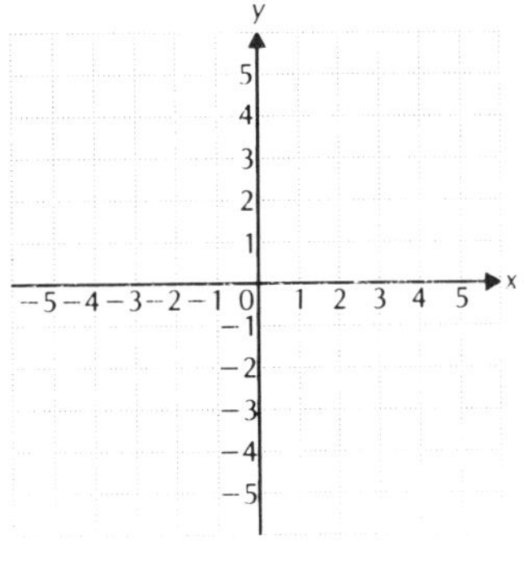

4. Find the x-intercepts of $f(x) = 3x^2 + 5x - 7$.

5. Find $\{1,4,7,8,16,20\} \cap \{1,7,9,11,18\}$.

6. Graph $\{x|x < -3\} \cup \{x|x > -1\}$.

Solve.

7. $-3x + 1 < 7$ and $2x + 3 < 9$

8. $|7 - x| > 9$

9. $|3x + 2| \le 5$

10. $|4x + 5| = 11$

11. $x^2 - 2x - 24 > 0$

12. $7x^2 - 27x - 4 < 0$

13. $\dfrac{2x - 1}{x + 7} \le 1$

ANSWERS

1. a) ________
   b) ________
   c) ________
2. a) ________
   b) ________
   c) ________
3. See graph.
4. ________
5. ________
6. See graph.
7. ________
8. ________
9. ________
10. ________
11. ________
12. ________
13. ________

NAME ______________________

TEST FORM D

ANSWERS

14. ____________

15. See graph.

16. See graph.

17. ____________

18. ____________

19. ____________

20. ____________

14. Of all the numbers whose difference is 22, find the two that have the minimum product.

Sketch a graph of the polynomial function.

15. $f(x) = x^4 - 5x^2$

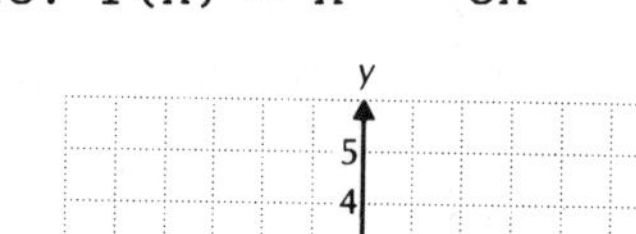

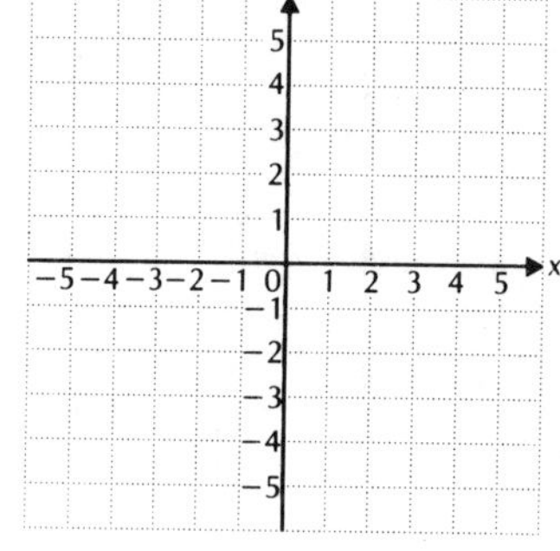

16. $f(x) = x^4 - 5x^2 + 4$

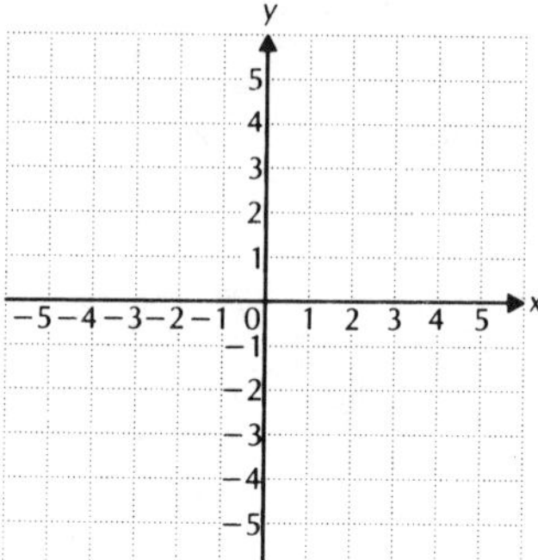

17. Find the remainder when $3x^4 - 2x^2 + x + 3$ is divided by $x + 3$.

18. Determine whether $x - 3$ is a factor of $x^3 - 9$.

19. Use synthetic division to find the quotient and the remainder. Show all your work.

$$(x^4 + 2x^3 - x + 4) \div (x + 3).$$

20. Use synthetic division to find $P(5)$:

$$P(x) = 4x^4 - 2x^3 + x - 9.$$

NAME ______________________

TEST FORM D

ANSWERS

21. Factor the polynomial P(x). Then solve the equation P(x) = 0.

$$P(x) = x^3 - 7x^2 + 7x + 15$$

21. ______________

22. Find a polynomial of degree 3 with roots 7, $2 + \sqrt{5}$, and $2 - \sqrt{5}$.

22. ______________

23. Find a polynomial of lowest degree having roots 1 and -2, and having -1 as a root of multiplicity 2 and 6 as a root of multiplicity 3.

23. ______________

24. Find a polynomial of lowest degree with rational coefficients that has $1 - \sqrt{5}$ and 4 as two of its roots.

24. ______________

25. List all possible rational roots of $8x^5 - 3x^3 + 2x - 4$.

25. ______________

26. What does Descartes' rule of signs tell you about the number of positive real roots and the number of negative real roots of the polynomial?

$$10x^6 - 5x^5 + 3x^4 - 2x^3 - x + 5$$

26. ______________

NAME ____________________

TEST FORM D

ANSWERS

27. ____________

28. ____________

29. See graph.

30. ____________

31. a) ____________

b) ____________

32. ____________

27. Find the smallest positive integer that is guaranteed by Theorem 15 to be an upper bound to the roots of

$$6x^5 - 3x^2 + 2x - 15.$$

Then find the largest negative integer that is guaranteed to be a lower bound.

28. Approximate the irrational roots of

$P(x) = 2x^3 + x^2 - 2.$

29. Graph: $f(x) = \dfrac{x^2 - 2x - 3}{x + 2}.$

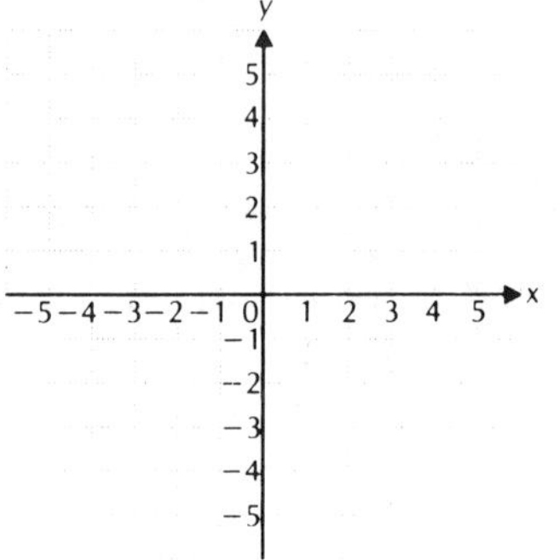

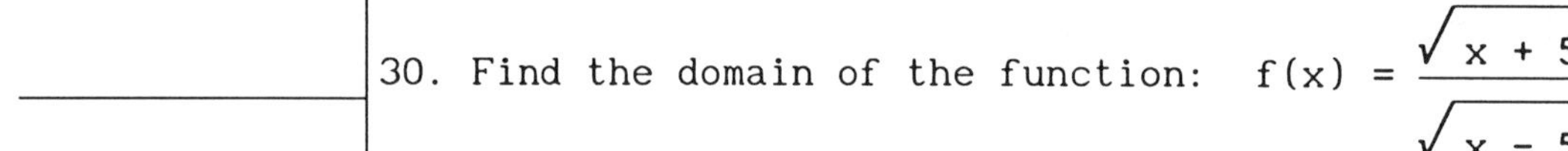

30. Find the domain of the function: $f(x) = \dfrac{\sqrt{x + 5}}{\sqrt{x - 5}}.$

31. A company has the following total-cost and total-revenue functions to use in producing and selling x units of a certain product.

$R(x) = 36x - x^2, \quad C(x) = 4x + 220$

a) Find the break-even values $(R(x) = C(x))$.

b) Find the values of x that produce a profit.

32. When $x^2 - 5x + 2k$ is divided by $x + 3$, the remainder is 10. Find the value of k.

CHAPTER 4 NAME ______

TEST FORM E CLASS ______ SCORE ______ GRADE ______

1. For the function $f(x) = 2x^2 - 10x - 5$,

a) find an equation of the type $f(x) = a(x - h)^2 + k$,

b) find the vertex; and

c) determine whether there is a maximum or minimum function value and find that value.

2. Graph: $f(x) = -x^2 + 3x - 1$. Also find the x-intercepts.

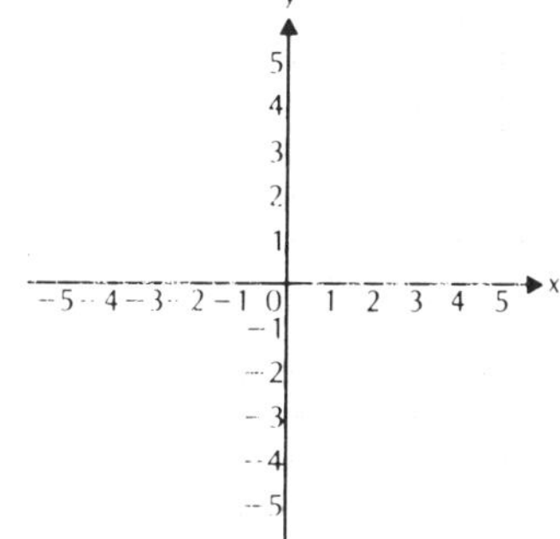

3. The sum of the length and width of a rectangle is 54. Find the dimensions for which the area is a maximum.

4. Find $\{b,d,f,h,j\} \cup \{d,h,\ell,p\}$.

5. Graph: $\{x|x < -\frac{3}{2}\} \cup \{x|x \geq -1\}$.

ANSWERS

1. a) ______

b) ______

c) ______

2. See graph.

3. ______

4. ______

5. See graph.

CHAPTER 4 *NAME* ____________________

TEST FORM E

ANSWERS

6. ________

7. ________

8. ________

9. ________

10. ________

11. ________

12. ________

13. See graph.

Matching: Solve and place the letter of the answer in the answer blank. Some letters may be used more than once, and some may not be used.

6. $|8 - 2x| = 4$

7. $|-3x + 7| \geq 19$

8. $x - 1 > -7$ and $-7x + 2 \geq 16$

9. $x^2 - 6x - 7 < 0$

10. $|x - 5| < 2$

11. $6x^2 - 17x + 10 \leq 0$

12. $\dfrac{x - 1}{2x + 3} > 1$

Solutions

a) $(3,7)$ b) $\left[\frac{5}{6}, 2\right]$

c) $(-\infty,-1) \cup (7,\infty)$

d) $\{-2,4\}$ e) $(-6,-2]$

f) $(-1,7)$ g) $\left(-4,-\frac{3}{2}\right)$

h) $(-\infty,-4] \cup \left[\frac{26}{3},\infty\right)$

i) $\{-6,-\frac{4}{3}\}$ j) $\{2,6\}$

k) $\left[4,\frac{26}{3}\right]$ ℓ) $\varnothing$

Sketch a graph of each polynomial function.

13. $f(x) = x^3 - 3x^2 + x + 1$

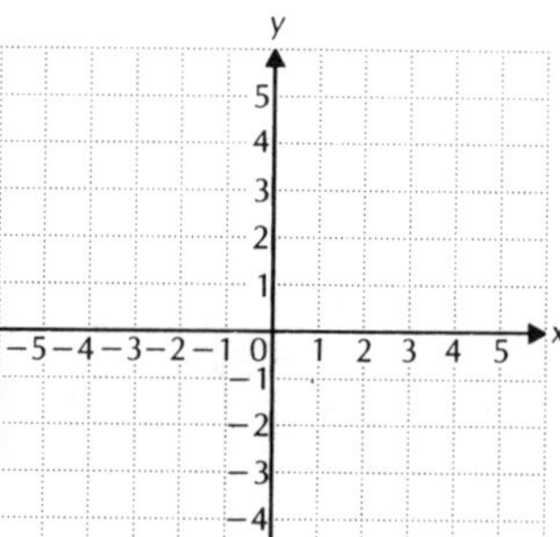

NAME ______________________

TEST FORM E

	ANSWERS
Sketch a graph of each polynomial function.	
14. $f(x) = 2x^4 + 3x^3 - 12x^2 - 7x + 6$	14. See graph.
15. Find the remainder when $P(x) = x^2 - 5ix + 4$ is divided by $x - i$.	15. ______
16. Use synthetic division to find the quotient and remainder. Show all your work. $(5x^5 - 2x^3 + x^2 - 2x + 3) \div (x + 2)$	16. ______
17. Use synthetic division to find $P(-5)$. $P(x) = x^4 - 2x^3 + 5x - 16$	17. ______
18. Factor the polynomial $P(x)$. Then solve the equation $P(x) = 0$. $P(x) = x^5 + 3x^4 - 26x^3 - 78x^2 + 25x + 75$	18. ______
19. Find a polynomial of degree 4 with roots $1 - \sqrt{3}$, $1 + \sqrt{3}$, $2i$, and $-2i$.	19. ______
20. Find a polynomial of lowest degree having roots 0 and -2, and having 5 as a root of multiplicity 4 and -7 as a root of multiplicity 3.	20. ______

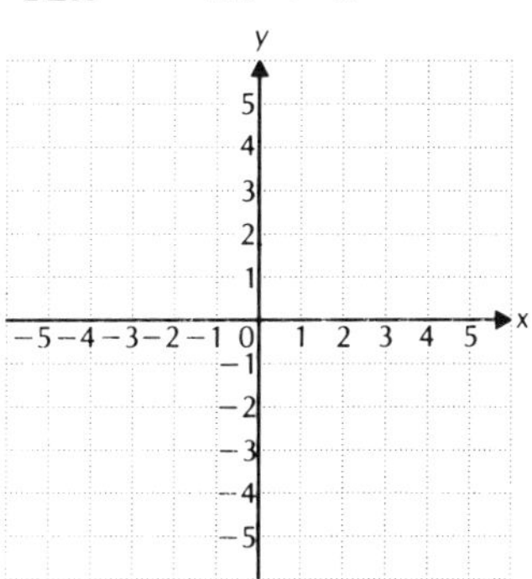

NAME ______________________

TEST FORM E

ANSWERS

21. ______

22. ______

23. ______

24. ______

25. ______

26. See graph.

21. Given that $x^3 - 64 = 0$ has 4 as a root, find the other roots.

22. A polynomial of degree 7 with rational coefficients has 4, -2i, $\sqrt{7}$, and 4 - 5i as roots. Find the other roots.

23. List all possible rational roots of $6x^3 - 5x + 4$.

24. What does Descartes' rule of signs tell you about the number of negative real roots of the polynomial?

$$-x^4 - 3x^3 + x^2 + 5x + 1$$

25. Find the smallest positive integer that is guaranteed by Theorem 15 to be an upper bound to the roots of

$$2x^5 - 3x^3 + x - 5.$$

Then find the largest negative integer that is guaranteed to be a lower bound.

26. Graph: $y = \dfrac{x^2 + 2x - 8}{x - 1}$.

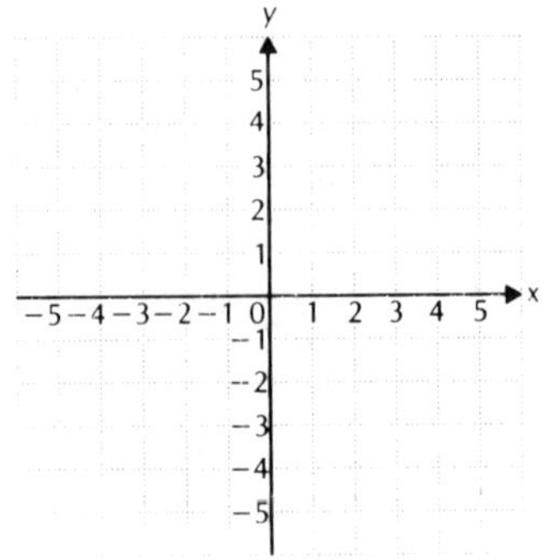

CHAPTER 4 NAME ______________________

TEST FORM F CLASS ________ SCORE ________ GRADE ______

ANSWERS

1. Determine whether there is a maximum or minimum function value of the function $f(x) = 4x^2 - 8x - 3$ and find that value.

   a) 7 is a minimum b) -2 is a maximum

   c) -2 is a maximum d) -7 is a minimum

1. ______________

2. Find the x-intercepts of $f(x) = x^2 - 5x + 4$.

   a) $(1 + \sqrt{2}, 0)$, $(0, 1 - \sqrt{2})$

   b) (1,0), (4,0)

   c) (5,0), (-1,0)

   d) No x-intercepts

2. ______________

3. Find $\{3,6,9,12,15,18\} \cap \{5,10,15,20,25\}$.

   a) {15} b) {3,6,9,10,12,15,20,25}

   c) {3,6,9,12,18} d) {15,20,25}

3. ______________

4. Graph: $\{x \mid 2x - 3 \geq 1\} \cup \{x \mid -x + 1 > -2\}$.

   a) 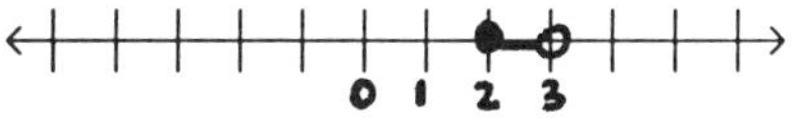

   b) 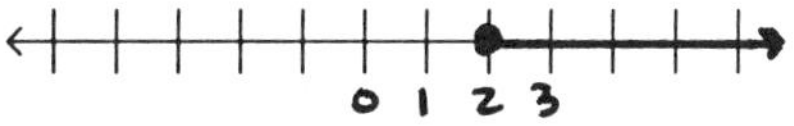

   c) 

   d) 

4. ______________

5. Solve: $|-4x + 3| > 15$.

   a) $\left(\frac{9}{2}, 13\right)$ b) $(-\infty, -3) \cup \left(\frac{9}{2}, \infty\right)$

   c) $\left(-3, \frac{9}{2}\right)$ d) $(-3, -2)$

5. ______________

6. Solve: $|x - 5| \leq 2$.

   a) [3,7] b) $(-\infty, 3] \cup [7, \infty)$

   c) [-3,7] d) [2,5]

6. ______________

ANSWERS	
7. ______	7. Solve: $\|6x + 3\| = 15$. a) $(-\infty,-3] \cup [2,\infty)$ b) $(-3,2]$ c) $\{-3,2\}$ d) $\{-5,1\}$
8. ______	8. Solve: $2x - 5 < -2x + 3$. a) $(-2,\infty)$ b) $(-2,2)$ c) $(-\infty,4)$ d) $(-\infty,2)$
9. ______	9. Solve: $x^2 - 5x - 14 > 0$. a) $(-3,9)$ b) $(-7,2)$ c) $(-\infty,-2) \cup (7,\infty)$ d) $(-2,7)$
10. ______	10. Solve: $3x^2 - 10x - 8 \le 0$. a) $\left(-\frac{3}{2},4\right)$ b) $\left[-\frac{2}{3},4\right]$ c) $\left(-\infty,-\frac{2}{3}\right) \cup (4,\infty)$ d) $(-\infty,-4) \cup \left[\frac{3}{2},\infty\right)$
11. ______	11. Solve: $\frac{x + 3}{x} > 2$. a) $(0,3)$ b) $(-\infty,0) \cup (3,\infty)$ c) $(2,3)$ d) $(-\infty,1) \cup (3,\infty)$
12. ______	12. Find one of two numbers whose sum is -12 and whose product is a maximum. a) -7 b) -10 c) -4 d) -6

NAME ______________________

TEST FORM F

ANSWERS

13. Find the remainder when $2x^5 - x^3 + 4x^2 - x + 4$ is divided by $x + 2$.

a) -42 b) 44 c) -34 d) 74

13. ______

14. Use synthetic division to find the quotient and remainder.

$$(x^4 + 3x^2 - 2x + 1) \div (x - 3)$$

a) Q: $x^2 + 9x + 25$, R: 76

b) Q: $x^3 + 3x^2 + 12x + 34$, R: 103

c) Q: $x^3 + 6x^2 + 18x + 52$, R: 157

d) Q: $x^3 - 3x^2 + 12x - 38$, R: 115

14. ______

15. Factor the polynomial P(x). Then solve the equation P(x) = 0.

$$P(x) = x^5 - x^4 - 8x^3 + 8x^2 - 9x + 9$$

Find the positive real roots.

a) 1 and 2 b) 1 and 4 c) 1 and 3 d) 3 only

15. ______

16. Find a polynomial of degree 4 with roots $\sqrt{5}$, $-\sqrt{5}$, $2 - i$, and $2 + i$.

a) $x^4 - 4x^3 + 20x - 25$

b) $x^4 + 2x^3 - 4x^2 + 15x + 5$

c) $x^4 - 25$

d) $x^4 - 4x^3 - 10x^2 + 20x - 25$

16. ______

NAME ______________________

TEST FORM F

ANSWERS

17. ____________

18. ____________

19. ____________

20. ____________

17. Find a polynomial of lowest degree having roots -3 and 1, and having 0 as a root of multiplicity 3 and -4 as a root of multiplicity 2.

a) $3x(x + 3)(x - 1)(x - 4)^2$

b) $x^3(x - 3)(x + 1)(x - 4)^2$

c) $(x + 3)(x - 1)(x + 4)$

d) $x^3(x + 3)(x - 1)(x + 4)^2$

18. A polynomial of degree 5 with rational coefficients has 4, $2 + \sqrt{7}$, and $-3 - 2i$ as roots. Find the other roots.

a) $-4, -2 - \sqrt{7}, 3 + 2i$ b) $2 - \sqrt{7}, -3 + 2i$

c) $-2 - \sqrt{7}, 3 + 2i$ d) $-2 + \sqrt{7}, 3 - 2i$

19. Given that $x^3 - 8$ has 2 as a root, find the other roots.

a) ± 2 b) $\pm 2i$ c) $-1 \pm \sqrt{3}i$ d) $-1 \pm \sqrt{3}$

20. List all possible rational roots of $4x^2 - 5x - 8$.

a) $\pm\left(\frac{1}{4}, \frac{1}{2}, 1, 2, 4, 8\right)$ b) $\pm (1, 2, 4, 8)$

c) $\pm\left(\frac{1}{4}, \frac{1}{2}, \frac{1}{8}, 1, 2, 4\right)$ d) $\pm (1, 2, 4)$

CHAPTER 5 NAME ______________

TEST FORM A CLASS ________ SCORE __________ GRADE ______

1. Find the inverse of the relation G given by

   G = {(2,-1), (5,-3), (7,1), (-4,2), (3.2,4.5)}.

2. Write an equation of the inverse of the relation

   $y = 3x - 4$.

3. Which of the following have inverses that are functions?

a) 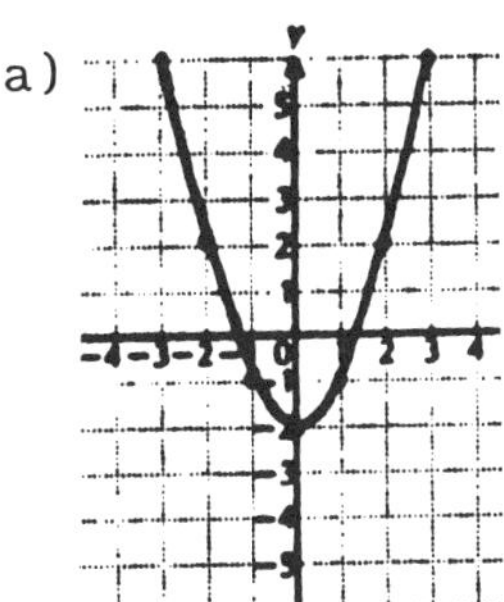

b) 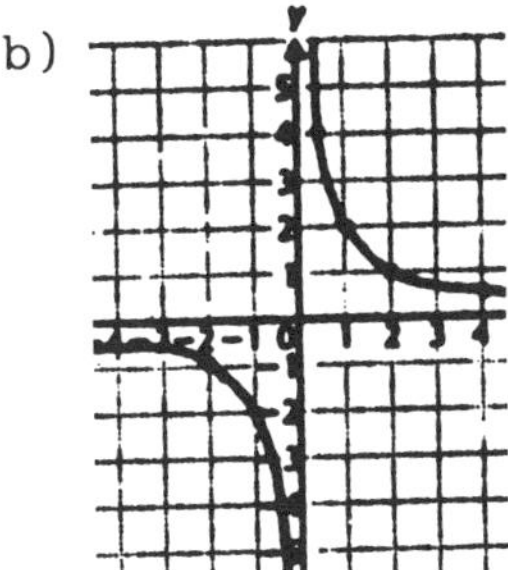

c) 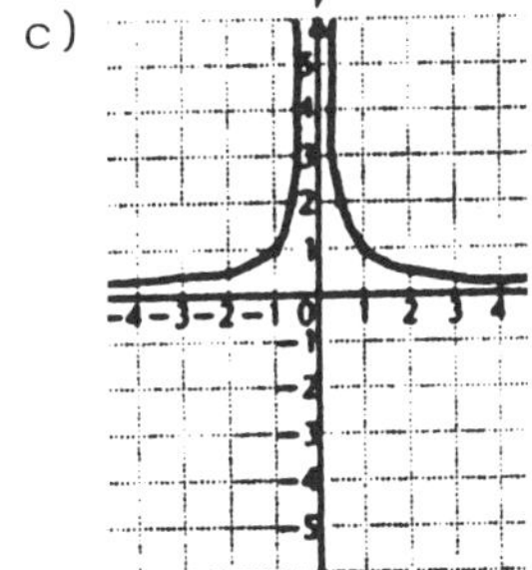

4. Find a formula for $f^{-1}(x)$: $f(x) = 6 - 2x$.

5. Find $h(h^{-1}(3))$: $h(x) = 35x^2 - 19x$.

6. Graph: $y = \log_8 x$.

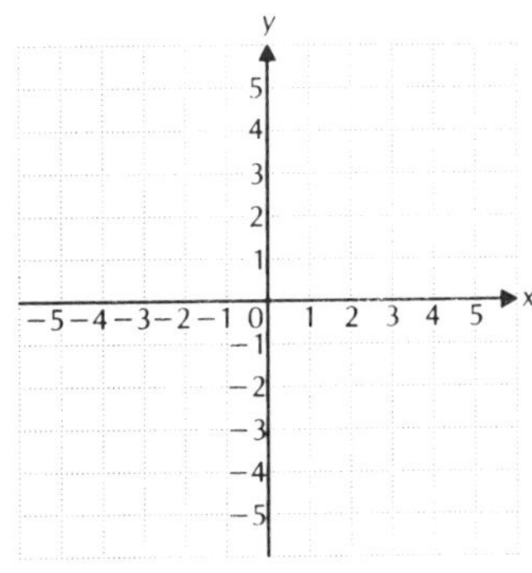

7. Graph: $f(x) = 3e^x$.

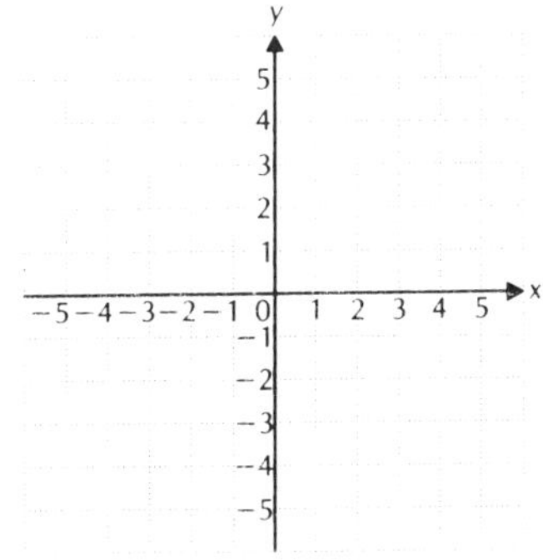

8. Find $\log_4 32$ using natural logarithms.

9. Write an exponential equation equivalent to $\log_{\sqrt{3}} 27 = 6$.

ANSWERS

1. ______________

2. ______________

3. ______________

4. ______________

5. ______________

6. See graph.

7. See graph.

8. ______________

9. ______________

NAME ______________________

TEST FORM A

ANSWERS

10. ______
11. ______
12. ______
13. ______
14. ______
15. ______
16. ______
17. ______
18. ______
19. ______
20. ______
21. ______

10. Write a logarithmic equation equivalent to

$$x^4 = 0.2401.$$

11. Simplify: $6^{\log_6 (x+2)}$.

12. Write an equivalent expression containing a single logarithm:

$$\frac{3}{4} \log_a x - \frac{1}{2} \log_a y + 5 \log_a z.$$

13. Express in terms of logarithms of a and b:

$$\log \frac{a^2}{b}.$$

Given that $\log_a 2 = 0.301$, $\log_a 7 = 0.845$, and $\log_a 6 = 0.778$, find each of the following.

14. $\log_a 3$

15. $\log_a 28$

16. $\log_a \sqrt[4]{6}$

17. Solve for x: $\log_d d^{3x^2} = 6$.

Solve.

18. $\log_4 x = 3$

19. $6^{2x+1} + 7 = 43$

20. $\log_5 (x + 4) + \log_5 x = 1$

21. $e^{-x} = 0.4$

TEST FORM A

Find using a calculator.

22. log (-1.04)

23. ln 0.000013

24. ln 2.05

25. log 0.04372

26. log 49,500

27. How many years will it take an investment of $3500 to double if interest is compounded annually at 10%?

28. An earthquake was measured at an intensity of $10^{6.34} \cdot I_0$. What was its magnitude on the Richter scale?

29. The population of a city was 60,000 in 1980 and 80,000 in 1990. Estimate the population in 2000.

30. The average walking speed R of people living in a city of population P, in thousands, is given by R = 0.37 ln P + 0.05, where R is in feet per second.

a) The population of Boston, Massachusetts, is 572,000. Find the average walking speed.

b) A city's population has an average walking speed of 3.1 ft/sec. Find the population.

ANSWERS

22. ______________

23. ______________

24. ______________

25. ______________

26. ______________

27. ______________

28. ______________

29. ______________

30. a) ______________

b) ______________

NAME ____________________

TEST FORM A

ANSWERS	
31. a) ______ b) ______ c) ______ d) ______	31. The cost of a Sweet ice cream bar in 1970 was 15 cents and was increasing at an exponential growth rate of 8.4%. a) Find an exponential function describing the growth of the cost of a Sweet bar. b) What will a Sweet bar cost in 1995? c) When will a Sweet bar cost \$3? d) What is the doubling time of the cost of a Sweet bar?
32. ______	32. The population of a city doubled in 20 years. What was the exponential growth rate?
33. ______	33. How long will it take an investment to double itself if it is invested at 6.8%, compounded continuously?
34. ______	34. How old is an animal bone that has lost 42% of its carbon-14?
35. ______	35. What is the loudness, in decibels, of a sound whose intensity is $550 \cdot I_0$?
36. ______	36. The hydrogen ion concentration of a substance is 2.4×10^{-5}. What is the pH?
37. ______	37. Determine whether $\pi^{4.2}$ or 4.2^{π} is larger?
38. ______	38. Solve: $5^{\log_5 (3x-7)} = 4$.

CHAPTER 5 *NAME* ____________________

TEST FORM B *CLASS* ________ *SCORE* ________ *GRADE* ________

1. Find the inverse of the relation H given by

   $H = \{(8,-2), (6,-7), (8,3), (-5,1), (5.6,-1.2)\}$.

2. Write an equation of the inverse of the relation

   $y = x^2 + 1$.

3. Which of the following have inverses that are functions?

a) 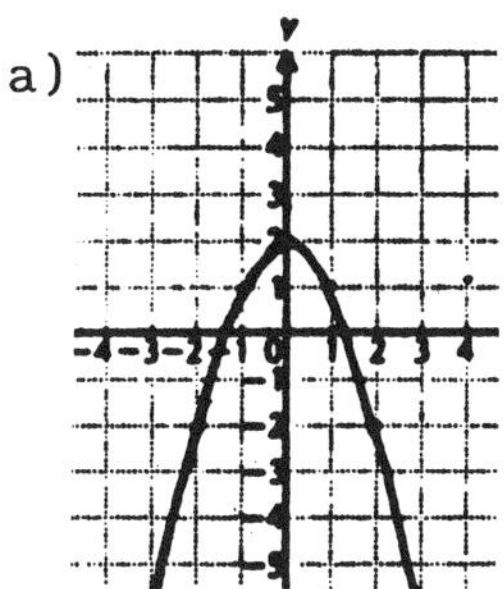

b) 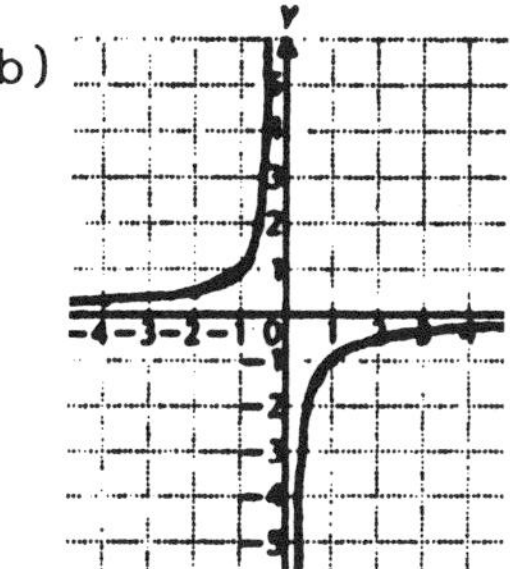

c) 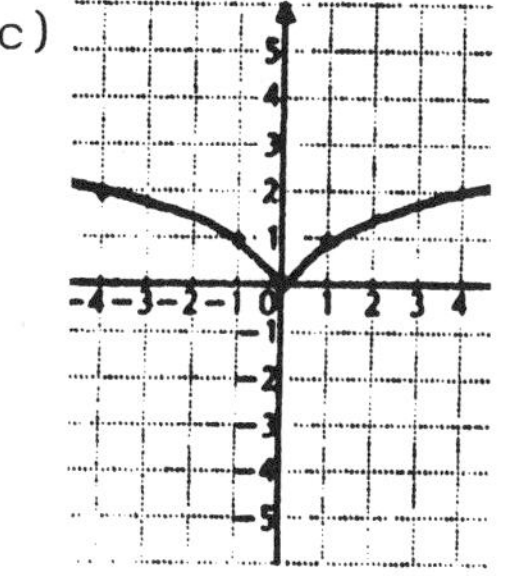

4. Find a formula for $f^{-1}(x)$: $f(x) = \dfrac{3}{x + 2}$.

5. Find $h(h^{-1}(-2))$: $h(x) = \dfrac{5x^2 - 1}{3}$.

6. Graph: $y = \log_2 x$.

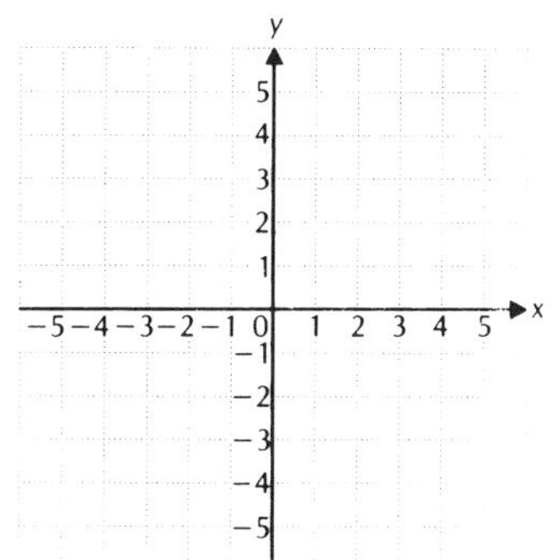

7. Graph: $f(x) = e^{x+2}$.

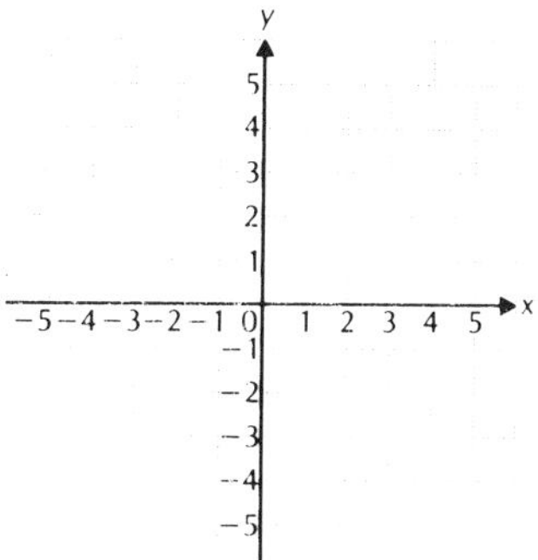

8. Find $\log_5 17$ using natural logarithms.

9. Write an exponential equation equivalent to $\log_{\sqrt{2}} 16 = 8$.

ANSWERS

1. ____________________

2. ____________________

3. ____________________

4. ____________________

5. ____________________

6. See graph.

7. See graph.

8. ____________________

9. ____________________

CHAPTER 5 NAME ______________________

TEST FORM B

ANSWERS	
10. ______	10. Write a logarithmic equation equivalent to $x^5 = 0.00243$.
11. ______	11. Simplify: $5^{\log_5 4x}$.
12. ______	12. Write an equivalent expression containing a single logarithm: $\frac{1}{2}\log_a x - 3\log_a y + 4\log_a z$.
13. ______	13. Express in terms of logarithms of x and y: $\log x^2\sqrt{y}$.
14. ______	
15. ______	Given that $\log_a 2 = 0.301$, $\log_a 5 = 0.699$, and $\log_a 8 = 0.903$, find each of the following.
16. ______	14. $\log_a 4$ 15. $\log_a 200$ 16. $\log_a \sqrt{2}$
17. ______	17. Solve for x: $\log_c c^{x^2+1} = 10$.
18. ______	Solve.
19. ______	18. $\log_6 x = 2$
20. ______	19. $5^{3x-1} - 2 = 23$
21. ______	20. $\log_2 x + \log_2 (x + 6) = 4$
	21. $e^x = 1.2$

NAME ______________________

TEST FORM B

Find using a calculator.

22. log 25,400

23. ln 4.92

24. ln 0.000051

25. log 0.00324

26. log (-13.62)

27. How many years will it take an investment of $5000 to double if interest is compounded annually at 6%?

28. What is the loudness, in decibels, of a sound whose intensity is $6{,}000 \cdot I_0$?

29. The population of a city was 50,000 in 1970 and 75,000 in 1985. Estimate the population in 2000.

30. A model for advertising response is given by $N(a) = 3000 + 200 \log a$, $a \geq 1$, where $N(a)$ = the number of units sold and a = the amount spent on advertising, in thousands of dollars.

    a) How many units were sold after spending $1000 (a = 1) on advertising?

    b) How many units were sold after spending $8000?

    c) How much would have to be spent in order to sell 4000 units?

ANSWERS

22. ______________

23. ______________

24. ______________

25. ______________

26. ______________

27. ______________

28. ______________

29. ______________

30. a) ______________

    b) ______________

    c) ______________

NAME ______________________

TEST FORM B

ANSWERS

31. a) ______
b) ______
c) ______

32. ______

33. ______

34. ______

35. ______

36. ______

37. a) ______
b) ______
c) ______
d) ______

38. ______

31. The population of Bosnia was 4.5 million in 1984. The exponential growth rate was 3.2% per year.

a) Find the exponential growth function.

b) Predict the population of Bosnia in the year 2000.

c) When will the population be 6.0 million?

32. The population of a city doubled in 34.2 years. What was the exponential growth rate?

33. How long will it take an investment to double itself if it is invested at 9.4%, compounded continuously?

34. How old is an animal bone that has lost 58% of its carbon-14?

35. An earthquake was measured at an intensity of $10^{2.54} \cdot I_0$. What was its magnitude on the Richter scale?

36. The pH of a substance is 6.1. What is the hydrogen ion concentration?

37. Approximate to six decimal places.

a) 5^3

b) $5^{3.14}$

c) $5^{3.141}$

d) $5^{3.1415}$

38. Simplify: $\dfrac{\log_8 16}{\log_8 2}$.

CHAPTER 5 NAME ______________________

TEST FORM C CLASS ________ SCORE ________ GRADE ________

1. Find the inverse of the relation J given by

   J = {(-1,3), (4,-7), (6,0), (2,5), (-1.3,7.6)}.

2. Write an equation of the inverse of the relation

   $y = |x + 3|$.

3. Which of the following have inverses that are functions?

   a) b) c)

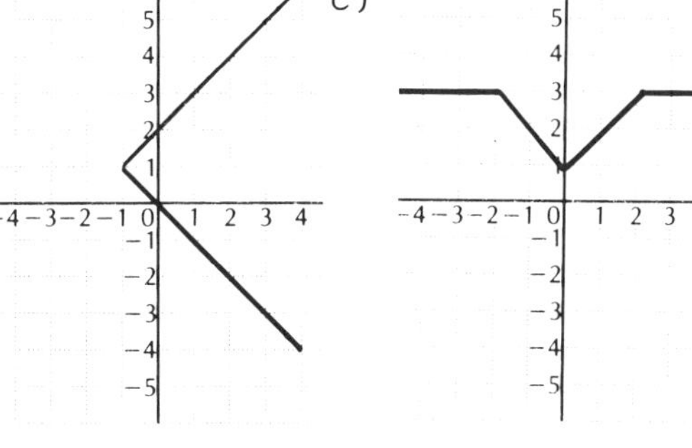

4. Find a formula for $f^{-1}(x)$: $f(x) = \dfrac{x + 2}{x - 1}$.

5. Find $h(h^{-1}(-2))$: $h(x) = \dfrac{25x - 7}{15}$.

6. Graph: $y = \log_5 x$.

7. Graph: $f(x) = 1.5e^x$.

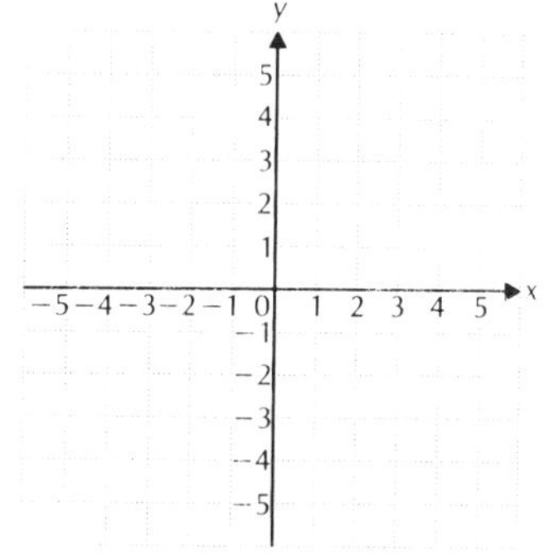

8. Find $\log_3 41$ using natural logarithms.

9. Write an exponential equation equivalent to $\log_{\sqrt{5}} 25 = 4$.

ANSWERS

1. ______________

2. ______________

3. ______________

4. ______________

5. ______________

6. See graph.

7. See graph.

8. ______________

9. ______________

NAME ____________________

TEST FORM C

ANSWERS

10. ____________
11. ____________
12. ____________
13. ____________
14. ____________
15. ____________
16. ____________
17. ____________
18. ____________
19. ____________
20. ____________
21. ____________

10. Write a logarithmic equation equivalent to

$$x^6 = 0.015625.$$

11. Simplify: $8^{\log_8 (x-1)}$.

12. Write an equivalent expression containing a single logarithm:

$$5 \log_a x + \frac{1}{3} \log_a y - 2 \log_a z.$$

13. Express in terms of logarithms of c and d:

$$\log 5cd^2.$$

Given that $\log_a 3 = 0.477$, $\log_a 4 = 0.602$, and $\log_a 6 = 0.778$, find each of the following.

14. $\log_a 2$

15. $\log_a 48$

16. $\log_a \sqrt[3]{6}$

17. Solve for x: $\log_b b^{x^2} = 4$.

Solve.

18. $\log_5 x = 3$

19. $4^{5x-1} - 1 = 63$

20. $\log_3 x + \log_3 (x - 8) = 2$

21. $e^x = 0.9$

NAME ______________________

TEST FORM C

Find using a calculator.

22. $\ln 0.0000021$

23. $\log 101{,}200$

24. $\log 0.13724$

25. $\log (-2.73)$

26. $\ln 5.12$

27. How many years will it take an investment of $2000 to double if interest is compounded annually at 8%?

28. An earthquake was measured at an intensity of $10^{5.17} \cdot I_0$. What was its magnitude on the Richter scale?

29. The population of a city was 100,000 in 1960 and 130,000 in 1970. Estimate the population in 2000.

30. It is known that $\frac{1}{4}$ of aluminum cans distributed will be recycled each year. A beverage company distributes 300,000 cans. The number still in use after time t, in years, is given by the function

$$N(t) = 300{,}000\left(\frac{1}{4}\right)^t.$$

a) After what amount of time will 50,000 still be in use?

b) After what amount of time will only 100 cans still be in use?

ANSWERS

22. ______

23. ______

24. ______

25. ______

26. ______

27. ______

28. ______

29. ______

30. a) ______

b) ______

NAME ____________________

TEST FORM C

ANSWERS	
31. a) ______ b) ______ c) ______	31. The cost of a first-class postage stamp became 3 cents in 1932, and the exponential growth rate of the cost was 3.8% per year. a) Find the exponential growth function. b) Predict the cost of a first-class postage stamp in the year 2010. c) When will the cost of the first-class postage stamp be \$1.50?
32. ______	32. The population of a city doubled in 18.4 years. What was the exponential growth rate?
33. ______	33. How long will it take an investment to double itself if it is invested at 10%, compounded continuously?
34. ______	34. How old is an animal bone that has lost 35% of its carbon-14?
35. ______	35. What is the loudness, in decibels, of a sound whose intensity is $4{,}350 \cdot I_0$?
36. ______	36. The pH of a substance is 4.9. What is the hydrogen ion concentration?
37. ______	37. Solve: $\log_{64} x = \frac{2}{3}$.
38. ______	38. Solve: $\log_4 \lvert x \rvert = 2$.

CHAPTER 5 NAME ____________________

TEST FORM D CLASS ________ SCORE ________ GRADE ______

1. Find the inverse of the relation K given by
   K = {(-4,3), (2,-5), (1,4), (6.2,4.3), (0,7)}.

2. Write an equation of the inverse of the relation
   xy = 3.

3. Which of the following have inverses that are functions?

a) 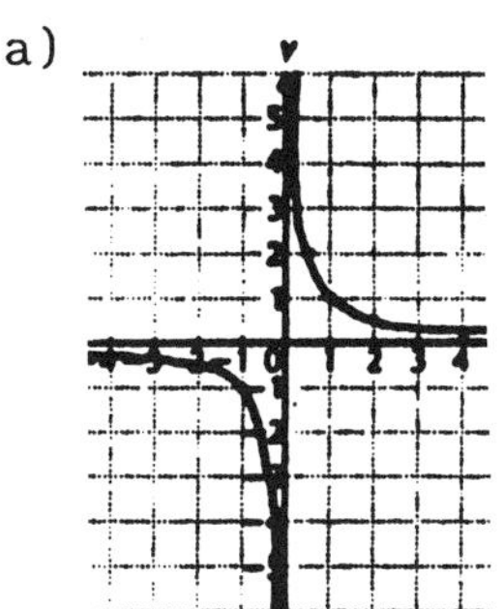

b) 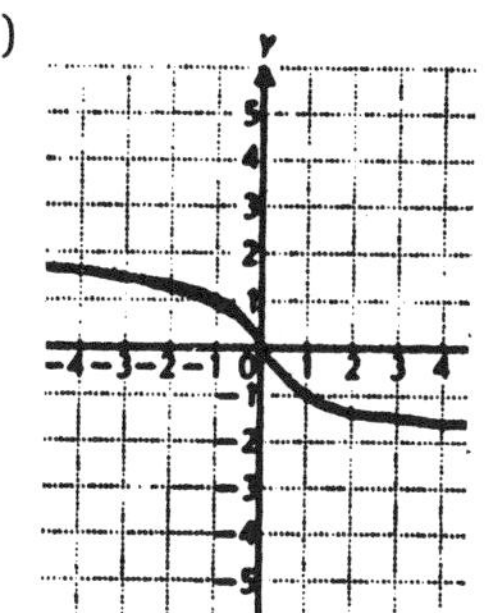

c) 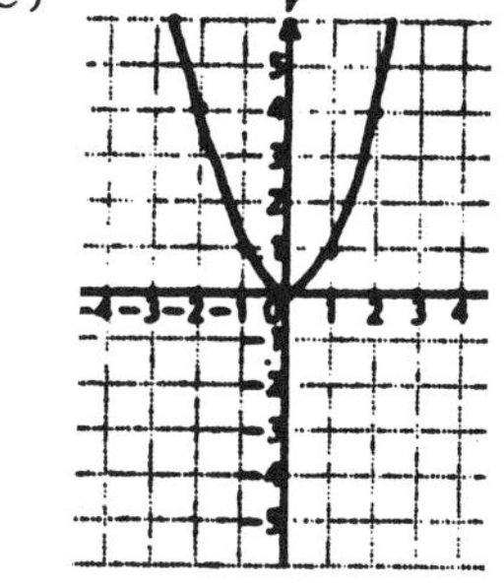

4. Find a formula for $f^{-1}(x)$: $f(x) = \sqrt{x + 3}$.

5. Find $h(h^{-1}(-7))$: $h(x) = \frac{x^2 - 7}{2x + 1}$.

6. Graph: $y = \log_9 x$.

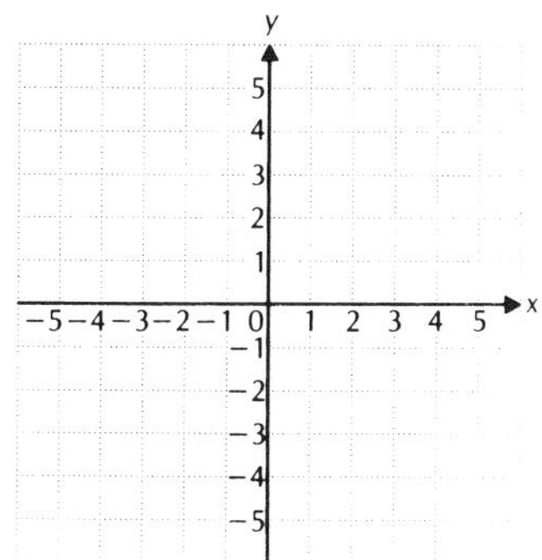

7. Graph: $f(x) = e^{2x-1}$.

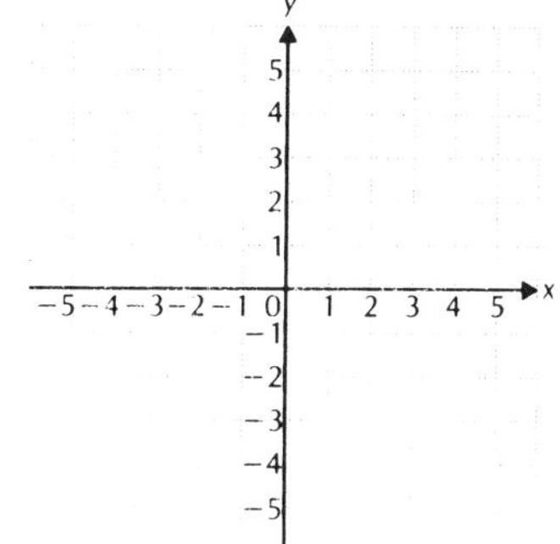

8. Find $\log_2 25$ using natural logarithms.

9. Write an exponential equation equivalent to $\log_{\sqrt{5}} 125 = 6$.

ANSWERS

1. ____________________

2. ____________________

3. ____________________

4. ____________________

5. ____________________

6. See graph.

7. See graph.

8. ____________________

9. ____________________

NAME ____________________

TEST FORM D

ANSWERS

10. ____________

11. ____________

12. ____________

13. ____________

14. ____________

15. ____________

16. ____________

17. ____________

18. ____________

19. ____________

20. ____________

21. ____________

10. Write a logarithmic equation equivalent to

$$x^5 = 0.00032.$$

11. Simplify: $7^{\log_7 5x}$.

12. Write an equivalent expression containing a single logarithm:

$$2 \log_a x + 4 \log_a y - \frac{2}{3} \log_a z.$$

13. Express in terms of logarithms of a and b:

$$\log \sqrt[3]{a^2 b}.$$

Given that $\log_a 4 = 0.602$, $\log_a 5 = 0.699$, and $\log_a 12 = 1.079$, find each of the following.

14. $\log_a 3$

15. $\log_a 100$

16. $\log_a \sqrt{12}$

17. Solve for x: $\log_a a^{3x^2} = 3$.

Solve.

18. $\log_2 x = 3$

19. $2^{3x-5} - 3 = 13$

20. $\log_4 x + \log_4 (x - 6) = 2$

21. $e^{-x} = 0.7$

NAME ______________________

TEST FORM D

Find using a calculator.

22. log 2.894

23. ln 6.23

24. log 17,350

25. log (-0.73)

26. ln 0.0000921

27. How many years will it take an investment of $8000 to double if interest is compounded annually at 11%?

28. What is the loudness, in decibels, of a sound whose intensity is $1{,}950I_0$?

29. The population of a city was 35,000 in 1960 and 42,000 in 1965. Estimate the population in 1990.

30. Students in a Spanish class took a final exam. They took equivalent forms of the exam in monthly intervals thereafter. The average score S(t), in percent, after t months was found to be given by

$$S(t) = 75 - 14 \log (t + 1), \quad t \geq 0.$$

a) What was the average score when they initially took the test, t = 0?

b) What was the average score after 6 months?

c) After what time t would the average score be 50?

ANSWERS

22. ______________

23. ______________

24. ______________

25. ______________

26. ______________

27. ______________

28. ______________

29. ______________

30. a) ______________

b) ______________

c) ______________

NAME ____________________

TEST FORM D

ANSWERS

31. a) ______

b) ______

c) ______

32. ______

33. ______

34. ______

35. ______

36. ______

37. ______

38. ______

31. Under ideal conditions, a population increase of rats has an exponential growth rate of 18.2% per day. Suppose one starts with a population of 100 rats.

a) Find the exponential growth function.

b) What will the population of rats be after 10 days?

c) What is the doubling time of the population of rats?

32. The population of a city doubled in 36 years. What was the exponential growth rate?

33. How long will it take an investment to double itself if it is invested at 8.7%, compounded continuously?

34. How old is an animal bone that has lost 24% of its carbon-14?

35. An earthquake was measured at an intensity of $10^{3.91} \cdot I_0$. What was its magnitude on the Richter scale?

36. The hydrogen ion concentration of a substance is 1.9×10^{-6}. What is the pH?

37. Solve: $\log_{\sqrt{7}} x = -3$.

38. Find the domain: $f(x) = \dfrac{1}{\sqrt{3 \ln x - 4}}$.

NAME ____________________

TEST FORM E

CLASS ______ SCORE ______ GRADE ______

1. Which of the following relations have inverses that are functions?

a)

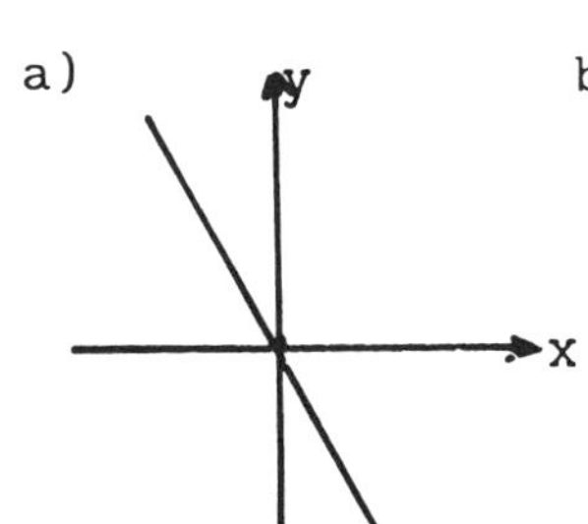

b)

c)

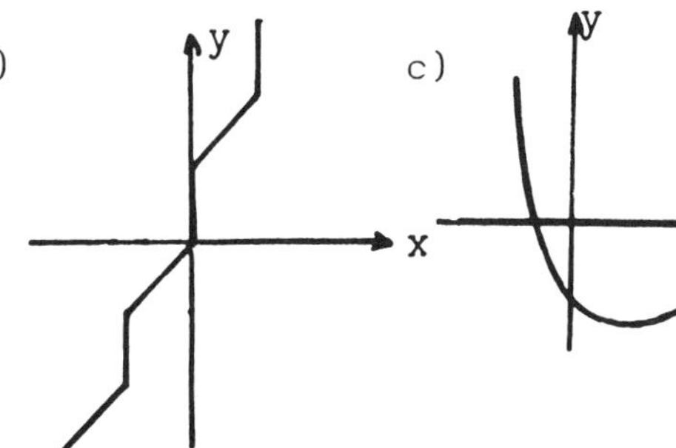

Find a formula for $f^{-1}(x)$.

2. $f(x) = x^2 + 3$

3. $f(x) = 5 + \frac{\sqrt{x}}{4}$

4. $h(x) = \frac{-5 - 2x}{4}$. Find $h^{-1}(h(6))$.

5. Graph: $y = \log_2 (3x - 1)$.

6. Graph: $f(x) = e^{-3x} - 2$ for nonnegative values of x.

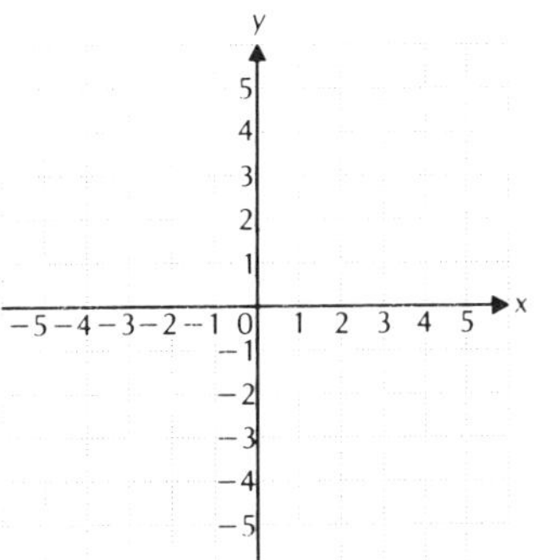

7. Write an exponential equation equivalent to $\log_{\sqrt{8}} 64 = 4$.

8. Write a logarithmic equation equivalent to $a^{-1/3} = 0.00143$.

Simplify.

9. $9^{\log_9 (5x+2)}$

10. $\log_7 7^{x^3}$

11. Find $\log_{12} 58$ using common logarithms.

ANSWERS

1. ____________________
2. ____________________
3. ____________________
4. ____________________
5. See graph.
6. See graph.
7. ____________________
8. ____________________
9. ____________________
10. ____________________
11. ____________________

CHAPTER 5 NAME ____________________

TEST FORM E

ANSWERS	
12. ______	12. Solve for x: $\log_a a^{x^2-2x} = 8$.
13. ______	13. Write an equivalent expression containing a single logarithm. $$7 \log_a x - \frac{1}{3} \log_a y + \frac{3}{4} \log_a z$$
14. ______	14. Find the domain: $f(x) = \log_8 (\log_5 x)$.
15. ______	Solve.
16. ______	15. $\log_{27} x = -\frac{1}{3}$ 16. $16^x = 64^{5x-2}$
17. ______	17. $\log_2 (x + 2) + \log_2 (x - 4) = 4$
18. ______	18. $\log_5 \sqrt{x^2 - 15} = 2$
19. ______	Find using a calculator.
20. ______	19. ln 5.14 20. ln 0.000149
21. ______	21. log (-2.74) 22. log 75,315
22. ______	Given that $\log_a 2 = 0.301$, $\log_a 12 = 1.079$, and $\log_a 7 = 0.845$, find each of the following.
23. ______	23. $\log_a \frac{7}{2}$ 24. $\log_a \sqrt{24}$ 25. $\log_a 42$
24. ______	26. An investment of $5000 doubles in 6 years. What is the interest rate if interest is compounded
25. ______	a) annually? b) continuously?
26. a) ______	27. The population of a city was 100,000 in 1960, and the exponential growth rate is 1.5% per year.
b) ______	a) What will the population be in 2000? b) When will the population be 250,000?
27. a) ______	28. How old is an animal bone which has lost 34% of its carbon-14?
b) ______	29. The average walking speed R of people living in a city of population P, in thousands, is given by $R = 0.37 \ln P + 0.05$, where R is in feet per second.
28. ______	a) The population of San Francisco is 705,300. Find the average walking speed.
29. a) ______	b) A city's population has an average walking speed of 1.7 ft/sec. Find the population.
b) ______	

CHAPTER 5 NAME ______________

TEST FORM F CLASS ________ SCORE ________ GRADE ______

Write the letter of your response on the answer blank.

ANSWERS

1. Which of the following relations have inverses that are functions?

a)

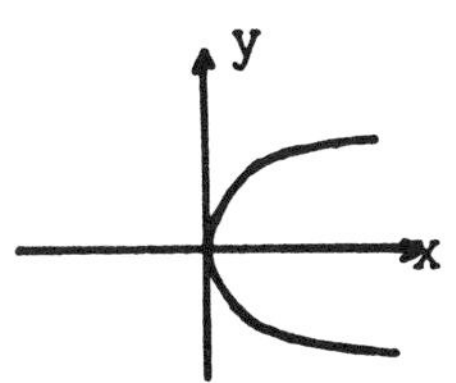

b)

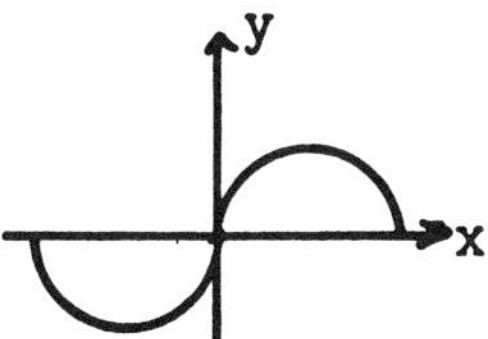

c)

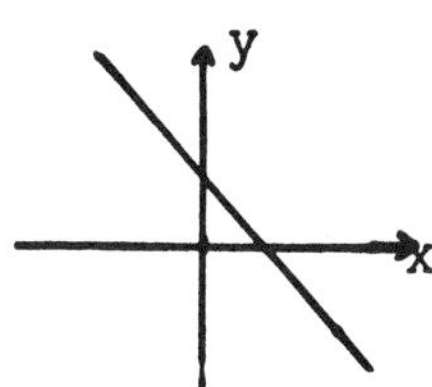

a) a, b b) a, c c) b, c d) b only

1. ______________

2. $h(x) = 3(x^2 + 7) - 2x$. Find $h(h^{-1}(9))$.

a) 2 b) 58 c) -9 d) 9

2. ______________

3. $f(x) = \sqrt{2x - 3}$. Find a formula for $f^{-1}(x)$.

a) $f^{-1}(x) = 4x^2 + 3$ b) $f^{-1}(x) = 2x - 3$

c) $f^{-1}(x) = \frac{x^2 + 3}{2}$ d) $f^{-1}(x) = \sqrt{2x + 3}$

3. ______________

CHAPTER 5 *NAME* ______________________________

TEST FORM F

ANSWERS

4. ____________

5. ____________

6. ____________

4. Graph $y = \log_2 (x + 1)$.

a)

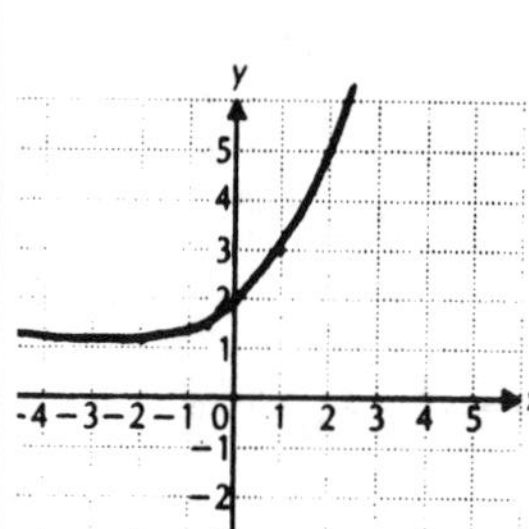

b)

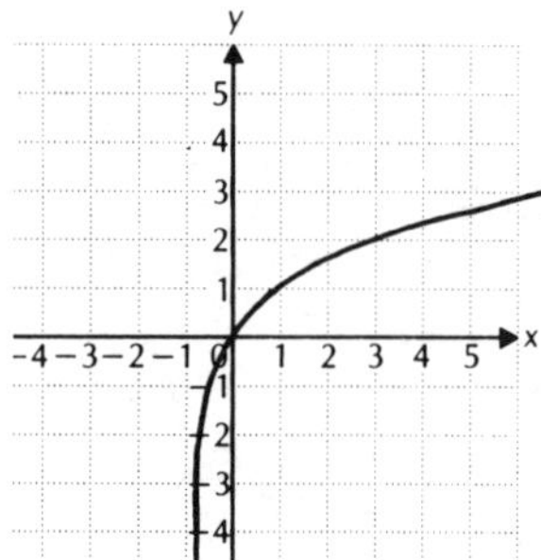

c)

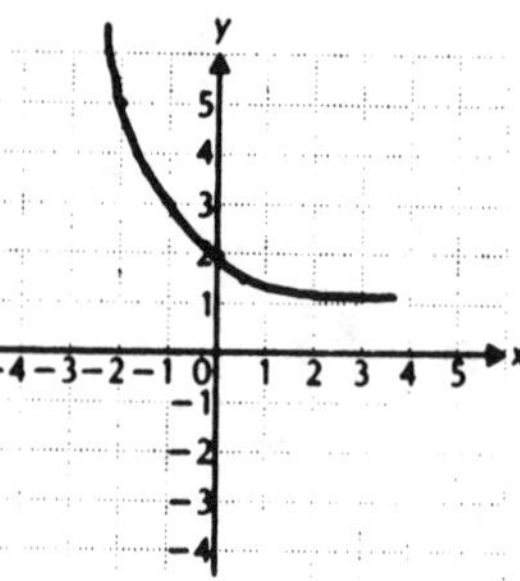

5. Graph $f(x) = 1 - e^{-0.5x}$, for nonnegative values of x.

a)

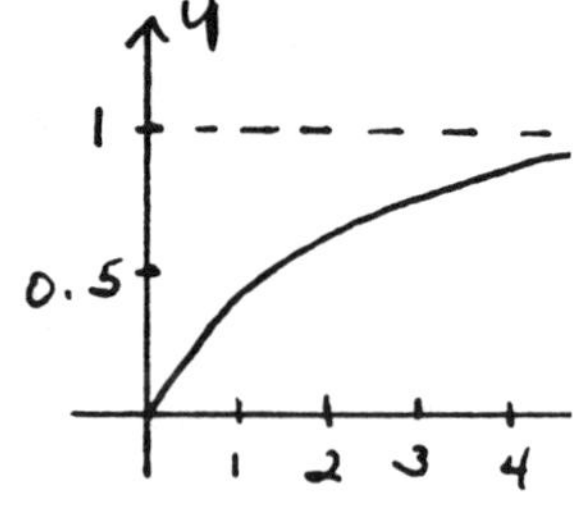

b)

c)

6. Find $\log_{0.5} 6$ using natural logarithms.

a) 1.5563 b) -1.4205 c) -2.5850 d) 0.3802

NAME ____________________

CLASS ______ SCORE ________ GRADE ______

ANSWERS

7. Write an exponential equation equivalent to $\log_{0.2} 125 = -3$.

a) $125^{0.2} = -3$ b) $10^{0.2} = 125^{-3}$

c) $125^{-3} = 0.2$ d) $(0.2)^{-3} = 125$

7. ____________

8. Write a logarithmic equation equivalent to $4.5^b = a$.

a) $\log_a b = 4.5$ b) $\log_{4.5} a = b$

c) $\log_{4.5} b = a$ d) $\log_b 4.5 = a$

8. ____________

9. Write an equivalent expression containing a single logarithm and simplify.

$$\log_a 4x + \log_a 3x - \log_a 3 - 2 \log_a 2x$$

a) $\dfrac{\log_a 6x}{\log_a (3 + x^2)}$ b) $\log_a \dfrac{6x}{3 + x^2}$

c) 0 d) 1

9. ____________

10. Simplify: $\log_6 6^{2x+3}$.

a) $-\frac{3}{2}$ b) 6 c) $2x + 3$ d) $\frac{7}{3}$

10. ____________

11. Simplify: $7^{\log_7 (x^2-9)}$.

a) $x^2 - 9$ b) 3 c) 7 d) 7^{x^2-9}

11. ____________

12. Solve for x: $\log_b b^{5x^2-2} = 9x$.

a) -2 b) $\frac{1}{5}$ c) $\frac{2}{5}$ d) $-\frac{1}{5}$, 2

12. ____________

13. Solve: $\log_{49} x = -\frac{1}{2}$.

a) 7 b) $\frac{1}{7}$ c) $-\frac{1}{7}$ d) -24.5

13. ____________

14. Solve: $3^{4x+1} - 5 = 22$.

a) $-\frac{1}{4}$ b) $\frac{3}{4}$ c) $\frac{1}{2}$ d) -2

14. ____________

NAME ______________________

TEST FORM F

ANSWERS

15. ________
16. ________
17. ________
18. ________
19. ________
20. ________
21. ________
22. ________
23. ________
24. ________

15. Solve: $\log_x (\log_3 27) = 2$.

a) 3 b) $\frac{3}{2}$ c) $\sqrt{2}$ d) $\sqrt{3}$

16. Solve: $\log_5 (x + 1) - \log_5 x = 2$.

a) 1 b) $\frac{1}{24}$ c) $\frac{2}{5}$ d) 5

17. If $\log_a x = 2$, $\log_a y = 3$, and $\log_a z = 4$, what is

$$\log_a \frac{\sqrt[3]{xz}}{\sqrt[3]{y^2z^{-2}}}?$$

a) $\frac{5}{2}$ b) $\frac{8}{3}$ c) $\frac{10}{3}$ d) $\frac{14}{3}$

18. Using a calculator, find log 0.0000517.

a) -4.2865 b) -4.7570 c) -4.1457 d) -3.2865

19. Using a calculator, find log 257,301.

a) 5.3751 b) 13.1755 c) 5.4104 d) 12.4580

20. Using a calculator, find ln 0.00025.

a) -3.6021 b) -8.2940 c) -3.2840 d) -5.9915

21. How many years will it take an investment of $20,000 to double if interest is compounded continuously at 12.5%? (Find answer to nearest tenth.)

a) 0.3 b) 4.8 c) 5.3 d) 5.5

22. What is the loudness, in decibels, of a sound whose intensity is 15,000 times I_0?

a) 42 b) 96 c) 4.2 d) 17,500

23. An earthquake had an intensity of $10^{6.45} \cdot I_0$. What was its magnitude on the Richter scale?

a) 0.8096 b) 1.8641 c) 6.45 d) 64.5

24. The hydrogen ion concentration of a substance is 3.7×10^{-4}. What is the pH?

a) 0.5 b) 3.4 c) 5.2 d) 1.9×10^{-4}

CHAPTER 6 NAME ______________________

TEST FORM A CLASS ________ SCORE __________ GRADE ______

ANSWERS

1. The point $\left(-\frac{2}{3}, \frac{\sqrt{5}}{3}\right)$ is on a unit circle. Find the coordinates of its reflection across the origin, the u-axis, and the v-axis.

1. ______________

2. On a unit circle, mark and label the points determined by $\frac{\pi}{3}$, $-\frac{9\pi}{2}$, and $\frac{11\pi}{4}$.

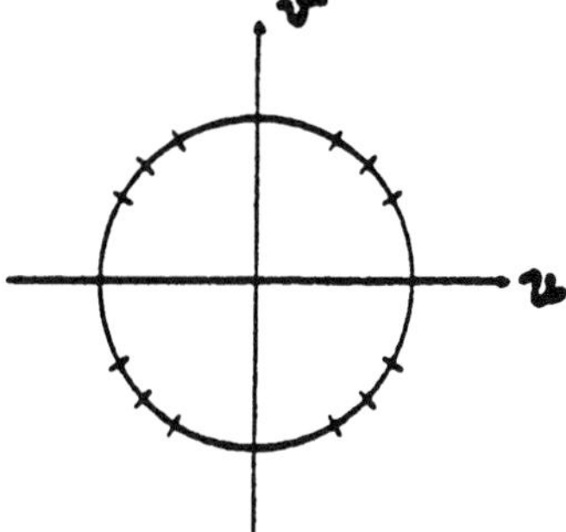

2. See circle.

3. Sketch a graph of y = sin x.

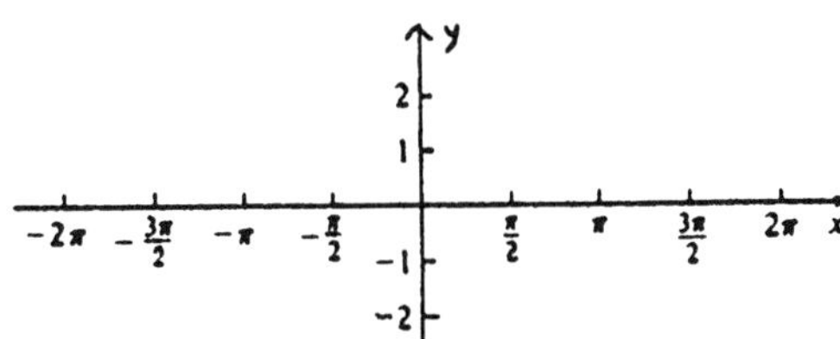

3. See graph.

4. What is the range of the sine function?

4. ______________

5. What is the period of the sine function?

5. ______________

6. What is the domain of the cosine function?

6. ______________

7. What is the amplitude of the cosine function?

7. ______________

8. Complete the following table.

x	$\frac{\pi}{3}$	$\frac{7\pi}{4}$	$-\frac{3\pi}{2}$
sin x			
cos x			

8. See table.

CHAPTER 6 *NAME* ______________________

TEST FORM A

ANSWERS

9. See graph.

10. ______

11. ______

12. ______

13. ______

14. ______

15. ______

16. ______

17. ______

18. ______

19. ______

20. ______

9. Sketch a graph of the tangent function.

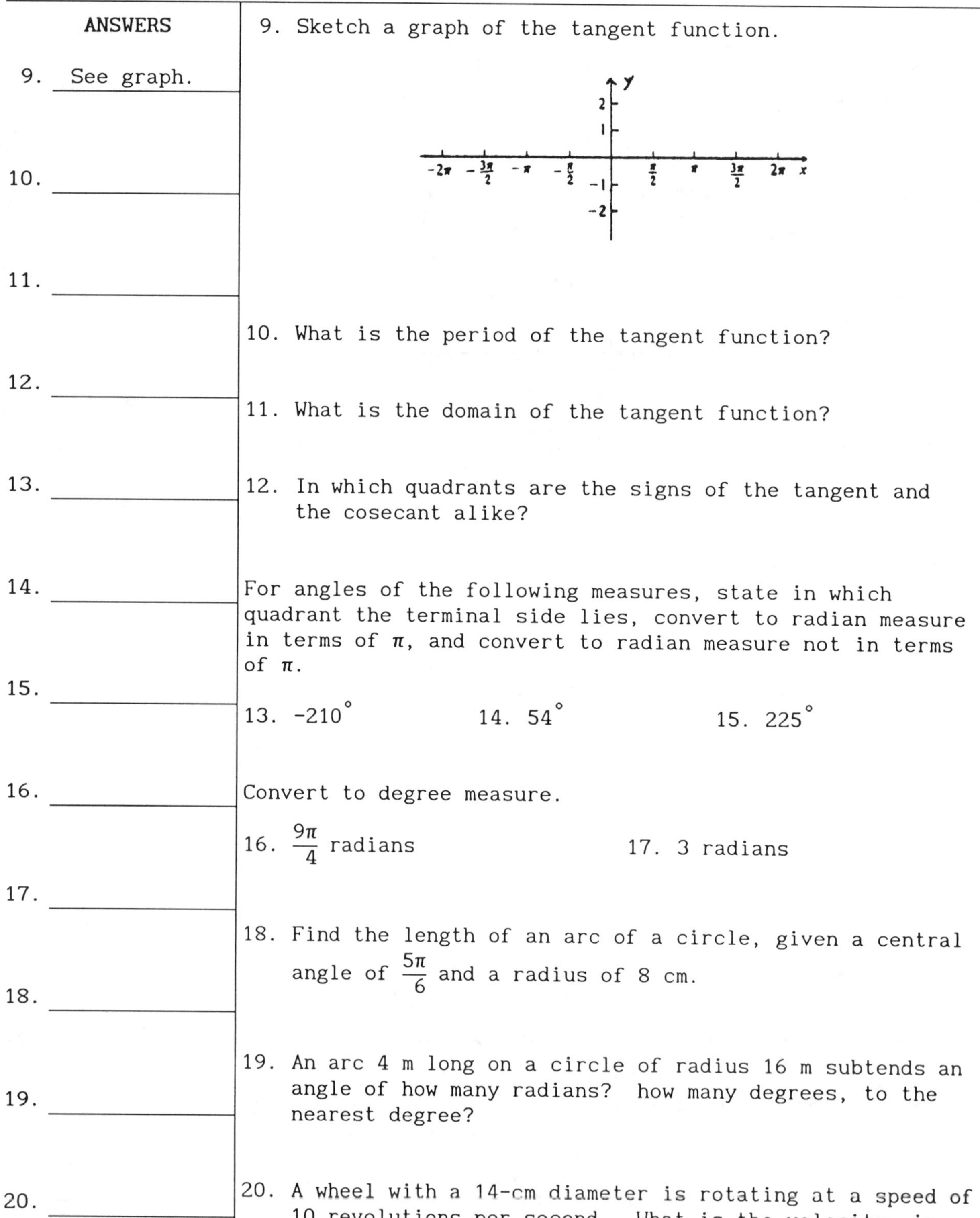

10. What is the period of the tangent function?

11. What is the domain of the tangent function?

12. In which quadrants are the signs of the tangent and the cosecant alike?

For angles of the following measures, state in which quadrant the terminal side lies, convert to radian measure in terms of π, and convert to radian measure not in terms of π.

13. -210° 14. 54° 15. 225°

Convert to degree measure.

16. $\frac{9\pi}{4}$ radians 17. 3 radians

18. Find the length of an arc of a circle, given a central angle of $\frac{5\pi}{6}$ and a radius of 8 cm.

19. An arc 4 m long on a circle of radius 16 m subtends an angle of how many radians? how many degrees, to the nearest degree?

20. A wheel with a 14-cm diameter is rotating at a speed of 10 revolutions per second. What is the velocity, in cm/sec, of a point on the rim?

CHAPTER 6 NAME ____________________

TEST FORM A CLASS ________ SCORE ________ GRADE ______

21. The old oaken bucket is being raised at a rate of 4 ft/sec. The radius of the drum is 6 in. What is the angular speed, in radians/sec, of the handle?

22. Find the six trigonometric function values for the angle θ shown.

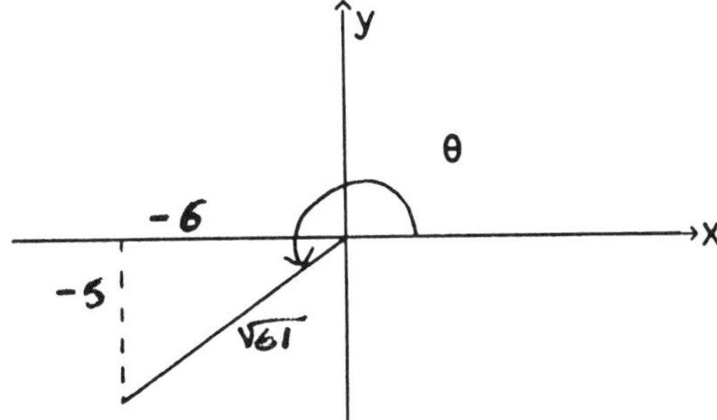

Find the following exactly. Do <u>not</u> use a calculator.

23. $\tan 480^\circ$ 24. $\sin(-120^\circ)$ 25. $\cos 240^\circ$

26. Given that $\sin\theta = \frac{1}{3}$ and that the terminal side is in quadrant II, find the other five function values.

27. Convert 28.35° to degrees and minutes.

28. Convert $89^\circ 36'$ to degrees and decimal parts of degrees.

Use a calculator to find each of the following.

29. $\sin(-153^\circ 15' 17'')$ 30. $\cos 1.115$

31. $\tan\left(-\frac{17\pi}{3}\right)$ 32. $\sec 19^\circ 15'$

33. $\csc 3.5124$ 34. $\cot(1573^\circ)$

ANSWERS

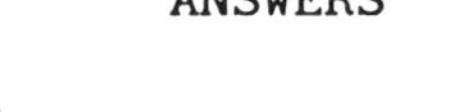

21. ____________

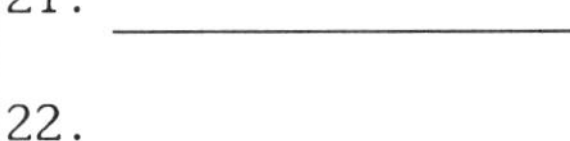

22. ____________

____

____

____

____

____

23. ____________

24. ____________

25. ____________

26. ____________

____

____

____

____

27. ____________

28. ____________

29. ____________

30. ____________

31. ____________

32. ____________

33. ____________

34. ____________

NAME ______________________________

TEST FORM A

ANSWERS

35. ______________

36. ______________

37. ______________

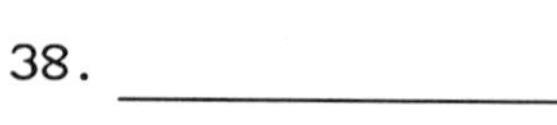

38. ______________

39. See graph.

40. ______________

41. ______________

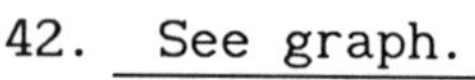

42. See graph.

43. ______________

44. ______________

Use a calculator to find the acute angle θ that is a solution of the equation. Give the answer in degrees and decimal parts of degrees, as well as in radians.

35. $\cos \theta = 0.5324$

36. $\csc \theta = 2.1538$

Use a calculator to find θ in the interval indicated.

37. $\sin \theta = 0.5932$, $(90^\circ, 180^\circ)$

38. $\tan \theta = -5.321$, $\left(-\frac{\pi}{2}, \frac{\pi}{2}\right)$

39. Sketch a graph of $y = 2 \sin (x + \pi)$.

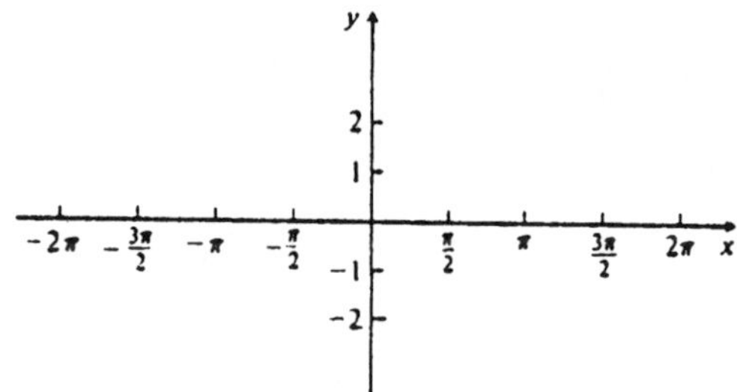

40. What is the period of the function in Question 39?

41. What is the phase shift of the function in Question 39?

42. Sketch the graph of $y = 2 \sin x + \cos x$ for values between 0 and 2π.

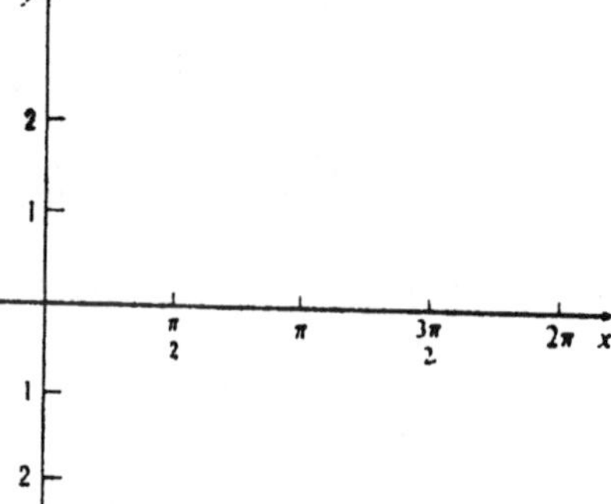

43. A point on the unit circle has a u-coordinate of $-\frac{1}{4}$. What is its v-coordinate?

44. Does $8 \cos x = 10$ have a solution for x? Why or why not?

NAME ______________________

TEST FORM B

CLASS ________ SCORE __________ GRADE ______

1. The point $\left(-\frac{1}{2}, -\frac{\sqrt{3}}{2}\right)$ is on a unit circle. Find the coordinates of its reflection across the origin, the u-axis, and the v-axis.

2. On a unit circle, mark and label the points determined by $\frac{3\pi}{4}$, $-\frac{5\pi}{2}$, and $\frac{13\pi}{6}$.

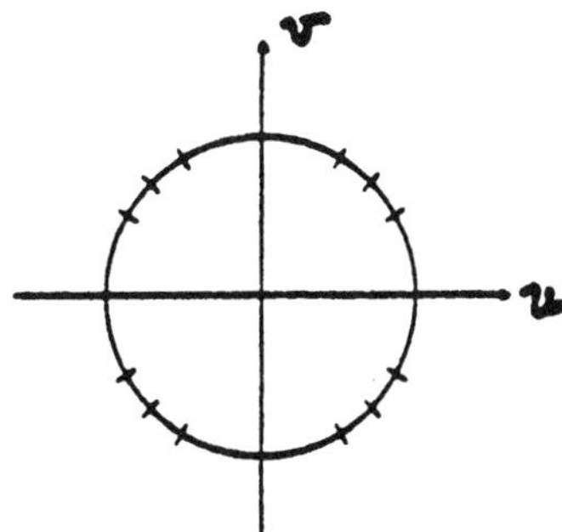

3. Sketch a graph of y = cos x.

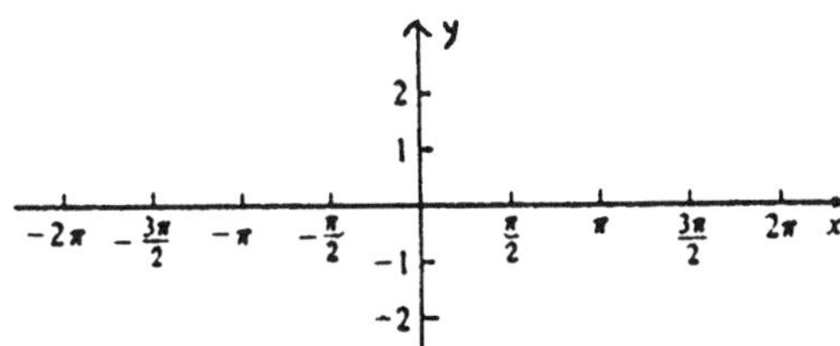

4. What is the range of the cosine function?

5. What is the period of the cosine function?

6. What is the domain of the sine function?

7. What is the amplitude of the sine function?

8. Complete the following table.

x	$\frac{3\pi}{4}$	$\frac{5\pi}{6}$	$-\frac{\pi}{2}$
sin x			
cos x			

ANSWERS

1. ______________

2. See circle.

3. See graph.

4. ______________

5. ______________

6. ______________

7. ______________

8. See table.

CHAPTER 6 *NAME* ____________________

TEST FORM B

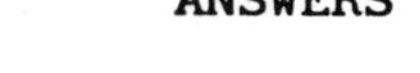

ANSWERS

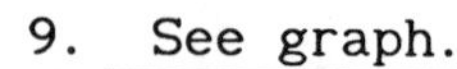

9. See graph.
10. __________
11. __________
12. __________
13. __________
14. __________
15. __________
16. __________
17. __________
18. __________
19. __________
20. __________

9. Sketch a graph of the cosecant function.

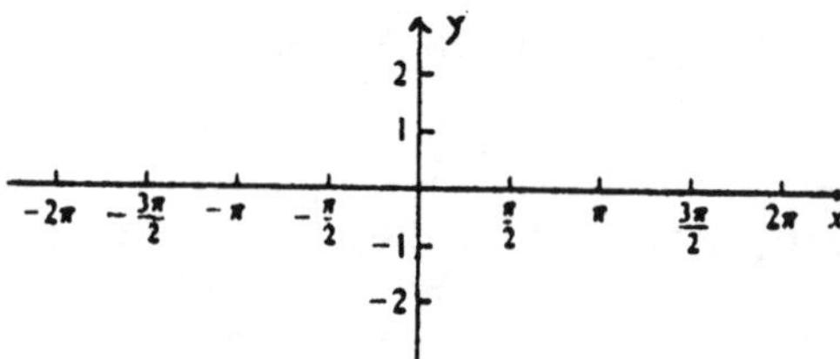

10. What is the period of the cosecant function?

11. What is the domain of the cosecant function?

12. In which quadrants are the signs of the cosine and the cotangent alike?

For angles of the following measures, state in which quadrant the terminal side lies, convert to radian measure in terms of π, and convert to radian measure not in terms of π.

13. 120° 14. 123° 15. 315°

Convert to degree measure.

16. $\frac{7\pi}{6}$ radians 17. 4 radians

18. Find the length of an arc of a circle, given a central angle of $\frac{3\pi}{4}$ and a radius of 10 cm.

19. An arc 10 cm long on a circle of radius 25 cm subtends an angle of how many radians? how many degrees, to the nearest degree?

20. A rock on a 5-ft string is rotated at 30 rpm. What is its linear speed, in ft/min?

CHAPTER 6 NAME ______________________

TEST FORM B CLASS ________ SCORE ________ GRADE ________

ANSWERS

21. ________

22. ________

____

____

____

____

____

23. ________

24. ________

25. ________

26. ________

____

____

____

____

27. ________

28. ________

29. ________

30. ________

31. ________

32. ________

33. ________

34. ________

21. A horse on a merry-go-round is 9 m from the center and travels at a speed of 12 km/h. What is its angular speed, in radians/h?

22. Find the six trigonometric function values for the angle θ shown.

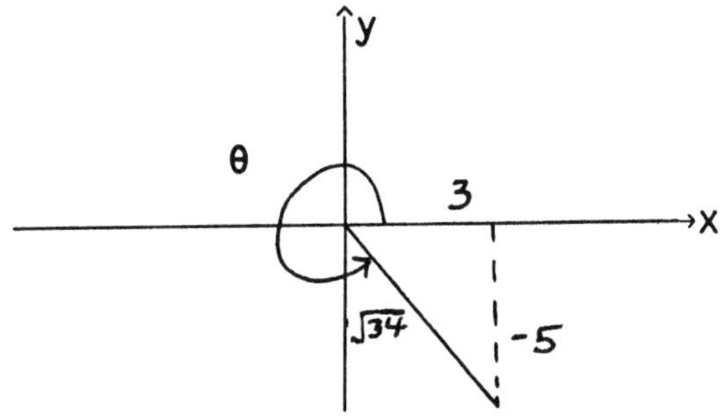

Find the following exactly. Do not use a calculator.

23. $\tan 510^\circ$ 24. $\sin(-225^\circ)$ 25. $\cos 330^\circ$

26. Given that $\cos\theta = \frac{1}{2}$ and that the terminal side is in quadrant IV, find the other five function values.

27. Convert 57.45° to degrees and minutes.

28. Convert $17^\circ 15'$ to degrees and decimal parts of degrees.

Use a calculator to find each of the following.

29. $\cos(89^\circ 12' 15'')$

30. $\tan 1.453$

31. $\sin\left(-\frac{14\pi}{5}\right)$

32. $\csc 37^\circ 12'$

33. $\cot 2.8915$

34. $\sec(2095^\circ)$

CHAPTER 6 *NAME* ______________________

TEST FORM B

ANSWERS

35. ______________

36. ______________

37. ______________

38. ______________

39. See graph.

40. ______________

41. ______________

42. See graph.

43. ______________

44. ______________

Use a calculator to find the acute angle θ that is a solution of the equation. Give the answer in degrees and decimal parts of degrees, as well as in radians.

35. $\tan \theta = 1.7502$

36. $\sec \theta = 5$

Use a calculator to find θ in the interval indicated.

37. $\cos \theta = 0.9735$, $(270^\circ, 360^\circ)$

38. $\sin \theta = 0.593$, $\left(\frac{\pi}{2}, \pi\right)$

39. Sketch a graph of $y = 3 \sin \left(2x + \frac{\pi}{2}\right)$.

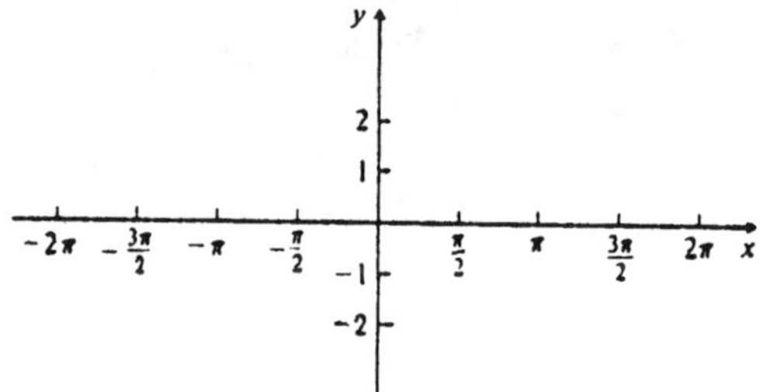

40. What is the period of the function in Question 39?

41. What is the phase shift of the function in Question 39?

42. Sketch the graph of $y = \cos x + \sin 2x$ for values between 0 and 2π.

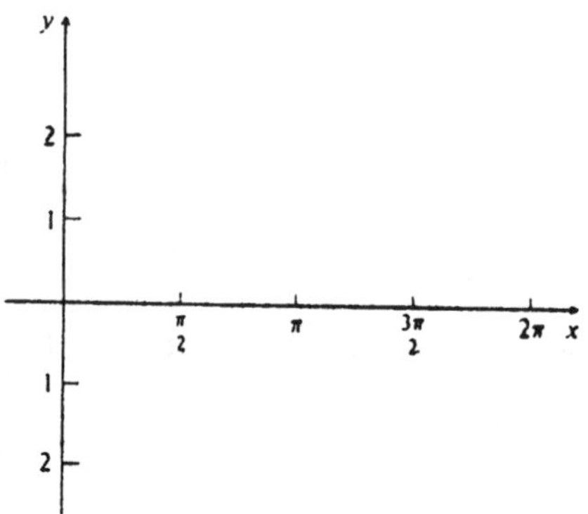

43. A grad is a unit of angle measure similar to a degree. A right angle has a measure of 100 grads. Convert 75° and $\frac{3\pi}{4}$ radians to grads.

44. Given that $\tan x = 3.2955$ and the terminal side is in quadrant III, find $\sin x$ and $\cos x$.

CHAPTER 6 NAME ______________________

TEST FORM C CLASS ________ SCORE ________ GRADE ______

ANSWERS

1. The point $\left(\frac{4}{7}, -\frac{\sqrt{33}}{7}\right)$ is on a unit circle. Find the coordinates of its reflection across the origin, the u-axis, and the v-axis.

2. On a unit circle, mark and label the points determined by $\frac{7\pi}{6}$, $-\frac{11\pi}{2}$, and $\frac{17\pi}{4}$.

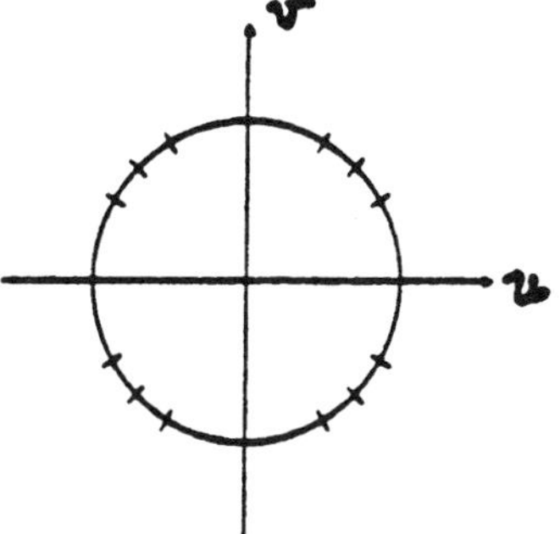

3. Sketch a graph of y = sin x.

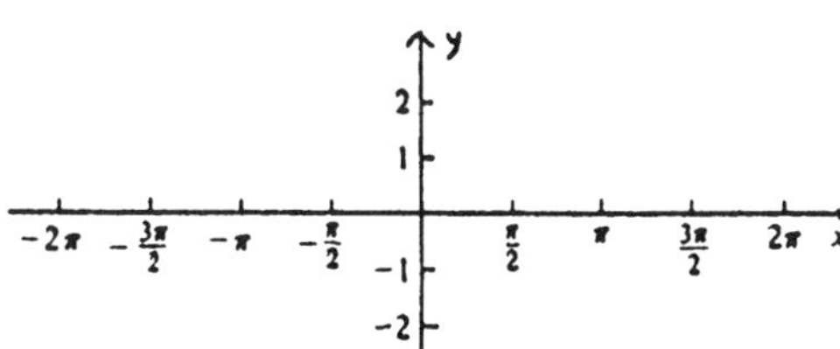

4. What is the range of the sine function?

5. What is the period of the sine function?

6. What is the domain of the cosine function?

7. What is the amplitude of the cosine function?

8. Complete the following table.

x	$\frac{\pi}{4}$	$\frac{7\pi}{6}$	$-\frac{5\pi}{2}$
sin x			
cos x			

1. ______________

2. See circle.

3. See graph.

4. ______________

5. ______________

6. ______________

7. ______________

8. See table.

NAME ______________________

TEST FORM C

ANSWERS	
9. See graph.	9. Sketch a graph of the cotangent function.
10. ______	10. What is the period of the cotangent function?
11. ______	11. What is the domain of the cotangent function?
12. ______	12. In which quadrants are the signs of the sine and the cosine alike?
13. ______	For angles of the following measures, state in which quadrant the terminal side lies, convert to radian measure in terms of π, and convert to radian measure not in terms of π.
14. ______	13. -300°
15. ______	14. 192° 15. 45°
16. ______	Convert to degree measure. 16. $\frac{5\pi}{3}$ radians
17. ______	17. 5 radians
18. ______	18. Find the length of an arc of a circle, given a central angle of $\frac{2\pi}{3}$ and a radius of 8 cm.
19. ______	19. An arc 12 ft long on a circle of radius 15 ft subtends an angle of how many radians? How many degrees, to the nearest degree?
20. ______	20. A 45-rpm record has a radius of 8 cm. What is the linear velocity of a point on the rim, in cm/min?

Graph axes for problem 9: y-axis marked 2, 1, −1, −2; x-axis marked -2π, $-\frac{3\pi}{2}$, $-\pi$, $-\frac{\pi}{2}$, $\frac{\pi}{2}$, π, $\frac{3\pi}{2}$, 2π.

CHAPTER 6 NAME ____________

TEST FORM C CLASS ______ SCORE ______ GRADE ______

ANSWERS

21. A wheel has a 36-cm diameter. The speed of a point on its rim is 11 m/sec. What is its angular speed, in radians/sec?

22. Find the six trigonometric function values for the angle θ shown.

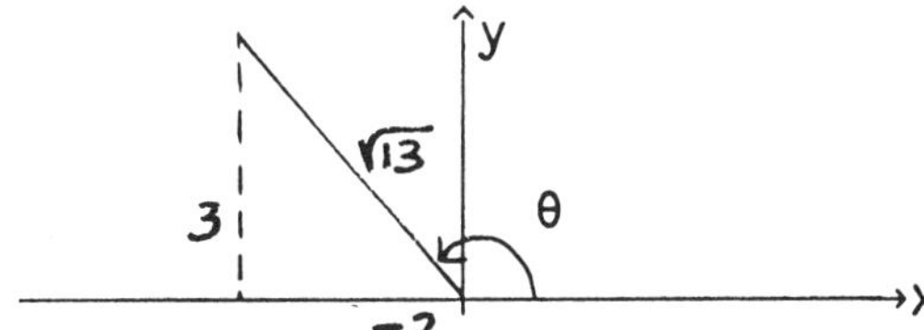

Find the following exactly. Do not use a calculator.

23. tan 405° 24. sin (-210°) 25. cos 135°

26. Given that $\tan \theta = \frac{3}{4}$ and that the terminal side is in quadrant III, find the other five function values.

27. Convert 101.75° to degrees and minutes.

28. Convert 124°54′ to degrees and decimal parts of degrees.

Use a calculator to find each of the following.

29. tan (-101°29′37″) 30. cos 1.832

31. $\sin\left(-\frac{15\pi}{4}\right)$ 32. cot 85°23′

33. sec 3.0726 34. csc (4310°)

21. ____
22. ____
23. ____
24. ____
25. ____
26. ____
27. ____
28. ____
29. ____
30. ____
31. ____
32. ____
33. ____
34. ____

NAME ______________________

TEST FORM C

ANSWERS

35. ____________

36. ____________

37. ____________

38. ____________

39. See graph.

40. ____________

41. ____________

42. See graph.

43. ____________

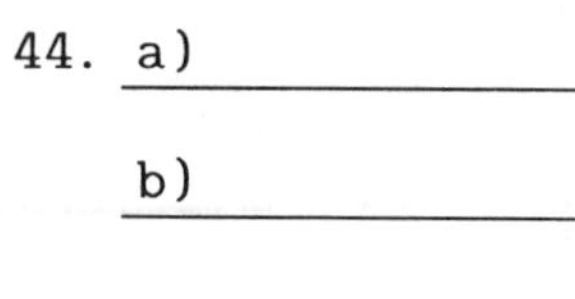

44. a) ____________

b) ____________

Use a calculator to find the acute angle θ that is a solution of the equation. Give the answer in degrees and decimal parts of degrees, as well as in radians.

35. $\csc \theta = 2.9514$

36. $\tan \theta = 8.9732$

Use a calculator to find θ in the interval indicated.

37. $\tan \theta = 3.5719$, $(180^\circ, 270^\circ)$

38. $\cos \theta = -0.972$, $\left(\pi, \frac{3\pi}{2}\right)$

39. Sketch a graph of $y = 2 \sin \left(x - \frac{\pi}{4}\right)$.

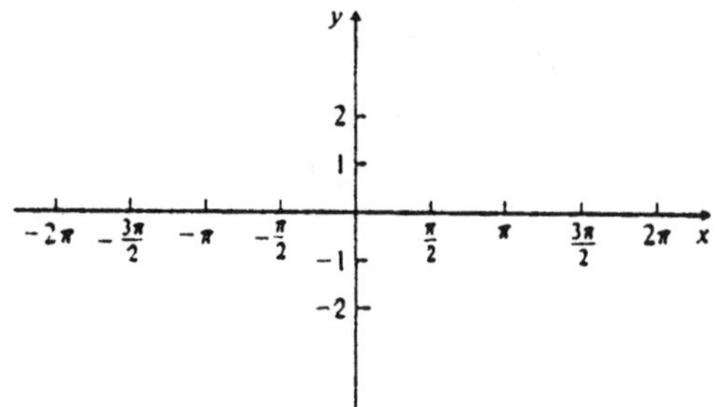

40. What is the period of the function in Question 39?

41. What is the phase shift of the function in Question 39?

42. Sketch the graph of $y = \sin x - \cos 2x$ for values between 0 and 2π.

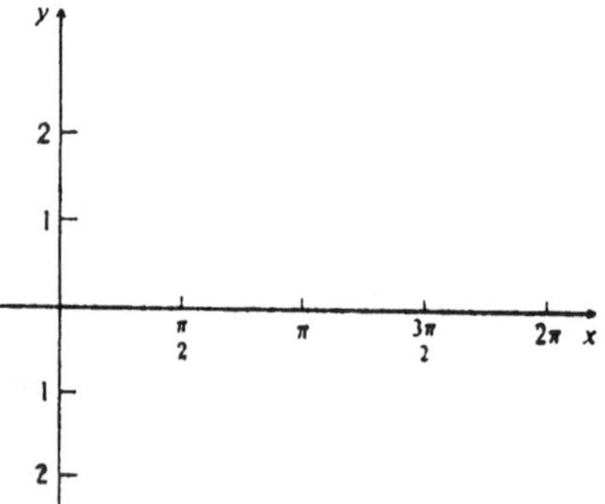

43. A point on the unit circle has u-coordinate $\frac{1}{5}$. What is the v-coordinate?

44. Given that $\cos \left(\frac{\pi}{16}\right) = 0.98079$, find each of the following.

a) $\sin \left(\frac{7\pi}{16}\right)$

b) $\sin \left(\frac{9\pi}{16}\right)$

CHAPTER 6 *NAME* ______________________

TEST FORM D *CLASS* ________ *SCORE* ________ *GRADE* ________

1. The point $\left(\frac{\sqrt{5}}{3}, \frac{2}{3}\right)$ is on a unit circle. Find the coordinates of its reflection across the origin, the u-axis, and the v-axis.

2. On a unit circle, mark and label the points determined by $\frac{2\pi}{3}$, $-\frac{7\pi}{4}$, and $\frac{9\pi}{2}$.

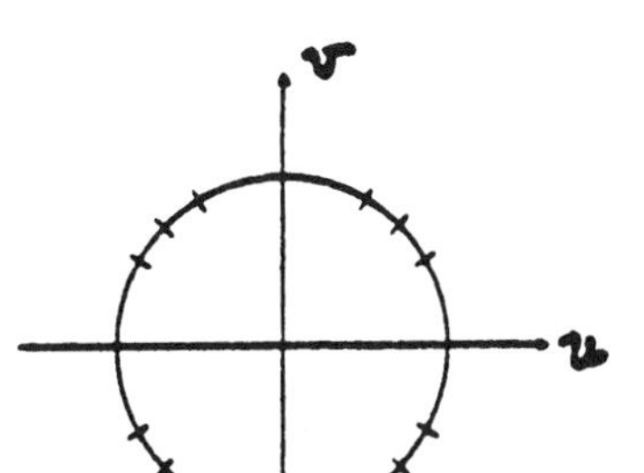

3. Sketch a graph of y = cos x.

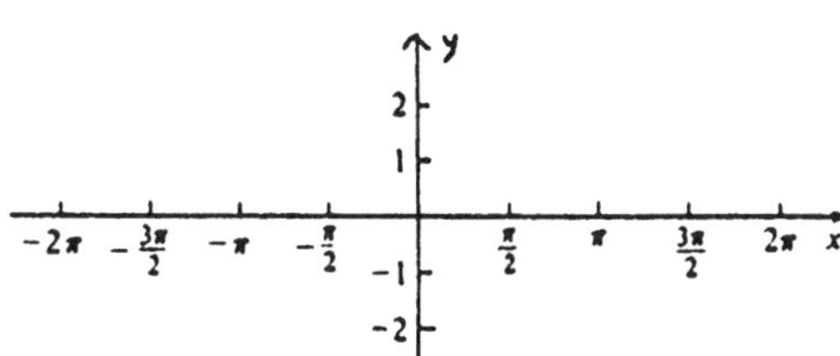

4. What is the range of the cosine function?

5. What is the period of the cosine function?

6. What is the domain of the sine function?

7. What is the amplitude of the sine function?

8. Complete the following table.

x	$\frac{2\pi}{3}$	$\frac{7\pi}{2}$	$-\frac{5\pi}{4}$
sin x			
cos x			

ANSWERS

1. ______________

2. See circle.

3. See graph.

4. ______________

5. ______________

6. ______________

7. ______________

8. See table.

CHAPTER 6 NAME ____________________

TEST FORM D

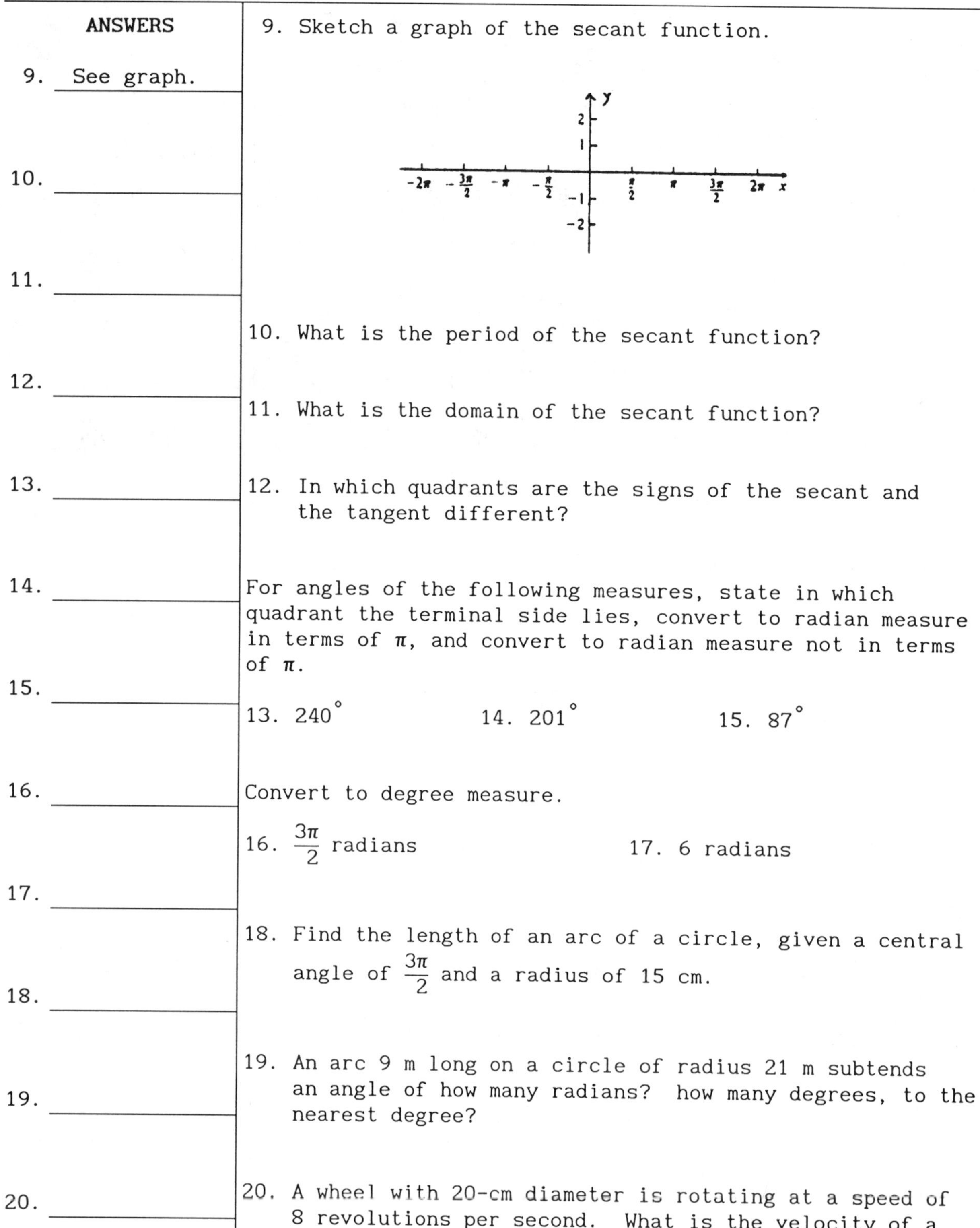

ANSWERS

9. See graph.

10. ____________

11. ____________

12. ____________

13. ____________

14. ____________

15. ____________

16. ____________

17. ____________

18. ____________

19. ____________

20. ____________

9. Sketch a graph of the secant function.

10. What is the period of the secant function?

11. What is the domain of the secant function?

12. In which quadrants are the signs of the secant and the tangent different?

For angles of the following measures, state in which quadrant the terminal side lies, convert to radian measure in terms of π, and convert to radian measure not in terms of π.

13. 240° 14. 201° 15. 87°

Convert to degree measure.

16. $\frac{3\pi}{2}$ radians 17. 6 radians

18. Find the length of an arc of a circle, given a central angle of $\frac{3\pi}{2}$ and a radius of 15 cm.

19. An arc 9 m long on a circle of radius 21 m subtends an angle of how many radians? how many degrees, to the nearest degree?

20. A wheel with 20-cm diameter is rotating at a speed of 8 revolutions per second. What is the velocity of a point on the rim, in cm/sec?

CHAPTER 6 NAME ____________________

TEST FORM D CLASS ________ SCORE ________ GRADE ______

21. A pulley belt runs, uncrossed, around two pulleys of radii 15 cm and 5 cm, respectively. A point on the belt travels at a rate of 40 m/sec. Find the angular speed, in radians/sec, of the larger pulley.

22. Find the six trigonometric function values for the angle θ shown.

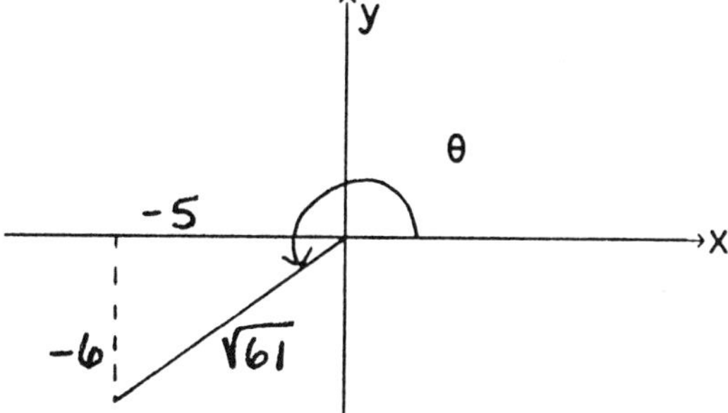

Find the following exactly. Do not use a calculator.

23. $\tan 540^\circ$ 24. $\sin(-45^\circ)$ 25. $\cos 390^\circ$

26. Given that $\cos \theta = \frac{2}{3}$ and that the terminal side is in quadrant IV, find the other five function values.

27. Convert 89.05° to degrees and minutes.

28. Convert $153^\circ 24'$ to degrees and decimal parts of degrees.

Use a calculator to find each of the following.

29. $\sin(124^\circ 17' 58'')$ 30. $\tan 1.502$

31. $\cos\left(-\frac{9\pi}{8}\right)$ 32. $\sec 100^\circ 18'$

33. $\cot 3.7849$ 34. $\csc(3216^\circ)$

ANSWERS

21. ____________

22. ____________

____

____

____

____

____

23. ____________

24. ____________

25. ____________

26. ____________

____

____

____

____

27. ____________

28. ____________

29. ____________

30. ____________

31. ____________

32. ____________

33. ____________

34. ____________

 NAME ____________________

TEST FORM D

ANSWERS

35. ____________________

36. ____________________

37. ____________________

38. ____________________

39. See graph.

40. ____________________

41. ____________________

42. See graph.

43. ____________________

44. ____________________

Use a calculator to find the acute angle θ that is a solution of the equation. Give the answer in degrees and decimal parts of degrees, as well as in radians.

35. $\cot \theta = 0.2184$

36. $\sin \theta = 0.9756$

Use a calculator to find θ in the interval indicated.

37. $\sin \theta = 0.3842$, $(0, 90^\circ)$

38. $\tan \theta = 9.176$, $(\pi, 2\pi)$

39. Sketch a graph of $y = -2 \cos \left(x - \frac{\pi}{2}\right)$.

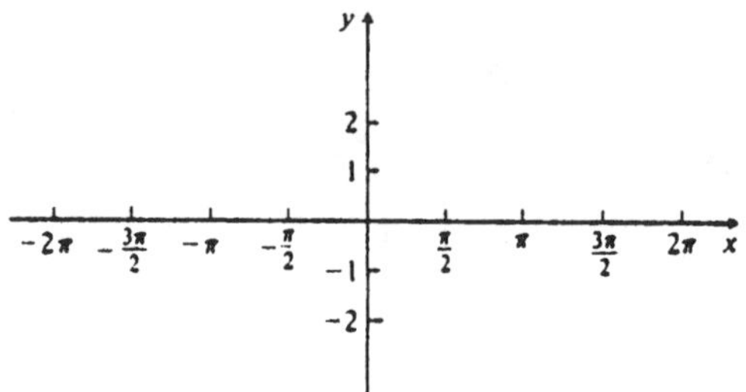

40. What is the period of the function in Question 39?

41. What is the phase shift of the function in Question 39?

42. Sketch the graph of $y = \cos x + 2 \sin x$ for values between 0 and 2π.

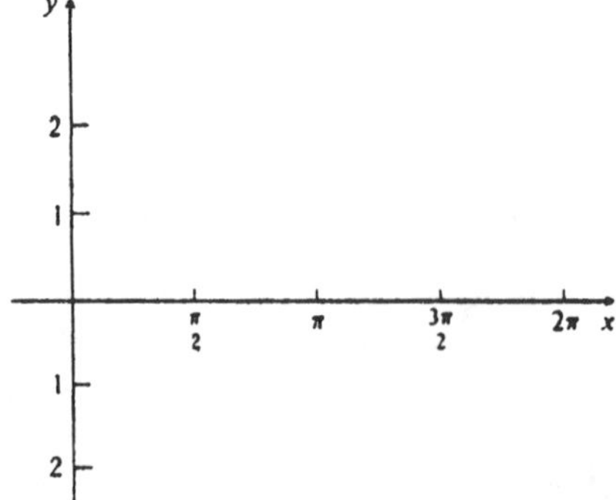

43. A grad is a unit of angle measure similar to a degree. A right angle has a measure of 100 grads. Convert 120° and $\frac{\pi}{6}$ radians to grads.

44. Given that $\sin x = 0.9372$ and the terminal side is in quadrant II, find $\cos x$ and $\tan x$.

CHAPTER 6 NAME ____________

TEST FORM E CLASS ______ SCORE ______ GRADE ______

1. Find two real numbers between -2π and 2π that determine point P.

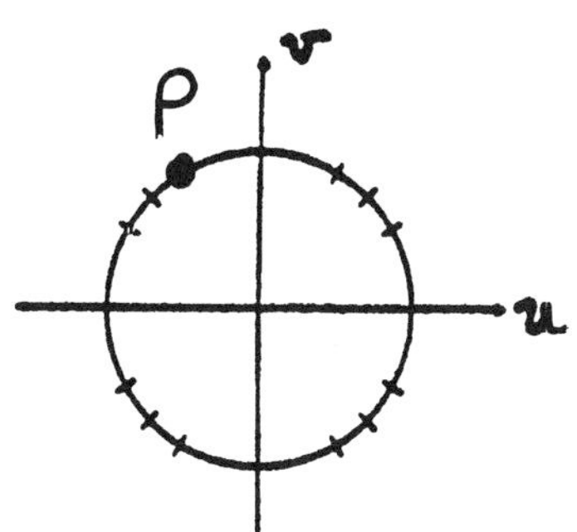

2. The number α determines a point on the unit circle with coordinates $\left(\frac{5}{8}, \frac{\sqrt{39}}{8}\right)$. What are the coordinates of the point determined by $-\alpha$?

3. Sketch a graph of $y = \sin x$.

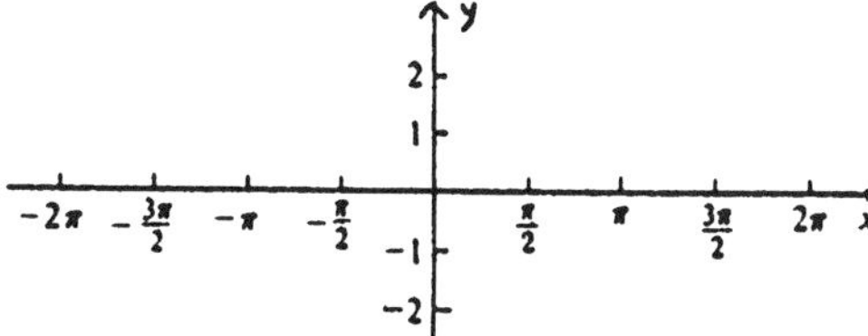

4. Sketch a graph of $y = \tan x$.

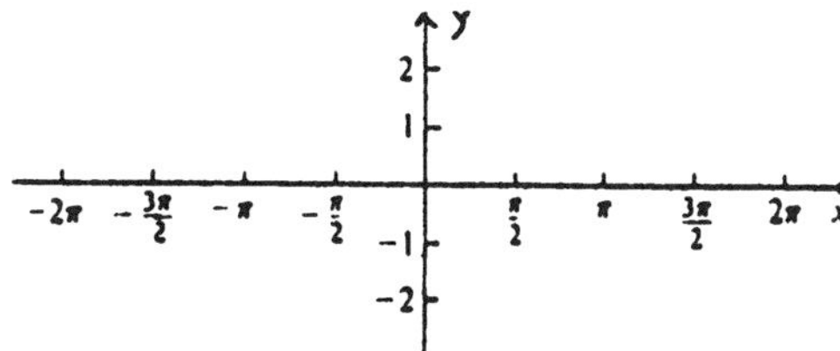

5. What is the domain of the sine function?

6. What is the range of the cosine function?

7. What is the period of the sine function?

8. What is the amplitude of the cosine function?

ANSWERS

1. ____________

2. ____________

3. See graph.

4. See graph.

5. ____________

6. ____________

7. ____________

8. ____________

NAME ______________________

TEST FORM E

ANSWERS	
9. ______	9. What is the domain of the cosecant function?
10. ______	10. What is the period of the cotangent function?
11. ______	11. What is the range of the secant function?
12. ______	12. What is the domain of the tangent function?
13. See table.	13. Complete the following table. Do not use a calculator.
14. ______	14. For an angle which measures -570°, state in which quadrant the terminal side lies and convert the measure to radian measure in terms of π.
15. ______	15. Convert $\frac{13\pi}{6}$ radians to degree measure.
16. ______	16. Convert 8 radians to degree measure (to the nearest degree).
17. ______	17. A wheel on a bicycle has a 50 cm-diameter. Through what angle (in radians) does the wheel turn while the bicycle travels 0.8 km?
18. ______	18. Two pulleys, 60 in. and 40 in. in diameter, respectively, are connected by a belt. The larger pulley makes 15 revolutions per minute. Find the angular speed of the smaller pulley, in radians/sec.

Table for 13:

x	$\frac{2\pi}{3}$	$\frac{-9\pi}{4}$	$\frac{5\pi}{6}$
sin x			
tan x			
sec x			
csc x			

CHAPTER 6 NAME ______________________

TEST FORM E CLASS ________ SCORE __________ GRADE ______

19. Find the six trigonometric function values for the angle θ as shown.

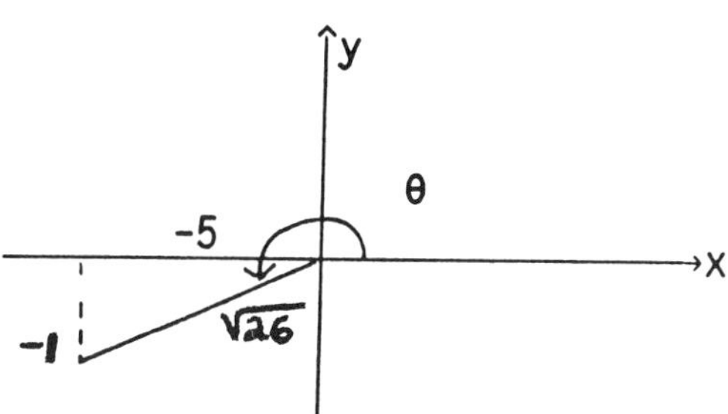

Find the following. Do not use a calculator.

20. $\cos(-330^\circ)$

21. $\csc \frac{3\pi}{2}$

22. $\sin \frac{14\pi}{3}$

23. $\tan 405^\circ$

24. Given that $\cot \theta = -3$ and that the terminal side is in quadrant IV, find the following.

a) $\sec \theta$ b) $\sin \theta$ c) $\tan \theta$

25. Convert 54.25° to degrees and minutes.

26. Convert $93^\circ 09'$ to degrees and a decimal part of a degree.

27. Use a calculator to find the following:

a) $\tan 18^\circ$ b) $\sin 25^\circ 15'$

c) $\sec 38^\circ 12'$ d) $\cot 105$

e) $\csc(-115^\circ\ 28')$ f) $\cos 13.9$

ANSWERS

19. ______________
____
____
____
____
____

20. ______________

21. ______________

22. ______________

23. ______________

24. a) ______________
b) ______________
c) ______________

25. ______________

26. ______________

27. a) ______________
b) ______________
c) ______________
d) ______________
e) ______________
f) ______________

ANSWERS

28. ____________

____

29. ____________

30. See graph.

31. See graph.

32. ____________

33. See graph.

28. Use a calculator to find the acute angle θ which is a solution of the equation. Give the answer in degrees and decimal parts of degrees, as well as in radians.

$\cos\theta = 0.7524$

29. Given that $\sin\theta = -0.7814$, find θ between 270° and 360°. Give the answer in degrees and decimal parts of degrees.

30. Sketch a graph of $y = -\cos x$.

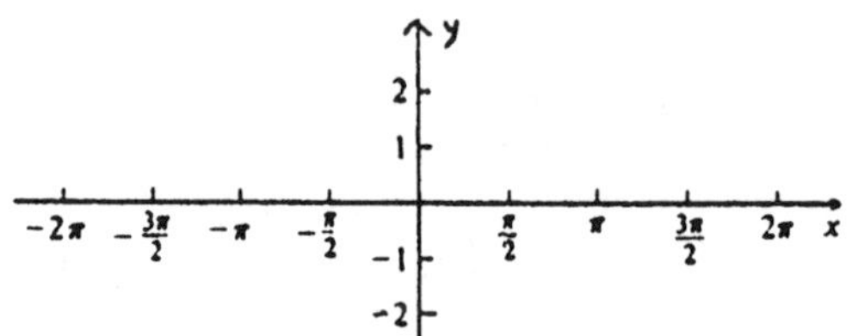

31. Sketch a graph of $y = 1 + \cos(2x - \pi)$.

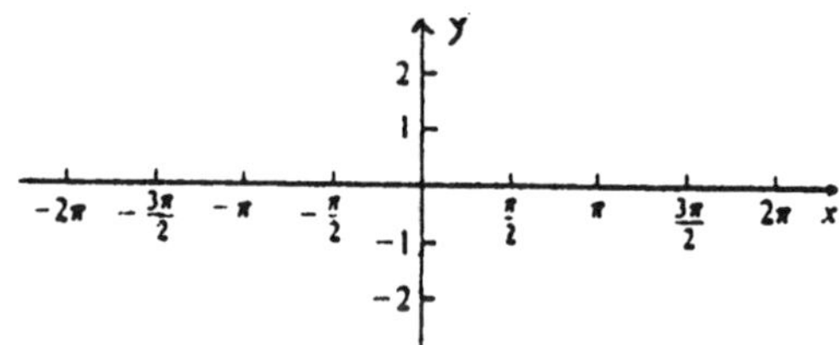

32. What is the phase shift of the function in Question 31?

33. Sketch a graph of $y = 2\sin x - \cos 4x$ for values between 0 and 2π.

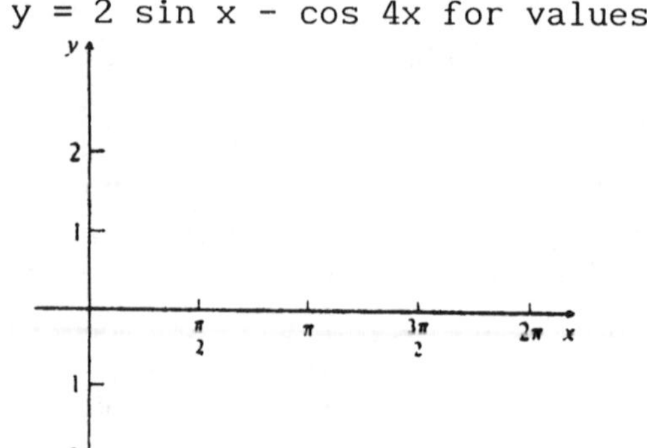

CHAPTER 6 NAME ______________

TEST FORM F CLASS ______ SCORE ______ GRADE ______

Write the letter of your response on the answer blank.

ANSWERS

1. The point $\left(-\frac{2}{5}, -\frac{\sqrt{21}}{5}\right)$ is on a unit circle. Find the coordinates of its reflection across the u-axis.

a) $\left(\frac{2}{5}, \frac{\sqrt{2}}{5}\right)$ b) $\left(-\frac{2}{5}, \frac{\sqrt{2}}{5}\right)$ c) $\left(\frac{2}{5}, -\frac{\sqrt{2}}{5}\right)$ d) $\left(-\frac{\sqrt{2}}{5}, -\frac{2}{3}\right)$

1. ______________

2. Find the coordinates of point P on the unit circle.

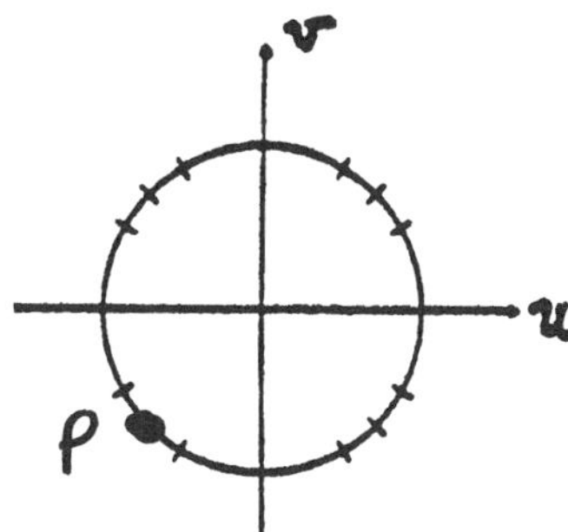

a) $\left(\frac{\sqrt{3}}{2}, -\frac{1}{2}\right)$ b) $\left(-\frac{\sqrt{2}}{2}, -\frac{\sqrt{2}}{2}\right)$

c) $\left(-\frac{1}{2}, -\frac{1}{2}\right)$ d) $\left(\frac{\sqrt{2}}{2}, -\frac{\sqrt{2}}{2}\right)$

2. ______________

3. Sketch a graph of $y = -\sin\theta$.

a)

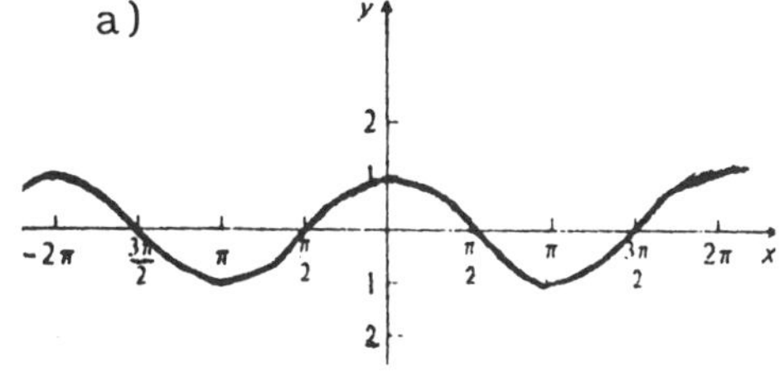

b)

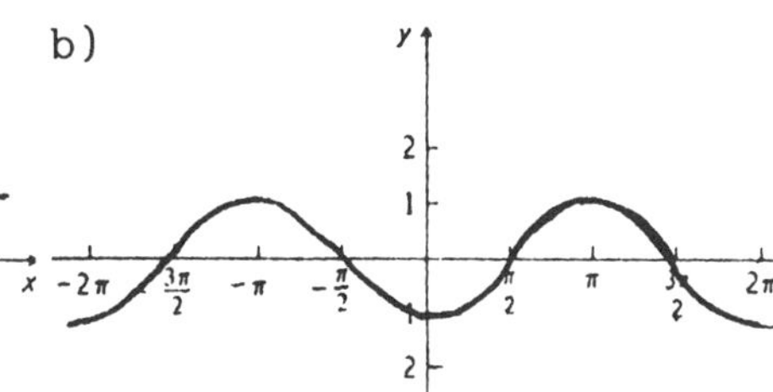

c)

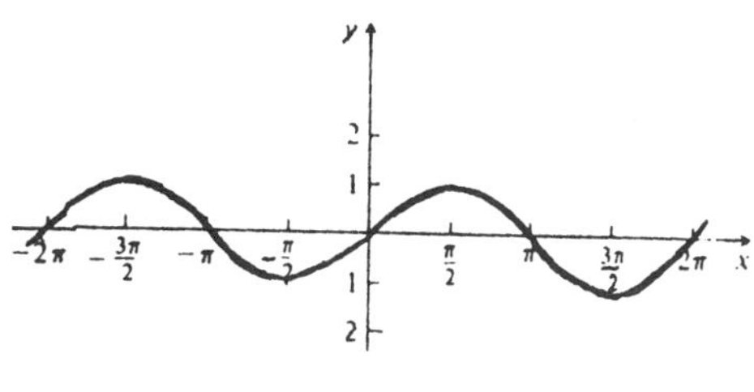

d)

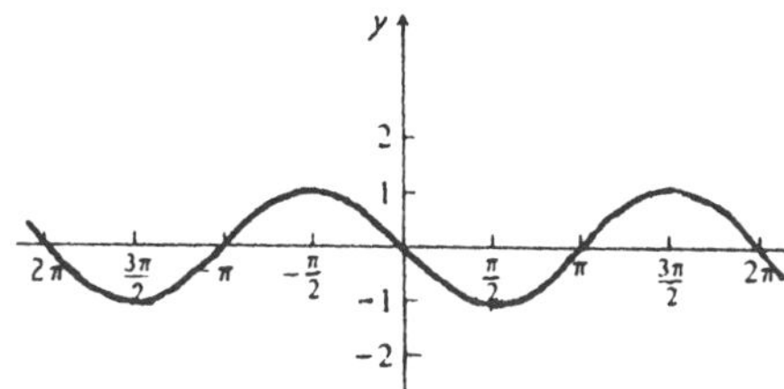

3. ______________

4. What is the period of the cosine function?

a) π b) 2π c) $\frac{\pi}{2}$ d) 4π

4. ______________

CHAPTER 6 NAME ____________________

TEST FORM F

ANSWERS

5. ______

6. ______

7. ______

8. ______

5. What is the range of the sine function?

a) All real numbers greater than 2π

b) All real numbers from -1 to 1, inclusive

c) All real numbers except -1 to 1

d) All real numbers

6. What is the amplitude of the sine function?

a) 1 b) -2 c) 2π d) $\frac{\pi}{2}$

7. Sketch a graph of the tangent function.

a)

b)

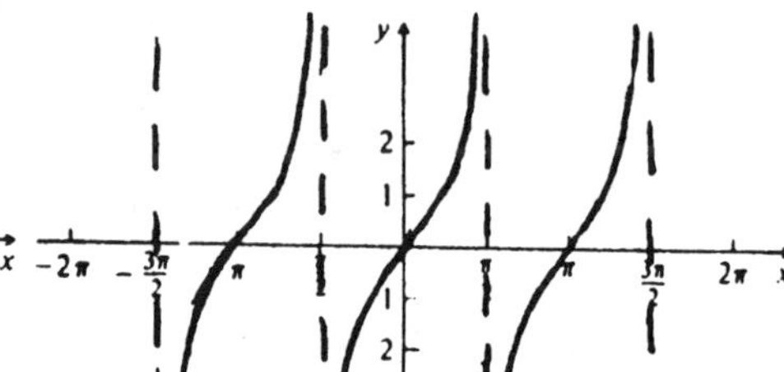

c)

d)

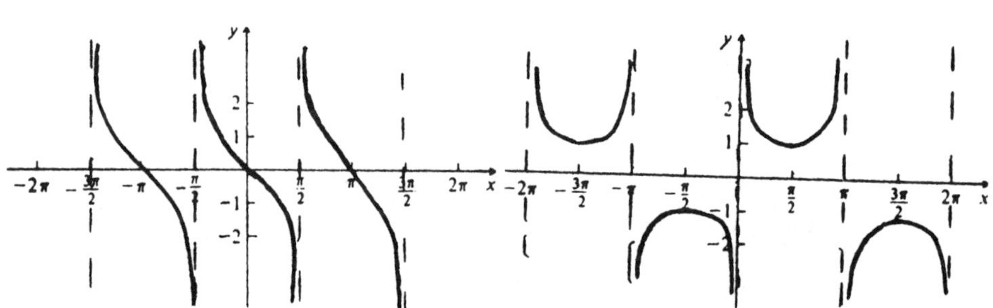

8. What is the domain of the tangent function?

a) All real numbers except between -1 and 1

b) All real numbers except $\frac{\pi}{2} + k\pi$

c) All real numbers except $k\pi$

d) All real numbers

ANSWERS

9. What is the period of the tangent function?

a) 4π b) $\frac{\pi}{2}$ c) 2π d) π

9. ____________

10. In which quadrants are the signs of the sine and cotangent the same?

a) I and II b) II and IV c) I and IV d) III and IV

10. ____________

11. For an angle which measures -150°, state in which quadrant the terminal side lies.

a) I b) II c) III d) IV

11. ____________

12. Convert 216° to radian measure in terms of π.

a) $\frac{5\pi}{3}$ b) $\frac{54{,}000}{\pi}$ c) $\frac{5\pi}{6}$ d) $\frac{6\pi}{5}$

12. ____________

13. Convert 5 radians to degree measure (to the nearest degree).

a) 900° b) 286° c) 1800° d) 19°

13. ____________

14. Convert $\frac{5\pi}{6}$ to degree measure (to the nearest degree).

a) 942° b) 300° c) 3° d) 150°

14. ____________

15. Find the length of an arc of a circle, given a central angle of $\frac{3\pi}{4}$ and a radius of 8 cm. Use 3.14 for π.

a) 0.393 cm b) 6 cm c) 18.84 cm d) 37.68 cm

15. ____________

16. An arc 4 in. long on a circle of radius 12 in. subtends an angle of how many radians?

a) 27 b) $\frac{1}{6}$ c) 3 d) $\frac{1}{3}$

16. ____________

17. A flywheel is rotating 6 radians/sec. It has an 8-cm diameter. What is the linear speed of a point on its rim, in cm/min?

a) 1440 b) 24 c) 720 d) 48

17. ____________

NAME ____________________

TEST FORM F

ANSWERS

18. ____________

19. ____________

20. ____________

21. ____________

22. ____________

18. A pulley belt runs uncrossed around two pulleys of radii 12 in. and 6 in. respectively. A point on the belt travels at a rate of 18 ft/sec. Find the angular speed, in radians per second, of the larger pulley.

a) 18 b) $\frac{3}{2}$ c) 2 d) 36

19. Find the six trigonometric function values for the angle θ as shown.

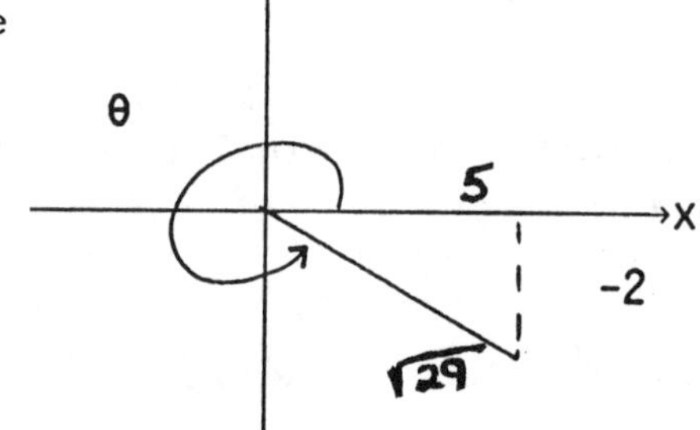

Find two of the six trigonometric function values.

a) $\sin\theta = \frac{5}{\sqrt{29}}$, $\cot\theta = -\frac{5}{2}$

b) $\cos\theta = -\frac{2}{\sqrt{29}}$, $\tan\theta = -\frac{2}{5}$

c) $\sec\theta = \frac{\sqrt{29}}{5}$, $\tan\theta = -\frac{2}{5}$

d) $\sin\theta = \frac{2}{\sqrt{29}}$, $\sec\theta = \frac{\sqrt{29}}{5}$

In Questions 20 - 24 find the exact values. Do not use a calculator.

20. Find $\tan\frac{5\pi}{6}$.

a) $-\sqrt{3}$ b) $-\frac{\sqrt{3}}{3}$ c) $-\frac{2\sqrt{3}}{3}$ d) $-\frac{1}{2}$

21. Find $\sin(-270^\circ)$.

a) 1 b) 0 c) Undefined d) -1

22. Find $\sec\left(-\frac{3\pi}{4}\right)$.

a) $-\frac{2\sqrt{3}}{3}$ b) $-\sqrt{2}$ c) 2 d) $-\frac{\sqrt{2}}{2}$

NAME ______________________

TEST FORM F

ANSWERS

23. Find sin 495°.

a) $\frac{\sqrt{2}}{2}$ b) $-\frac{\sqrt{3}}{2}$ c) $-\frac{\sqrt{2}}{2}$ d) $\frac{1}{2}$

23. ______________

24. Find cot 360°.

a) 0 b) Undefined c) -1 d) 1

24. ______________

25. Given that $\sin\theta = -\frac{3}{4}$ and that the terminal side is in quadrant III, find sec θ.

a) $\frac{5}{4}$ b) $-\frac{3\sqrt{13}}{13}$ c) $-\frac{3\sqrt{7}}{7}$ d) $-\frac{4\sqrt{7}}{7}$

25. ______________

26. Given that $\tan\theta = \sqrt{10}$ and that the terminal side is in quadrant I, find cos θ.

a) $\frac{\sqrt{11}}{11}$ b) $\frac{3\sqrt{10}}{10}$ c) $\frac{\sqrt{110}}{11}$ d) $\frac{\sqrt{5}}{5}$

26. ______________

27. Convert 14.8° to degrees and minutes.

a) 14°8′ b) 14°80′ c) 4°10′ d) 14°48′

27. ______________

28. Convert 74°12′ to degrees and a decimal part of a degree.

a) 74.2° b) 74.12° c) 74.72° d) $74.41\overline{6}°$

28. ______________

In Questions 29 - 31, use a calculator.

29. Find cos 29°18′.

a) 0.4894 b) -0.0401 c) 0.8721 d) 0.7683

29. ______________

30. Find tan 2.0712.

a) -0.5468 b) 0.8774 c) 64.23° d) -1.8287

30. ______________

31. Find sin 19°17′.

a) 0.9439 b) 0.3302 c) 0.4099 d) 0.9968

31. ______________

NAME ______________________

TEST FORM F

ANSWERS

32. ____________

33. ____________

34. ____________

35. ____________

36. ____________

37. ____________

In Questions 32 and 33, use a calculator to find the acute angle θ which is a solution of the equation.

32. $\sin \theta = 0.5937$

a) 30.70° b) 53.58° c) 36.42° d) 47.38°

33. $\sec \theta = 2.5$

a) 21.80° b) 66.42° c) 10.1° d) 23.58°

34. Given that $\cot \theta = -0.4383$, find θ between 270° and 360°.

a) $293^\circ 40'$ b) $336^\circ 20'$ c) $23^\circ 40'$ d) $113^\circ 40'$

35. Given that $\csc \theta = 7.132$, find θ between 0 and $\frac{\pi}{2}$.

a) 3.252 b) 1.7115 c) 0.1393 d) 0.1407

36. Sketch a graph of $y = \frac{1}{2} \sin \left(x + \frac{\pi}{2}\right)$.

a) b)

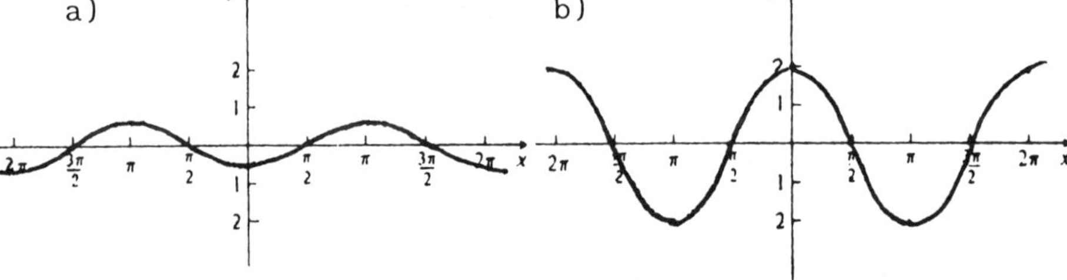

c) d)

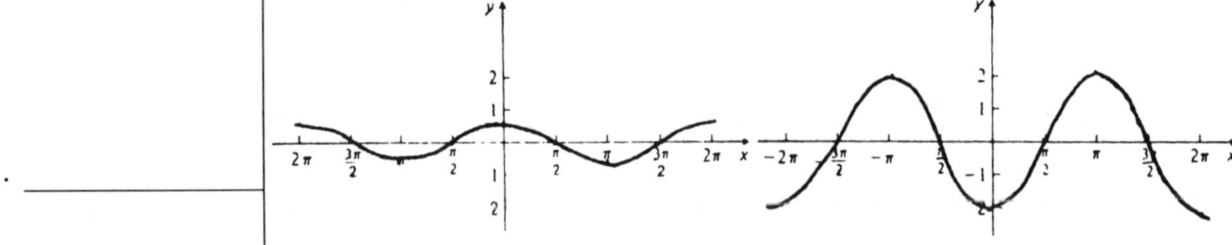

37. What is the phase shift of the function in Question 36?

a) π b) $-\frac{\pi}{2}$ c) $\frac{\pi}{2}$ d) $\frac{\pi}{4}$

CHAPTER 7 NAME ______

TEST FORM A CLASS ______ SCORE ______ GRADE ______

Use the sum and difference formulas to write equivalent expressions. You need not simplify.

1. $\cos(\pi + x)$

2. $\tan(19^\circ + 35^\circ)$

3. Simplify: $\sin 85^\circ \cos 15^\circ - \cos 85^\circ \sin 15^\circ$.

4. Find $\cos 75^\circ$ exactly.

5. Given that $\cos \alpha = \frac{\sqrt{3}}{2}$ and $\sin \beta = \frac{\sqrt{2}}{2}$ and that α and β are between 0 and $\frac{\pi}{2}$, evaluate $\tan(\alpha - \beta)$ exactly.

6. Find an identity for $\csc(x - 540^\circ)$.

Complete these identities.

7. $\sin^2 x + \cos^2 x =$ ______

8. $1 + \tan^2 x =$ ______

9. $\sin\left(x - \frac{\pi}{2}\right) =$ ______

10. $\csc\left(\frac{\pi}{2} - x\right) =$ ______

11. $\cot\left(\frac{\pi}{2} - x\right) =$ ______

12. Express $\tan x$ in terms of $\sec x$.

Simplify.

13. $\dfrac{\sqrt{\csc^2 x - 1}}{\cos x}$

14. $\dfrac{\sin^2 x \cot^2 x - 1}{\tan x \sin^2 x \cot x}$

15. Rationalize the denominator: $\sqrt{\dfrac{\tan x}{\cos x}}$.

ANSWERS

1. ______

2. ______

3. ______

4. ______

5. ______

6. ______

7. ______

8. ______

9. ______

10. ______

11. ______

12. ______

13. ______

14. ______

15. ______

CHAPTER 7 NAME ____________________

TEST FORM A

ANSWERS	
16. ______	16. Find $\sin 2\theta$, $\cos 2\theta$, and $\tan 2\theta$ and the quadrant in which 2θ lies if $\sin \theta = \frac{3}{5}$ and θ is in quadrant II.
______	
______	17. Find $\sin \frac{7\pi}{12}$ exactly.
______	
17. ______	18. Simplify: $\frac{\sec \theta \csc \theta}{2}$.
18. ______	
19. See work.	19. Prove the identity $\sec^2 \theta + \csc^2 \theta = \sec^2 \theta \csc^2 \theta$.
20. ______	20. Find, in radians, $\arccos \left(\frac{1}{2}\right)$.
21. ______	
22. ______	21. Use a calculator to find, in degrees, $\sin^{-1} (0.6137)$.
23. ______	
24. ______	22. Find $\tan^{-1} \sqrt{3}$. 23. Find $\sec^{-1} (2)$.
25. ______	
26. ______	24. Find $\sin \left(\cos^{-1} \frac{\sqrt{3}}{2}\right)$. 25. Find $\sin \left[\arcsin (-1)\right]$.
27. ______	
28. ______	Find all solutions of the following equations in $[0, 2\pi)$.
29. ______	26. $2 \cos^2 x - 11 \cos x = 6$
______	
______	27. $4 \sin^3 x - \sin x = 0$
______	
______	28. $2 \tan^2 x + 5 \tan x + 3 = 0$
______	
______	29. Given that $\sin \theta = 0.7431$, $\cos \theta = 0.6691$, $\tan \theta = 1.111$, $\cot \theta = 0.9004$, $\sec \theta = 1.495$, $\csc \theta = 1.346$, find the six function values for $90^\circ - \theta$.
30. ______	30. Prove the identity $\sin \left(x + \frac{5\pi}{2}\right) = \cos x$.

CHAPTER 7 *NAME* __________

TEST FORM B *CLASS* ______ *SCORE* ______ *GRADE* ______

Use the sum and difference formulas to write equivalent expressions. You need not simplify.

1. $\sin(\pi - x)$

2. $\tan(27^\circ - 34^\circ)$

3. Simplify: $\cos 74^\circ \cos 16^\circ - \sin 74^\circ \sin 16^\circ$.

4. Find $\sin 105^\circ$ exactly.

5. Given that $\cos \alpha = \frac{1}{2}$ and $\sin \beta = \frac{\sqrt{2}}{2}$ and that α and β are between 0 and $\frac{\pi}{2}$, evaluate $\tan(\alpha - \beta)$ exactly.

6. Find an identity for $\sec(x + 630^\circ)$.

Complete these identities.

7. $\sin^2 x + \cos^2 x =$ ______

8. $1 + \cot^2 x =$ ______

9. $\sin\left(\frac{\pi}{2} - x\right) =$ ______

10. $\cot\left(\frac{\pi}{2} - x\right) =$ ______

11. $\cos\left(x + \frac{\pi}{2}\right) =$ ______

12. Express $\cot x$ in terms of $\csc x$.

Simplify.

13. $\frac{\sqrt{1 + \tan^2 x}}{\tan x}$

14. $\frac{\sec^2 x \sin^2 x + 1}{\sin x \tan^2 x \csc x}$

15. Rationalize the denominator: $\sqrt{\frac{\sin x}{\tan x}}$.

ANSWERS

1. ______
2. ______
3. ______
4. ______
5. ______
6. ______
7. ______
8. ______
9. ______
10. ______
11. ______
12. ______
13. ______
14. ______
15. ______

NAME ______________________

TEST FORM B

ANSWERS	
16. ______	16. Find $\sin 2\theta$, $\cos 2\theta$, and $\tan 2\theta$ and the quadrant in which 2θ lies if $\tan \theta = -\frac{3}{4}$ and θ is in quadrant II.
______	
______	17. Find $\sin \frac{5\pi}{12}$ exactly.
______	
17. ______	18. Simplify: $\frac{\cot \theta - \tan \theta}{\cot \theta + \tan \theta}$.
18. ______	
19. See work.	19. Prove the identity: $(\tan \theta)(1 + \cot^2 \theta) = (\cot \theta)(1 + \tan^2 \theta)$.
20. ______	20. Find, in radians, $\arctan \left(-\frac{\sqrt{3}}{3}\right)$.
21. ______	21. Use a calculator to find, in degrees, $\cos^{-1}(0.1237)$.
22. ______	22. Find $\sin^{-1} \frac{\sqrt{2}}{2}$.
23. ______	23. Find $\csc^{-1}(-1)$.
24. ______	24. Find $\tan \left[\sin^{-1}\left(-\frac{1}{2}\right)\right]$.
25. ______	25. Find $\arctan \left(\tan \frac{\pi}{6}\right)$.
	Find all solutions of the following equations in $[0, 2\pi)$.
26. ______	26. $2 \sin^2 x - \sin x = 1$
27. ______	27. $4 \cos^3 x - \cos x = 0$
28. ______	28. $\tan^2 x + 5 \tan x + 3 = 0$
29. ______	29. Given that $\sin \theta = 0.8290$, $\cos \theta = 0.5592$, $\tan \theta = 1.483$, $\cot \theta = 0.6745$, $\sec \theta = 1.788$, $\csc \theta = 1.206$, find the six function values for $90^\circ - \theta$.
______	
______	
______	
______	
______	
30. ______	30. Simplify: $\sin\left(x + \frac{\pi}{2}\right)\left[\sec x - \cos x\right]$.

CHAPTER 7 — *NAME* ____________________

TEST FORM C — *CLASS* ______ *SCORE* ______ *GRADE* ______

Use the sum and difference formulas to write equivalent expressions. You need not simplify.

1. $\sin(\pi + x)$

2. $\tan(37^\circ + 45^\circ)$

3. Simplify: $\sin 37^\circ \cos 18^\circ + \cos 37^\circ \sin 18^\circ$.

4. Find $\cos 15^\circ$ exactly.

5. Given that $\cos\alpha = \frac{\sqrt{2}}{2}$ and $\sin\beta = \frac{1}{2}$ and that α and β are between 0 and $\frac{\pi}{2}$, evaluate $\tan(\alpha - \beta)$ exactly.

6. Find an identity for $\csc(x + 810^\circ)$.

Complete these identities.

7. $\sin^2 x + \cos^2 x =$ ______

8. $1 + \tan^2 x =$ ______

9. $\tan\left(\frac{\pi}{2} - x\right) =$ ______

10. $\sin\left(\frac{\pi}{2} - x\right) =$ ______

11. $\sec\left(\frac{\pi}{2} - x\right) =$ ______

12. Express $\sec x$ in terms of $\tan x$.

Simplify.

13. $\dfrac{\sqrt{\sec^2 x - 1}}{\sec x}$

14. $\dfrac{\cos^2 x \tan^2 x - 1}{\sin x \cos^2 x \csc x}$

15. Rationalize the denominator: $\sqrt{\dfrac{\sec x}{\cot x}}$.

ANSWERS

1. ______
2. ______
3. ______
4. ______
5. ______
6. ______
7. ______
8. ______
9. ______
10. ______
11. ______
12. ______
13. ______
14. ______
15. ______

CHAPTER 7 NAME ______________________

TEST FORM C

ANSWERS

16. ______________________

____

____

____

17. ______________________

18. ______________________

19. See work.

20. ______________________

21. ______________________

22. ______________________

23. ______________________

24. ______________________

25. ______________________

26. ______________________

27. ______________________

28. ______________________

29. ______________________

____

____

____

____

____

30. ______________________

16. Find $\sin 2\theta$, $\cos 2\theta$, and $\tan 2\theta$ and the quadrant in which 2θ lies if $\tan \theta = -\frac{5}{12}$ and θ is in quadrant IV.

17. Find $\sin \frac{3\pi}{8}$ exactly.

18. Simplify: $2\cos 2y \sin 2y$.

19. Prove the identity $\sin 2x = \frac{2 \cot x}{\csc^2 x}$.

20. Find, in radians, $\arcsin (-1)$.

21. Use a calculator to find, in degrees, $\tan^{-1} (0.7321)$.

22. Find $\cos^{-1} \frac{1}{2}$.

23. Find $\csc^{-1} \left(\frac{2\sqrt{3}}{3}\right)$.

24. Find $\cos \left(\sin^{-1} \frac{\sqrt{2}}{2}\right)$.

25. Find $\tan \left[\arctan \left(-\frac{\sqrt{3}}{3}\right)\right]$.

Find all solutions of the following equations in $[0,2\pi)$.

26. $4 \cos^2 x + 1 = 4 \cos x$

27. $2 \sin^3 x - \sin x = 0$

28. $\tan^2 x + 1 = 3 \tan x$

29. Given that $\sin \theta = 0.9613$, $\cos \theta = 0.2756$, $\tan \theta = 3.487$, $\cot \theta = 0.2867$, $\sec \theta = 3.628$, $\csc \theta = 1.040$, find the six function values for $90^\circ - \theta$.

30. Find $\sin \theta$, $\cos \theta$, and $\tan \theta$ given that $\sin 2\theta = \frac{2}{3}$, $\frac{\pi}{2} \le 2\theta \le \pi$.

CHAPTER 7 NAME ____________________

TEST FORM D CLASS ________ SCORE ________ GRADE ______

Use the sum and difference formulas to write equivalent expressions. You need not simplify.

1. $\cos\left(\frac{\pi}{2} - x\right)$

2. $\tan(92^\circ - 18^\circ)$

3. Simplify: $\cos 92^\circ \cos 15^\circ + \sin 92^\circ \sin 15^\circ$.

4. Find $\sin 15^\circ$ exactly.

5. Given that $\cos \alpha = \frac{1}{2}$ and $\sin \beta = \frac{1}{2}$ and that α and β are between 0 and $\frac{\pi}{2}$, evaluate $\tan(\alpha - \beta)$ exactly.

6. Find an identity for $\sec(x - 720^\circ)$.

Complete these identities.

7. $\sin^2 x + \cos^2 x =$ ________

8. $1 + \cot^2 x =$ ________

9. $\sec\left(\frac{\pi}{2} - x\right) =$ ________

10. $\tan\left(\frac{\pi}{2} - x\right) =$ ________

11. $\cos\left(x - \frac{\pi}{2}\right) =$ ________

12. Express $\cos x$ in terms of $\sin x$.

Simplify.

13. $\frac{\sqrt{1 - \sin^2 x}}{\sin x}$

14. $\frac{\csc^2 x \cos^2 x + 1}{\sec x \cot^2 x \cos x}$

15. Rationalize the denominator: $\sqrt{\frac{\csc x}{\tan x}}$.

ANSWERS

1. ____________
2. ____________
3. ____________
4. ____________
5. ____________
6. ____________
7. ____________
8. ____________
9. ____________
10. ____________
11. ____________
12. ____________
13. ____________
14. ____________
15. ____________

NAME ______________________

TEST FORM D

ANSWERS	
16. ______	16. Find $\sin 2\theta$, $\cos 2\theta$, and $\tan 2\theta$ and the quadrant in which 2θ lies if $\tan \theta = -\frac{8}{15}$ and θ is in quadrant II.
______	
______	17. Find $\cos \frac{7\pi}{8}$ exactly.
______	
17. ______	18. Simplify: $2 \sin x \cos^3 x - 2 \sin^3 x \cos x$.
18. ______	19. Prove the identity $\sin x = \dfrac{\tan x}{\sqrt{1 + \tan^2 x}}$.
19. See work.	
20. ______	20. Find, in radians, $\arccos \left(\frac{\sqrt{3}}{2}\right)$.
21. ______	21. Use a calculator to find, in degrees, $\cos^{-1} (0.3456)$.
22. ______	22. Find $\sin^{-1} \left(\frac{1}{2}\right)$.
23. ______	23. Find $\cot^{-1} \left(\frac{\sqrt{3}}{3}\right)$.
24. ______	24. Find $\cos \left[\tan^{-1} \left(-\frac{\sqrt{3}}{3}\right)\right]$.
25. ______	25. Find $\arccos \left(\cos \frac{2\pi}{3}\right)$.
26. ______	Find all solutions of the following equations in $[0,2\pi)$.
	26. $6 \sin^2 x + 3 \sin x = 3$
27. ______	27. $\sin x - 2 \sin^3 x = 0$
28. ______	28. $\tan^2 x + 2 = 4 \tan x$
29. ______	29. Given that $\sin \theta = 0.6293$, $\cos \theta = 0.7771$, $\tan \theta = 0.809$, $\cot \theta = 1.235$, $\sec \theta = 1.287$, $\csc \theta = 1.589$, find the six function values for $90^\circ - \theta$.
______	
______	
______	
______	
______	
30. ______	30. Solve: $\lvert \cos x \rvert = \frac{\sqrt{3}}{2}$. (Restrict solutions to $[0,360^\circ)$).

CHAPTER 7 NAME ______________________

TEST FORM E CLASS ________ SCORE __________ GRADE ______

ANSWERS

1. Using a sum or a difference formula write an equivalent expression for $\tan(75^\circ - 13^\circ)$.

2. Simplify: $\cos x \cos \pi + \sin x \sin \pi$.

3. Find $\cos 165^\circ$.

4. Find $\tan\left(-\frac{5\pi}{12}\right)$.

5. Given $\sin \alpha = \frac{2}{3}$ and $\tan \beta = 1$, and α and β are between 0 and $\frac{\pi}{2}$, evaluate $\cos(\alpha + \beta)$ exactly.

6. Find an identity for $\csc\left(\frac{\pi}{2} + \theta\right)$.

Complete these identities:

7. $1 + \cot^2 \theta =$ ______

8. $\sin\left(\theta + \frac{\pi}{2}\right) =$ ______

9. $1 + \tan^2 \theta =$ ______

10. $\cos\left(\theta - \frac{\pi}{2}\right) =$ ______

11. $\sin^2 \theta + \cos^2 \theta =$ ______

12. $\sin\left(\frac{\pi}{2} - \theta\right) =$ ______

Simplify.

13. $\sec x \cdot \cot x \cdot \sin x$

14. $\frac{\sqrt{1 - \sin^2 x}}{\sin x}$

15. $\frac{\cos x - \sec x}{\sec x} + \cos^2 x \tan^2 x$

1. ______________
2. ______________
3. ______________
4. ______________
5. ______________
6. ______________
7. ______________
8. ______________
9. ______________
10. ______________
11. ______________
12. ______________
13. ______________
14. ______________
15. ______________

NAME ______________________

TEST FORM E

ANSWERS

16. ______

17. ______

______

______

______

18. ______

19. See work.

20. ______

21. ______

22. ______

23. ______

24. ______

25. ______

26. ______

27. ______

28. ______

29. ______

16. Rationalize the denominator: $\sqrt{\dfrac{\csc x}{\tan x}}$.

17. Find sin 2θ, cos 2θ, and tan 2θ and the quadrant in which 2θ lies if $\cos\theta = -\dfrac{3}{\sqrt{23}}$ and θ is in quadrant II.

Simplify.

18. $2\cos^2\dfrac{\theta}{2} - 1$

19. $\tan 2\theta\,(1 - \tan^2\theta)$

20. Prove the identity $\dfrac{2\cos^2\theta - 1}{\sin\theta\cos\theta} = \cot\theta - \tan\theta$.
Show your work below.

21. Use a calculator to find $\sin^{-1} 0.7984$ in degrees.

22. Use a caluclator to find $\cot^{-1} 3.4097$ in degrees.

Find each of the following:

23. $\tan^{-1}\left(-\sqrt{3}\right)$

24. $\arcsin\dfrac{\sqrt{2}}{2}$

25. $\cos\left(\sin^{-1} 0\right)$

26. $\arcsin\sin\dfrac{3\pi}{4}$

Find all solutions of the following equations in $[0,2\pi)$ or $[0^\circ,360^\circ)$.

27. $|\cos 2x| = \dfrac{1}{2}$

28. $4\sin^3 x - \sin x = 0$

29. $\ln\cot x = 1$

CHAPTER 7 NAME ____________________

TEST FORM F CLASS ________ SCORE ________ GRADE ______

Write the letter of your response on the answer blank.

ANSWERS

1. Using a sum or a difference formula write an equivalent expression for $\sin\left(x + \frac{\pi}{2}\right)$.

   a) $\cos x \cos \frac{\pi}{2} + \sin x \sin \frac{\pi}{2}$

   b) $\sin x \cos \frac{\pi}{2} - \cos x \sin \frac{\pi}{2}$

   c) $\cos x \cos \frac{\pi}{2} - \sin x \sin \frac{\pi}{2}$

   d) $\sin x \cos \frac{\pi}{2} + \cos x \sin \frac{\pi}{2}$

1. ____________

2. Simplify: $\sin 74^\circ \cos 17^\circ + \cos 74^\circ \sin 17^\circ$.

   a) $\cos 91^\circ$ b) $\sin 91^\circ$ c) $\cos 57^\circ$ d) $\sin 91^\circ$

2. ____________

3. Find $\cos 195^\circ$ exactly.

   a) $-\frac{\sqrt{6} + \sqrt{2}}{2}$ b) $-\frac{\sqrt{6} + \sqrt{2}}{4}$

   c) $\frac{\sqrt{2} - \sqrt{6}}{4}$ d) $\frac{\sqrt{6} - \sqrt{2}}{4}$

3. ____________

4. Given $\tan \alpha = \frac{\sqrt{3}}{3}$, $\sin \beta = \frac{\sqrt{2}}{2}$, and α and β are between 0 and $\frac{\pi}{2}$, evaluate $\cos(\alpha + \beta)$ exactly.

   a) $\frac{\sqrt{6} - \sqrt{2}}{4}$ b) 0

   c) $\frac{\sqrt{3}}{2}$ d) $\frac{\sqrt{2} + \sqrt{6}}{2}$

4. 

5. Complete this Pythagorean identity: $1 + \cot^2 \theta =$ ____.

   a) $\sec^2 \theta$ b) $\csc \theta$ c) $\sec \theta$ d) $\csc^2 \theta$

5. ____________

6. Complete this ccfunction identity: $\sin\left(\theta - \frac{\pi}{2}\right) =$ ____.

   a) $\sin \theta$ b) $-\sin \theta$ c) $-\cos \theta$ d) $\cos \theta$

6. ____________

NAME ______________________

TEST FORM F

ANSWERS

7. ______

8. ______

9. ______

10. ______

11. ______

12. ______

7. Complete this cofunction identity: $\cos\left(\frac{\pi}{2} - \theta\right) = $ ____.

a) $-\cos\theta$ b) $\sin\theta$ c) $-\sin\theta$ d) $\cos\theta$

8. Express $\sec\theta$ in terms of $\tan\theta$.

a) $\sec\theta = \pm\sqrt{1 + \tan^2\theta}$ b) $\sec\theta = \pm\sqrt{1 + \cot^2\theta}$

c) $\sec\theta = \pm\sqrt{1 - \tan^2\theta}$ d) $\sec\theta = \pm\sqrt{\cos^2\theta - 1}$

9. Simplify: $\frac{1}{\sec x}(\tan x + \cot x)$.

a) $\cos x$ b) $\frac{1}{\sin x \cdot \cos x}$

c) $\csc x$ d) $\sin x$

10. Simplify: $\frac{\sqrt{1 + \cot^2 x}}{\cot x}$.

a) $\frac{\sin x}{\cos^2 x}$ b) $\sec x$

c) $\sqrt{\tan x + \cot x}$ d) $\sqrt{1 + \cot x}$

11. Rationalize the denominator: $\sqrt{\frac{\cos x}{\tan x}}$.

a) $\cot x$ b) $\frac{\sqrt{\cos x}}{\tan x}$

c) $\frac{\sqrt{\sin x}}{\csc x}$ d) $\frac{\sqrt{\sin x}}{\tan x}$

12. Find $\cos 2\theta$ if $\tan\theta = -\frac{4}{3}$ and θ is in quadrant IV.

a) $-\frac{7}{25}$ b) 1 c) $\frac{7}{25}$ d) $\frac{7}{5}$

CHAPTER 7 NAME ______________________

TEST FORM F

ANSWERS

13. Find $\sin \frac{7\pi}{8}$ exactly. 13. ________

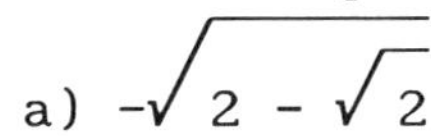

a) $-\sqrt{2 - \sqrt{2}}$ b) $\sqrt{2 - \sqrt{2}}$

c) $\frac{\sqrt{2 - \sqrt{2}}}{2}$ d) $-\sqrt{1 - \sqrt{2}}$

14. Simplify: $\frac{\sin 2x}{\sin^2 x}$. 14. ________

a) $2 \cot x$ b) $2 \tan x$

c) $2 - \cos^2 x$ d) $- \cos^2 x$

15. Use a calculator to find arcsin (-0.3842). 15. ________

a) 23° b) 203° c) 67° d) -23°

16. Find $\operatorname{arccot} \frac{\sqrt{3}}{3}$. 16. ________

a) $\frac{\pi}{6}$ b) $\frac{\pi}{3}$ c) $\frac{5\pi}{6}$ d) $\frac{2\pi}{3}$

17. Find $\sin \cos^{-1} \left(-\frac{\sqrt{2}}{2}\right)$. 17. ________

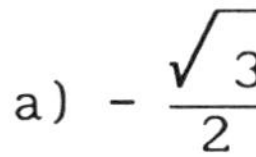

a) $-\frac{\sqrt{3}}{2}$ b) $-\frac{1}{2}$ c) $-\frac{\sqrt{2}}{2}$ d) $\frac{\sqrt{2}}{2}$

18. Find $\operatorname{Arccos} \cos \frac{\pi}{6}$.

a) $\frac{\pi}{6}$ b) $\frac{1}{2}$ c) $\frac{5\pi}{6}$ d) $\frac{\sqrt{3}}{2}$ 18. ________

NAME ______________________

TEST FORM F

ANSWERS

19. ______

20. ______

21. ______

22. ______

23. See work.

In questions 19 and 20 use the following information:

$\sin \theta = 0.6691$	$\tan \theta = 0.9004$	$\sec \theta = 1.346$
$\cos \theta = 0.7431$	$\cot \theta = 1.111$	$\csc \theta = 1.495$

19. Find $\sec (90^\circ - \theta)$.

    a) 1.111 b) 0.7431 c) 0.6691 d) 1.495

20. Find $\cot (90^\circ - \theta)$.

    a) 0.6691 b) 0.9004 c) 1.111 d) 1.346

21. Solve: $2 \sin^2 x - \sqrt{2} \sin x = 0$. Restrict solutions to $[0, 2\pi)$.

    a) $\frac{\pi}{2}, \pi$ b) $0, \frac{\pi}{3}, \frac{2\pi}{3}, \pi$

    c) $0, \pi, \frac{4\pi}{3}, \frac{5\pi}{3}$ d) $0, \frac{\pi}{4}, \frac{3\pi}{4}, \pi$

22. Solve: $\ln \sin x = 0$. Restrict solutions to $[0, 2\pi)$.
    The solution set

    a) is empty. b) contains only one solution.

    c) contains only two solutions.

    d) contains only four solutions

23. Prove the identity $\csc 2\theta + \cot 2\theta = \cot \theta$.
    Show your work below.

CHAPTER 8 *NAME* ______

TEST FORM A *CLASS* ______ *SCORE* ______ *GRADE* ______

Solve each of the following right triangles. Standard lettering has been used.

1. $A = 39.6^\circ$, $b = 21.5$

2. $a = 16$, $c = 21$

3. An observer stands on level ground, 150 m from the base of a tower, and looks up at an angle of 30.5° to see the top of the tower. How high is the tower above the observer's eye level?

4. Solve triangle ABC, where $A = 57^\circ$, $C = 22^\circ$, and $a = 5.24$.

5. In triangle ABC, $b = 15$, $c = 12$, and $A = 42^\circ$. Find a.

6. In an isosceles triangle, the base angles each measure 47.5° and the base is 250 cm long. Find the lengths of the other two sides.

7. Find the area of triangle ABC if $a = 5.1$ m, $c = 7.3$ m, and $B = 79.4^\circ$.

8. Two forces of 8 newtons and 14 newtons act on an object at right angles. Find the magnitude of the resultant and the angle that it makes with the smaller force.

9. Find rectangular notation for the vector $(15, 210^\circ)$.

10. Find polar notation for the vector $\langle 6,8 \rangle$.

ANSWERS

1. ______

2. ______

3. ______

4. ______

5. ______

6. ______

7. ______

8. ______

9. ______

10. ______

NAME ______________________

TEST FORM A

ANSWERS	
11. a) ________ b) ________	11. Do the following calculations for $\mathbf{u} = \langle 3,-1\rangle$ and $\mathbf{v} = \langle 4,9\rangle$: (a) $\mathbf{u} - \mathbf{v}$; (b) $\lvert 3\mathbf{u} + \mathbf{v}\rvert$.
12. a) ________ b) ________	12. Vector **u** has a westerly component of 5 and a northerly component of 7. Vector **v** has an easterly component of 9 and a southerly component of 2. Find: a) the components of $\mathbf{u} + \mathbf{v}$; b) the magnitude and the direction of $\mathbf{u} + \mathbf{v}$.
13. ________	13. Find Cartesian coordinates of the point $(-7, 150^\circ)$.
14. ________	14. Convert $r = 9$ to a rectangular equation.
15. ________	15. Solve $x^2 = 3i$.
16. See graph.	16. Graph the pair of complex numbers $6 - 3i$ and $-1 + 2i$ and their sum.

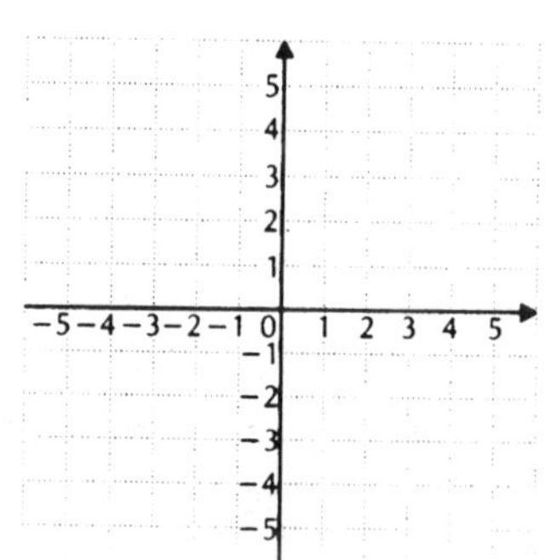

 NAME ____________________

ANSWERS

17. Find rectangular notation for $2 \text{ cis } 45^\circ$.

17. ____________

18. Find polar notation for $-1 + i$.

18. ____________

19. Find the product of $5 \text{ cis } 30^\circ$ and $2 \text{ cis } 40^\circ$.

19. ____________

20. Divide $4 \text{ cis } 80^\circ$ by $3 \text{ cis } 15^\circ$.

20. ____________

21. Find $(1 - i)^5$. Write polar notation for the answer.

21. ____________

22. Find the cube roots of $-i$.

22. ____________

NAME ____________________

ANSWERS

23. See graph.

24. See graph.

25. ____________________

23. Graph: $r = -4\cos\theta$.

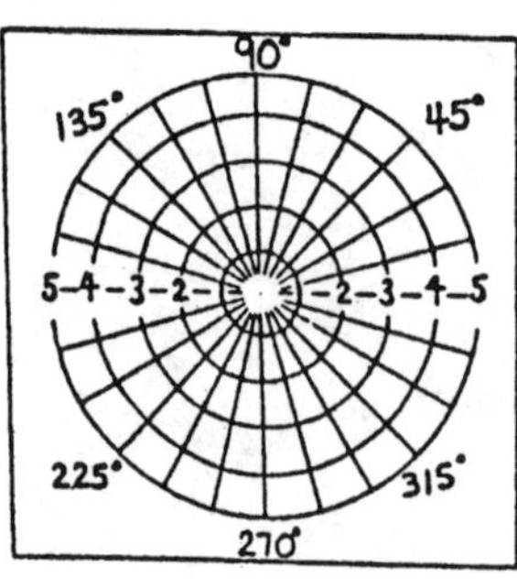

24. Graph: $r = \dfrac{3}{1 + \cos\theta}$.

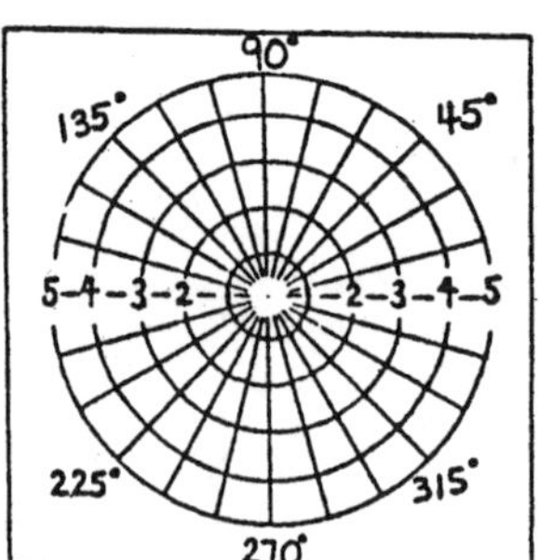

25. Find h to the nearest tenth.

CHAPTER 8 *NAME* ______

TEST FORM B *CLASS* ______ *SCORE* ______ *GRADE* ______

Solve each of the following right triangles. Standard lettering has been used.

1. $B = 19.5^\circ$, $a = 32$

2. $a = 12$, $b = 8$

3. A building 90 m high casts a shadow 120 m long. What is the angle of elevation of the sun?

4. Solve triangle ABC, where $A = 92^\circ$, $B = 51^\circ$, and $a = 10.4$.

5. In triangle ABC, $a = 9$, $b = 10$, and $C = 75^\circ$. Find c.

6. A triangular flower bed measures 6 ft on one side, and that side makes angles of 29.4° and 60° with the other sides. How long are the other two sides?

7. Find the area of triangle ABC if $b = 10.2$ cm, $c = 4.6$ cm, and $A = 39.5^\circ$.

8. Two forces of 10 newtons and 15 newtons act on an object at right angles. Find the magnitude of the resultant and the angle that it makes with the larger force.

9. Find rectangular notation for the vector $(18, 135^\circ)$.

10. Find polar notation for the vector $\langle 9, -15 \rangle$.

ANSWERS

1. ______

2. ______

3. ______

4. ______

5. ______

6. ______

7. ______

8. ______

9. ______

10. ______

NAME ______________________

TEST FORM B

ANSWERS

11. a) ______________

b) ______________

12. a) ______________

b) ______________

13. ______________

14. ______________

15. ______________

16. 

11. Do the following calculations for **u** = <5,2> and **v** = <-7,-3>: (a) **u** + **v**; (b) $|2\mathbf{u} - \mathbf{v}|$.

12. Vector **u** has a easterly component of 15 and a northerly component of 10. Vector **v** has an westerly component of 3 and a southerly component of 9. Find:

a) the components of **u** + **v**;

b) the magnitude and the direction of **u** + **v**.

13. Find Cartesian coordinates of the point $(-2, 45^\circ)$.

14. Convert r = 11 to a rectangular equation.

15. Solve $x^2 = 12i$.

16. Graph the pair of complex numbers -2 + 5i and -1 - 2i and their sum.

Imaginary

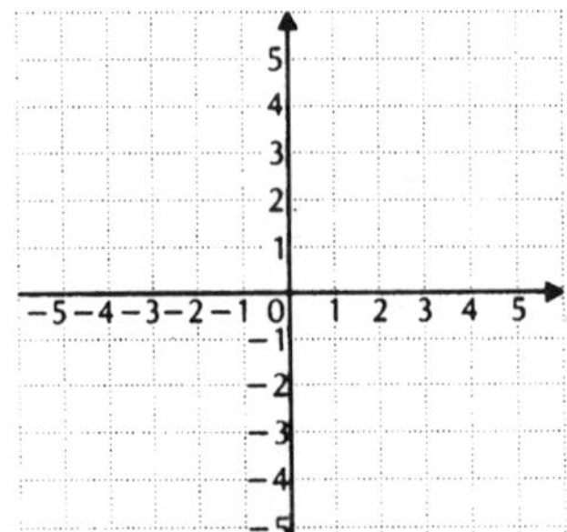

Real

NAME ______________________

ANSWERS

17. Find rectangular notation for 5 cis (-60°).

17. __________

18. Find polar notation for $\sqrt{3} - i$.

18. __________

19. Find the product of 4 cis 20° and 8 cis 15°.

19. __________

20. Divide 9 cis 100° by 3 cis 25°.

20. __________

21. Find $(2 + 2i)^8$. Write polar notation for the answer.

21. __________

22. Find the cube roots of 1.

22. __________

NAME ______________________

ANSWERS

23. See graph.

24. See graph.

25. ______________

23. Graph: $r = 2 + 2\cos\theta$.

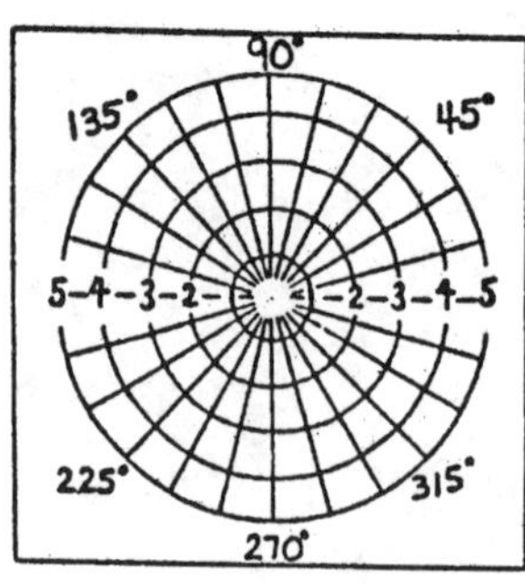

24. Graph: $r = \dfrac{4}{1 + \sin\theta}$.

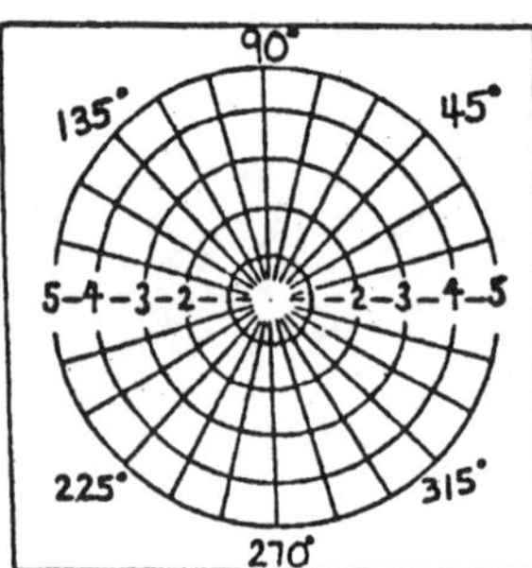

25. A parallelogram has sides of lengths 5.12 and 9.34. Its area is 23.35. Find the sizes of its angles.

CHAPTER 8

TEST FORM C

NAME ____________________

CLASS ________ SCORE __________ GRADE ______

Solve each of the following right triangles. Standard lettering has been used.

1. $A = 61.4^\circ$, $c = 11.5$

2. $b = 9$, $c = 19$

3. A guy wire is 21.8 m long and is fastened from the ground to a pole 8.2 m above the ground. What angle does the wire make with the ground?

4. Solve triangle ABC, where $A = 65^\circ$, $B = 54^\circ$, and $b = 11.2$.

5. In triangle ABC, $a = 11$, $c = 15$, and $B = 58^\circ$. Find b.

6. In an isosceles triangle the base angles each measure 31.4° and the base is 115 cm long. Find the lengths of the other two sides.

7. Find the area of triangle ABC if $a = 4.9$ in., $b = 7.1$ in., and $C = 92.4^\circ$.

8. Two forces of 8 newtons and 15 newtons act on an object at right angles. Find the magnitude of the resultant and the angle that it makes with the smaller force.

9. Find rectangular notation for the vector $(10,120^\circ)$.

10. Find polar notation for the vector $\langle -4,7 \rangle$.

ANSWERS

1. ____________

2. ____________

3. ____________

4. ____________

5. ____________

6. ____________

7. ____________

8. ____________

9. ____________

10. ____________

11. Do the following calculations for **u** = <-7,3> and **v** = <2,-5>: (a) **u** − **v**; (b) $|\mathbf{u} + 2\mathbf{v}|$.

12. Vector **u** has a westerly component of 7 and a northerly component of 3. Vector **v** has an westerly component of 2 and a southerly component of 7. Find:

    a) the components of **u** + **v**;

    b) the magnitude and the direction of **u** + **v**.

13. Find Cartesian coordinates of the point $(-4, 60^\circ)$.

14. Convert r = 12 to a rectangular equation.

15. Solve $x^2 = 9i$.

16. Graph the pair of complex numbers 3 - 4i and 2 - i and their sum.

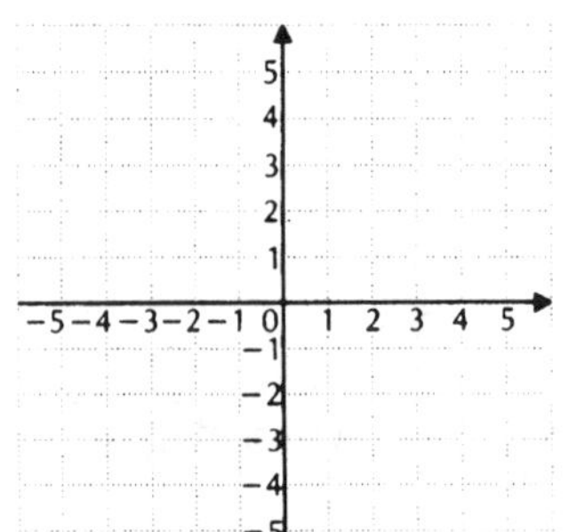

ANSWERS

11. a) ____________

    b) ____________

12. a) ____________

    b) ____________

13. ____________

14. ____________

15. ____________

16. See graph.

ANSWERS

17. Find rectangular notation for 6 cis (210°).

17. ______

18. Find polar notation for $-1 - i$.

18. ______

19. Find the product of 7 cis 29° and 3 cis 18°.

19. ______

20. Divide 8 cis 150° by 4 cis 40°.

20. ______

21. Find $(\sqrt{3} - i)^5$. Write polar notation for the answer.

21. ______

22. Find the cube roots of -27.

22. ______

NAME ______________________

ANSWERS

23. See graph.

24. See graph.

25. ______________

23. Graph: $r = 5 \sin 3\theta$.

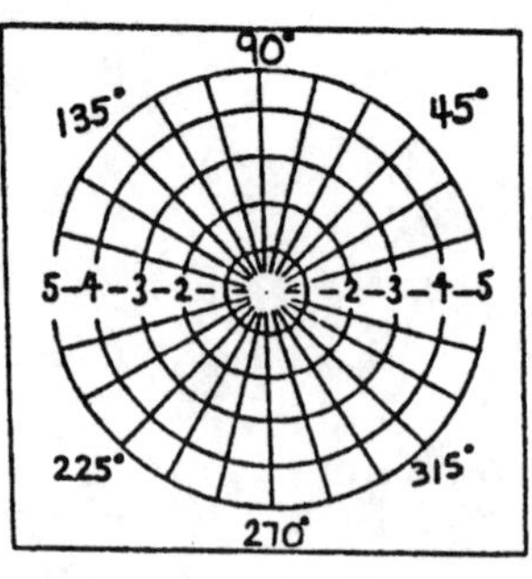

24. Graph: $r = \dfrac{4}{1 + \cos \theta}$.

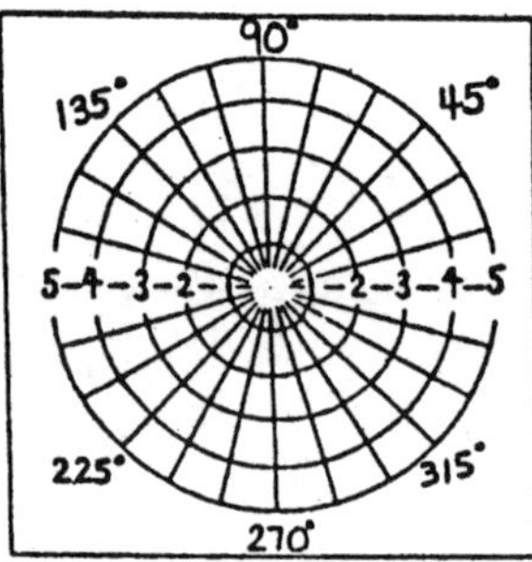

25. Let **u** = $\langle 4,3 \rangle$. Find a vector that has the same direction as **u** but length 2.

CHAPTER 8 NAME ____________________

TEST FORM D CLASS ________ SCORE ________ GRADE ______

Solve each of the following right triangles. Standard lettering has been used.

1. $B = 58.2^\circ$, $c = 21.4$

2. $a = 3$, $b = 8$

3. A kite string makes an angle of 40.5° with the (level) ground, and 475 ft of string is out. How high is the kite?

4. Solve triangle ABC, where $B = 85^\circ$, $C = 43^\circ$, and $c = 25.4$.

5. In triangle ABC, $a = 5$, $b = 19$, and $C = 104^\circ$. Find c.

6. A triangular flower bed measures 10 ft on one side, and that side makes angles of 42.8° and 51° with the other sides. How long are the other two sides?

7. Find the area of triangle ABC if $b = 20.1$ ft, $c = 15.2$ ft, and $A = 100.2^\circ$.

8. Two forces of 20 newtons and 14 newtons act on an object at right angles. Find the magnitude of the resultant and the angle that it makes with the larger force.

9. Find rectangular notation for the vector $(11, 315^\circ)$.

10. Find polar notation for the vector $\langle -6, -11 \rangle$.

ANSWERS

1. ____________

2. ____________

3. ____________

4. ____________

5. ____________

6. ____________

7. ____________

8. ____________

9. ____________

10. ____________

NAME ______________________

TEST FORM D

ANSWERS

11. a) ____________

b) ____________

12. a) ____________

b) ____________

13. ____________

14. ____________

15. ____________

16. See graph.

11. Do the following calculations for **u** = <-1,-6> and **v** = <3,-2>: (a) **u** + **v**; (b) |**u** - 3**v**|.

12. Vector **u** has a westerly component of 3 and a northerly component of 5. Vector **v** has an easterly component of 6 and a northerly component of 4. Find:

a) the components of **u** + **v**;

b) the magnitude and the direction of **u** + **v**.

13. Find Cartesian coordinates of the point $(-5, 240^\circ)$.

14. Convert r = 13 to a rectangular equation.

15. Solve $x^2 = 15i$.

16. Graph the pair of complex numbers -3 + 2i and -1 + 3i and their sum.

NAME ______________________

17. Find rectangular notation for $4 \text{ cis } (-45^\circ)$.

18. Find polar notation for $8\sqrt{3} + 8i$.

19. Find the product of $5 \text{ cis } 17^\circ$ and $4 \text{ cis } 16^\circ$.

20. Divide $5 \text{ cis } 165^\circ$ by $3 \text{ cis } 100^\circ$.

21. Find $(\sqrt{3} + i)^5$. Write polar notation for the answer.

22. Find the cube roots of $-8i$.

ANSWERS

17. ______________

18. ______________

19. ______________

20. ______________

21. ______________

22. ______________

ANSWERS	
23. See graph.	23. Graph: $r = 4 \sin 2\theta$.
24. See graph.	24. Graph: $r = \frac{3}{1 + \sin \theta}$.
25. ______	25. A parallelogram has sides of lengths 5.4 m and 9.2 m. One of its angles is 112°. Find the area of the parallelogram.

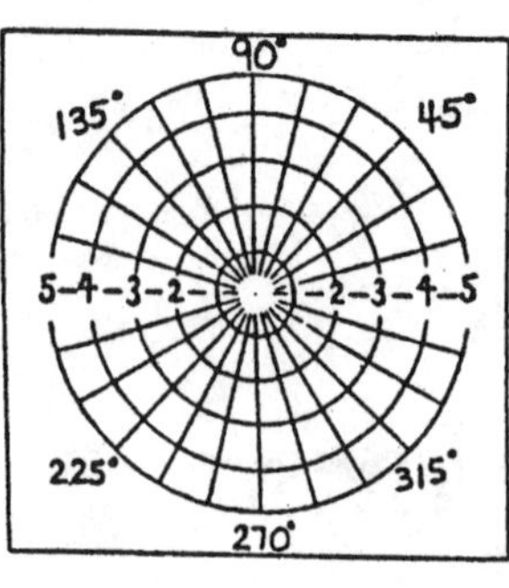

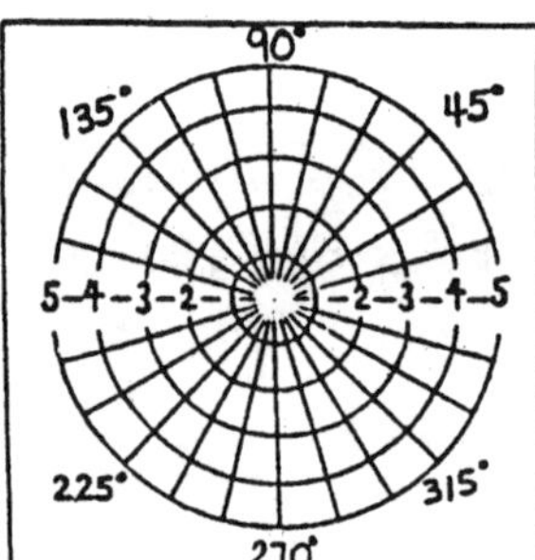

CHAPTER 8 *NAME* ____________________

TEST FORM E *CLASS* ________ *SCORE* ________ *GRADE* ______

Solve each of the following right triangles. Standard lettering has been used.

1. $B = 15.2^\circ$, $a = 34.2$

2. $a = 7.5$, $b = 3.9$

3. From a fire tower 91 ft high, a small fire is seen. The angle of depression to the fire is 41.6°. How far is the fire from the foot of the fire tower?

4. Solve triangle ABC, where $B = 115.9^\circ$, $C = 12.3^\circ$, and $c = 0.125$.

5. In triangle ABC, $a = 5$, $b = 8$, and $c = 12$. Find B.

6. A guy wire to a pole makes a 45.8° angle with level ground. At a point 6.1 m farther from the pole than the guy wire, the angle of elevation of the top of the pole is 31.7°. How long is the guy wire?

7. Two forces of 22 newtons and 53 newtons act on an object. The angle between the forces is 50°. Find the magnitude of the resultant and the angle that it makes with the smaller force.

8. Find polar notation for the vector $\langle -3,15 \rangle$.

9. Find rectangular notation for the vector $(24,108^\circ)$.

ANSWERS

1. ____________
2. ____________
3. ____________
4. ____________
5. ____________
6. ____________
7. ____________
8. ____________
9. ____________

CHAPTER 8 NAME ______________________

TEST FORM E

ANSWERS	
10. a) ______	10. Do the following calculations for $\mathbf{u} = \langle 2,-1\rangle$ and $\mathbf{v} = \langle 5,-3\rangle$: (a) $\lvert\mathbf{u}\rvert + \lvert 3\mathbf{v}\rvert$; (b) $2\mathbf{u} - \mathbf{v}$.
b) ______	
	11. Find the area of triangle ABC if $a = 7.2$ m, $c = 4.9$ m, and $B = 110.5^\circ$.
	12. Find polar coordinates of the point $(7,-7)$.
11. ______	
	13. Convert to a polar equation: $x^2 + y^2 = 400$.
	14. Convert to a rectangular equation: $r + r \sin \theta = 5$.
12. ______	
	15. Solve: $x^2 = 2 + 2\sqrt{3}i$.
13. ______	16. Graph the pair of complex numbers $-4 - 2i$ and $5 + 6i$ and their sum.
14. ______	
15. ______	
16. See graph.	

Imaginary

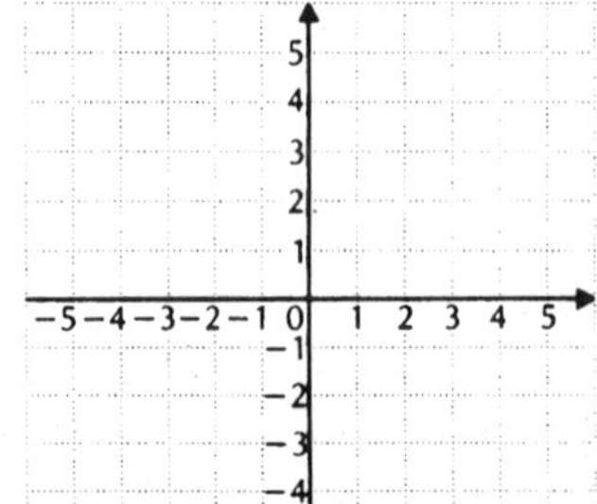

Real

ANSWERS

17. Find rectangular notation for $4 \text{ cis } \left(-\frac{\pi}{2}\right)$.

17. ______________

18. Find the product of $16 \text{ cis } 190^\circ$ and $3 \text{ cis } 30^\circ$.

18. ______________

19. Convert $\dfrac{3 - \sqrt{3}i}{\sqrt{3} + i}$ to polar notation and divide.

19. ______________

20. Find $(\sqrt{3} - i)^5$. Write polar notation for the answer.

20. ______________

21. Find the sixth roots of 64.

21. ______________

NAME ______________________

TEST FORM E

ANSWERS

22. See graph.

23. See graph.

24. ______________

22. Graph: $r = 4 \sin \theta$.

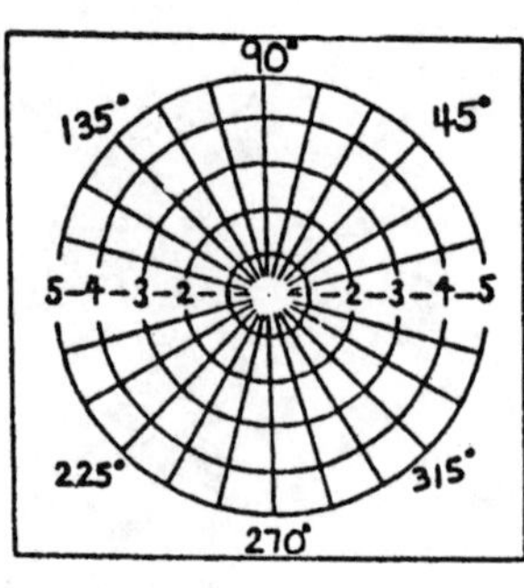

23. Graph: $r^2 = 4 \cos 2\theta$.

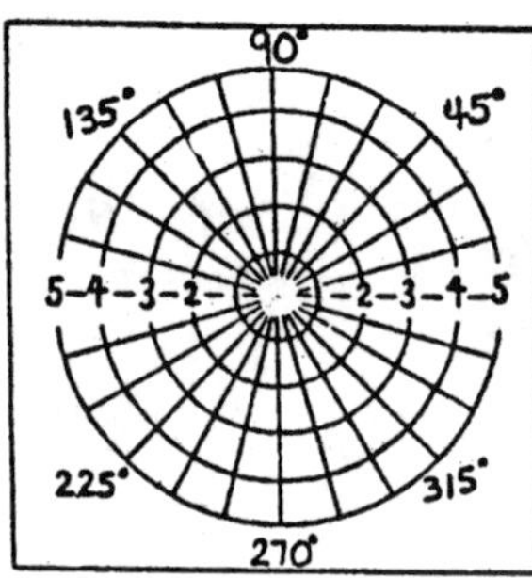

24. A balloon is rising 11 ft/sec while a wind is blowing 9 ft/sec. Find the speed of the balloon and the angle it makes with the horizontal.

CHAPTER 8 *NAME* ______

TEST FORM F *CLASS* ______ *SCORE* ______ *GRADE* ______

Write the letter of your response on the answer blank.

ANSWERS

1. Solve this right triangle. Standard lettering has been used.

   $A = 42.8^\circ$, $a = 308$

   Find b.

   a) 285 b) 209 c) 333 d) 22

1. ______

2. Solve this right triangle. Standard lettering has been used.

   $b = 5.2$, $c = 9.1$

   Find the measure of angle A to the nearest degree.

   a) 35° b) 99° c) 60° d) 55°

2. ______

3. A guy wire from the top of a pole is 10 m long and makes an angle of 75° with the ground. How tall is the pole?

   a) 13.02 m b) 9.66 m c) 9.93 m d) 2.59 m

3. ______

4. Solve triangle ABC, where $A = 32^\circ$, $B = 51^\circ$, and $c = 20$. Find b.

   a) 15.66 b) 10.68 c) 19.85 d) 15.54

4. ______

5. In triangle ABC, $a = 5$, $c = 3$, and $B = 45^\circ$. Find b.

   a) 4.7 b) 5.2 c) 6.3 d) 3.6

5. ______

ANSWERS

6. ____________

6. In an isosceles triangle, the base angles each measure 29.5° and the base measures 12.2 in. Find the lengths of the other two sides.

a) 21.24 in. b) 7.01 in. c) 6.17 in. d) 99.37 in.

7. ____________

7. Two forces of 10 newtons and 20 newtons act on an object at right angles. Find the magnitude of the resultant.

a) 26.6 newtons b) 29.5 newtons

c) 22.4 newtons d) 49.1 newtons

8. ____________

8. Find polar notation for the vector <-5,-4>.

a) (6.4, 218.7°) b) (3.0,218.7°)

c) (6.1, 321.3°) d) 6.4, 38.7°)

9. ____________

9. Find rectangular notation for the vector (6,240°).

a) <-5.20,-3> b) <-3.00,-5.20>

c) <4.05,6.37> d) <10.4,3.46>

ANSWERS

10. If **u** = <2,-4> and **v** = <1,-2>, find $|3\mathbf{v} - 4\mathbf{u}|$.

a) <6,-12> b) 4.24 c) <2,-4) d) 4.47

10. ________

11. Vector **u** has a westerly component of 5 and a northerly component of 8. Vector **v** has a westerly component of 15 and a southerly component of 2. Find the direction of **u** + **v**.

a) S 73.3° W b) N 73.3° W

c) N 59° W d) N 16.7° W

11. ________

12. Convert to a polar equation: $3x - 2y = 8$.

a) $3r \sin \theta + 2r \cos \theta = 8$ b) $2r \sin \theta - 3r \cos \theta = 8$

c) $3r \cos \theta - 2r \sin \theta = 8$ d) $3r \cos \theta + 2r \sin \theta = 8$

12. ________

13. Convert to a rectangular equation: $r = 2 \sin \theta$.

a) $x^2 - 2x = -y^2$ b) $x^2 + y^2 = 2y$

c) $x^2 = 3y^2$ d) $r^2 = 2y$

13. ________

NAME ____________________

TEST FORM F

ANSWERS

14. ____________________

15. ____________________

16. ____________________

17. ____________________

14. Solve: $x^2 = 8i$.

a) $2 + 4i, -2 - 4i$

b) $2 + 2i, -2 - 2i$

c) $4i, -4i$

d) $2i, -2i$

15. Graph the pair of complex numbers $2 + 4i$, $3 - 3i$ and their sum.

a)

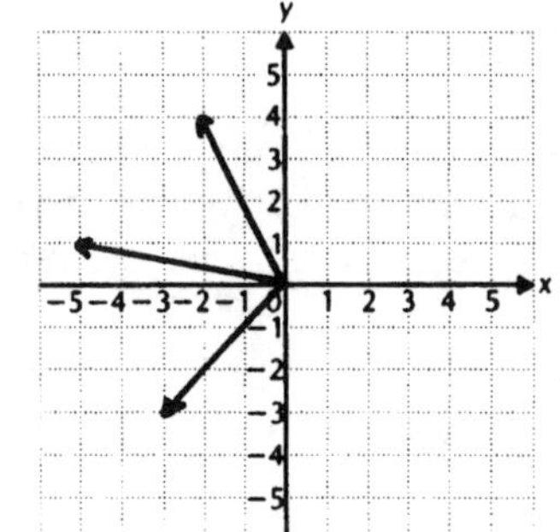

b)

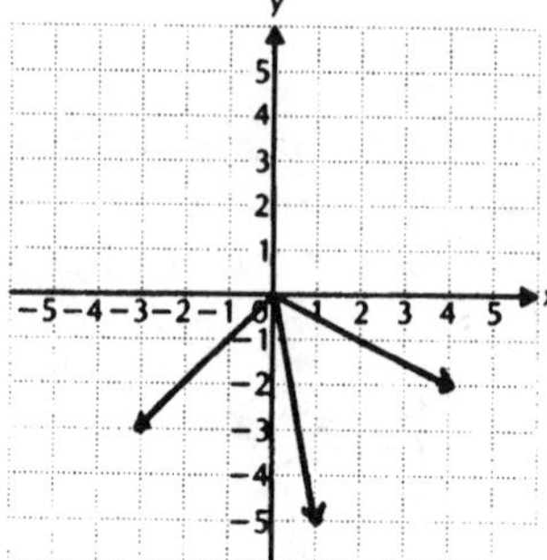

c)

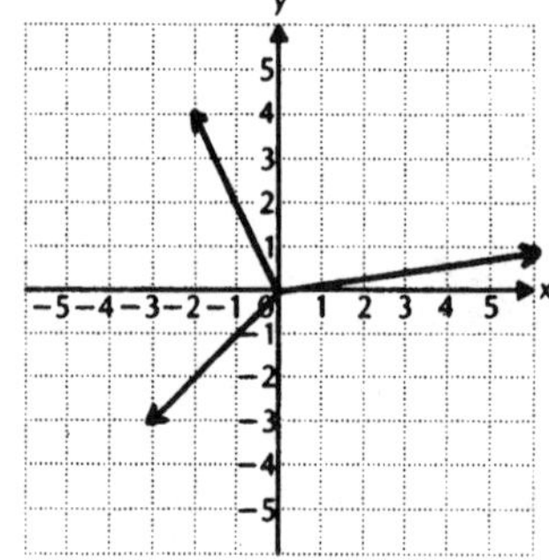

d)

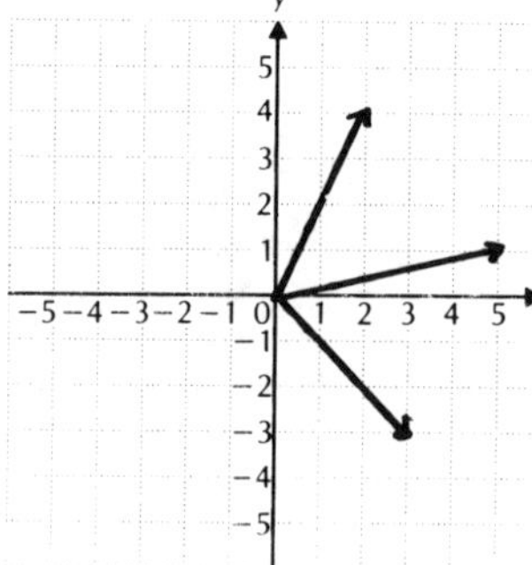

16. Find the rectangular notation for $12 \text{ cis } 300^\circ$.

a) $6 - 6i$

b) $-6 + 6\sqrt{3}i$

c) $6 - 6\sqrt{3}i$

d) $12 - 12\sqrt{3}i$

17. Find polar notation for $-6\sqrt{2} - 6\sqrt{2}i$.

a) $12 \text{ cis } 225^\circ$

b) $6 \text{ cis } 225^\circ$

c) $12 \text{ cis } 45^\circ$

d) $6 \text{ cis } 45^\circ$

NAME ____________________

TEST FORM F

ANSWERS

18. Find the product of 4 cis 60° and 3 cis 40°.

a) 12 cis 240°   b) 7 cis 20°

c) 12 cis 20°   d) 12 cis 100°

18. ____________

19. Divide 20 cis 240° by 4 cis 80°.

a) 5 cis 3°   b) 5 cis 160°

c) 16 cis 160°   d) 5 cis 320°

19. ____________

20. Find $(1 - i)^8$. Write polar notation for the answer.

a) 16 cis 0°   b) 16 cis 135°

c) $\sqrt{2}$ cis 0°   d) $\sqrt{2}$ cis 315°

20. ____________

21. Find the cube roots of -64.

a) 16 cis 45°, 16 cis 165°, 16 cis 285°

b) 64 cis 60°, 64 cis 180°, 64 cis 300°

c) 4 cis 60°, 4 cis 180°, 4 cis 300°

d) 4 cis 30°, 4 cis 150°, 4 cis 270°

21. ____________

22. Find the area of triangle ABC if b = 15.4 m, c = 9.3 m, and A = 59.7°.

a) 123.6 m   b) 36.1 m   c) 21.3 m   d) 61.8 m

22. ____________

ANSWERS

23. ______

24. ______

23. Graph $r = 5 \sin \theta$.

a)

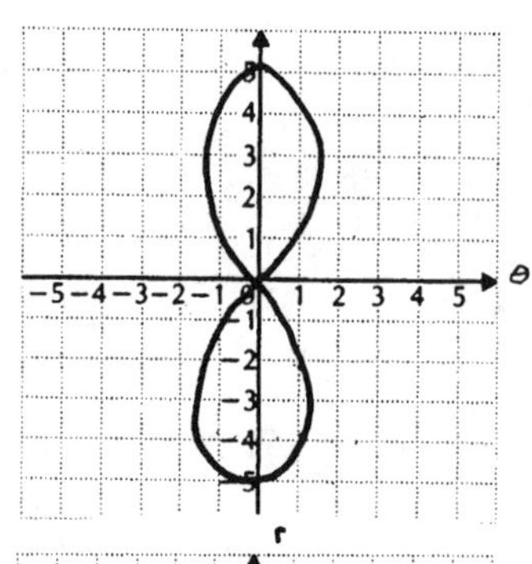

b)

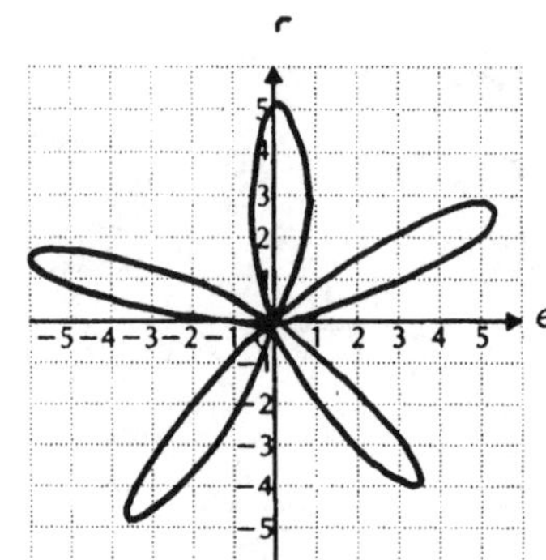

c)

d)

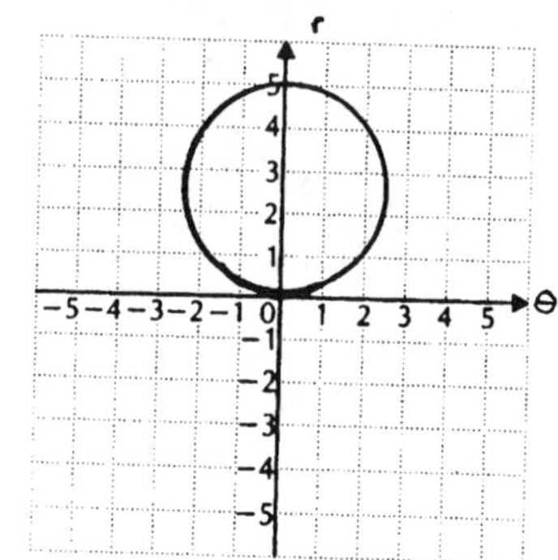

24. Graph $r = 3 - 3 \cos \theta$.

a)

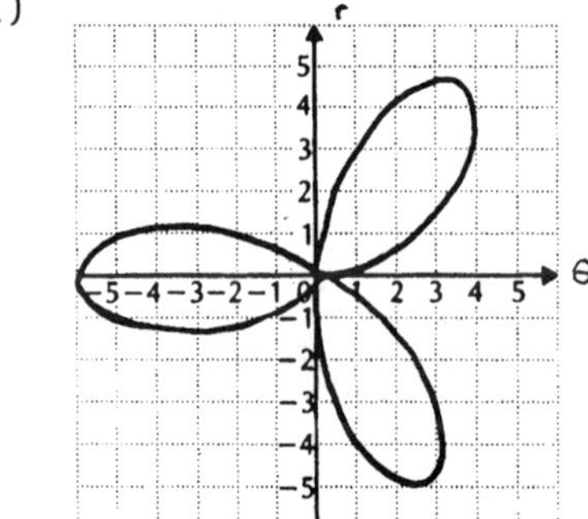

b)

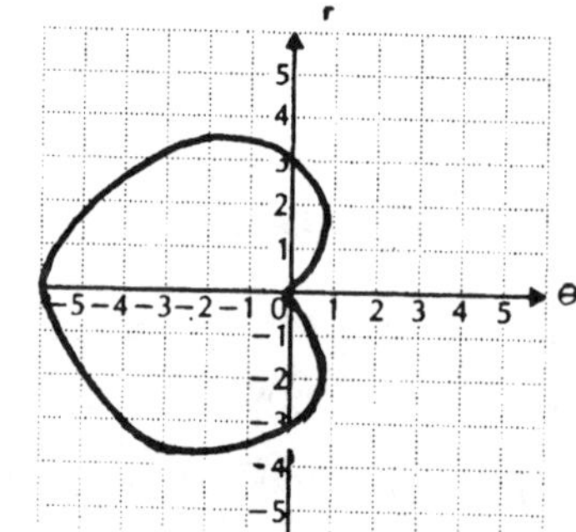

c)

d)

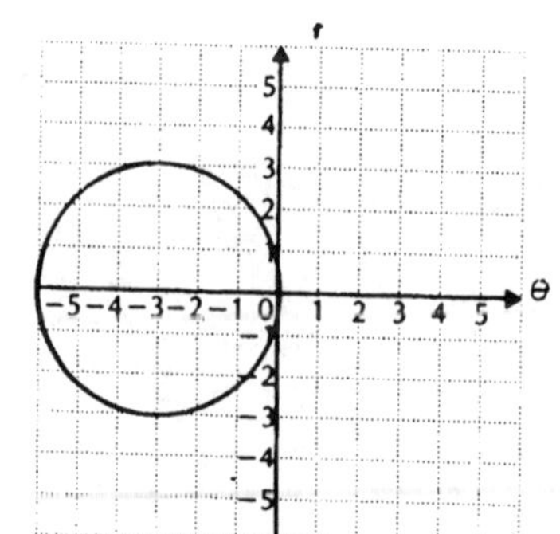

CHAPTER 9 NAME ______________

TEST FORM A CLASS ______ SCORE ______ GRADE ______

Solve.

1. $0.3x - 0.2y = 1$,
   $0.01x + 0.1y = 0.14$.

2. A boat travels 50 km downstream in 2 hr. It travels 51 km upstream in 3 hr. Find the speed of the boat and the speed of the stream.

3. One week, a business sold 35 sweatshirts. Black ones cost $20.50 and red ones cost $19.75. In all, $707 worth of sweatshirts were sold. How many of each color were sold?

4. The sum of three numbers is 29. The third number is twice the second minus three. The second number is four more than the first. Find the numbers.

Solve using matrices. If there is more than one solution, list three of them.

5. $8x - 4y = 5$,
   $10x + 8y = 3$

6. $2x - y - 2z = 4$,
   $x - 2y + z = -1$,
   $6x - 6y - 2z = 6$

7. $3x - y + z = 2$,
   $x - 2y + z = 1$,
   $4x + y + 2z = 7$

Classify as consistent or inconsistent, dependent or independent.

8. $x - y = -5$,
   $-x - y = 5$

9. $x - 5y + z = 2$,
   $4x + y - z = 3$,
   $3x + 6y - 2z = 1$

10. Find numbers a, b, and c such that the function $f(x) = ax^2 + bx + c$ fits the data points (0,3), (1,4), and (-2,-5). Then write the equation for the function.

Evaluate.

11. $\begin{vmatrix} \frac{1}{2} & 4 \\ -2 & 6 \end{vmatrix}$

12. $\begin{vmatrix} 1 & 3 & -1 \\ 0 & 4 & 1 \\ 1 & 0 & 5 \end{vmatrix}$

ANSWERS

1. ______________
2. ______________
3. ______________
4. ______________
5. ______________
6. ______________
7. ______________
8. ______________
9. ______________
10. ______________
11. ______________
12. ______________

NAME ______________________

TEST FORM A

ANSWERS
13. ______
14. ______
15. ______
16. ______
17. ______
18. ______
19. ______
20. ______
21. ______
22. ______
23. ______
24. ______
25. ______
26. ______
27. ______

Solve using Cramer's rule.

13. $3x + y = 7,$
$2x - 4y = 14$

14. $x + y - z = -4,$
$2x + y + 3z = 9,$
$x + 3y - 2z = -11$

For Exercises 15 - 24, let

$$A = \begin{bmatrix} -1 & 4 \\ 3 & 2 \end{bmatrix} \quad B = \begin{bmatrix} -1 & 0 & 2 \\ 4 & 2 & 5 \end{bmatrix} \quad C = \begin{bmatrix} 1 & 0 \\ 2 & -1 \\ 3 & 4 \end{bmatrix}$$

$$D = \begin{bmatrix} 1 & 2 & -3 \\ 4 & 1 & 5 \\ 0 & 2 & 1 \end{bmatrix} \quad E = \begin{bmatrix} 3 & -1 & 2 \\ 0 & 1 & 4 \\ 5 & 2 & 6 \end{bmatrix} \quad F = \begin{bmatrix} 1 & 0 & -4 \end{bmatrix}$$

$$G = \begin{bmatrix} -1 & 5 \\ 2 & -3 \end{bmatrix} \quad O = \begin{bmatrix} 0 & 0 \\ 0 & 0 \end{bmatrix} \quad I = \begin{bmatrix} 1 & 0 \\ 0 & 1 \end{bmatrix}.$$

Find each of the following, if possible.

15. GC
16. O + A
17. A + I
18. D + E
19. A - G
20. BC
21. B + C
22. -F
23. 2D - E
24. GI

Find A^{-1}, if it exists.

25. $A = \begin{bmatrix} 3 & 2 \\ -1 & 1 \end{bmatrix}$

26. $A = \begin{bmatrix} -1 & 2 & 1 \\ 3 & -4 & 2 \\ 2 & -4 & -2 \end{bmatrix}$

27. $A = \begin{bmatrix} 2 & -1 & 0 \\ 3 & -2 & 1 \\ 4 & 1 & -3 \end{bmatrix}$

NAME ______________________

TEST FORM A

28. Let $A = \begin{bmatrix} 5 & 2 & -1 \\ 3 & 4 & 2 \\ -1 & 3 & 0 \end{bmatrix}$. Find a_{12}, M_{12}, and A_{12}.

Evaluate.

29. $\begin{vmatrix} -1 & 2 & 1 \\ 3 & -1 & 2 \\ 1 & 2 & 1 \end{vmatrix}$

30. $\begin{vmatrix} -3 & 2 & -2 \\ 1 & 1 & 1 \\ 6 & -4 & 4 \end{vmatrix}$

31. $\begin{vmatrix} 2 & 1 & -2 & 3 \\ -1 & -3 & 2 & -1 \\ 2 & 0 & 0 & 0 \\ 3 & 4 & -2 & 1 \end{vmatrix}$

32. $\begin{vmatrix} -2 & 2 & -2 \\ 3 & 1 & 2 \\ 5 & 1 & 0 \end{vmatrix}$

33. Factor: $\begin{vmatrix} bc & ac & ab \\ a & b & c \\ 1 & 1 & 1 \end{vmatrix}$.

34. Write a matrix equation equivalent to this system of equations and use the inverse of the coefficient matrix to solve the system. Show all your work.

$$3x - 2y = -11,$$
$$x + 3y = 11$$

35. Graph: $2x - 4y \leq 8$.

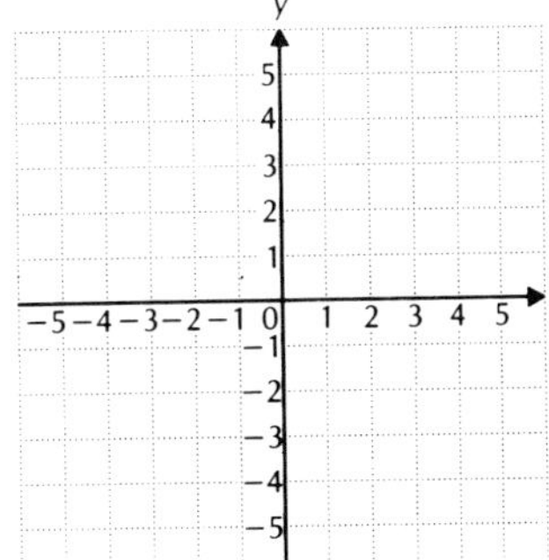

ANSWERS

28. ______________

29. ______________

30. ______________

31. ______________

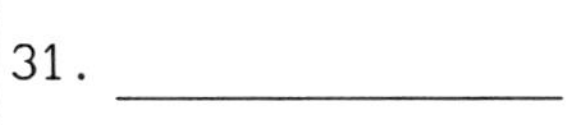

32. ______________

33. ______________

34. ______________

35. See graph.

NAME ______________________

ANSWERS

36. ______________

37. ______________

38. ______________

39. ______________

36. Maximize and minimize $T = 10x + 5y$ subject to

$$x + 3y \le 15,$$
$$2 \le x \le 4,$$
$$y \ge 0.$$

37. A bakery can produce either bread or cake. In a given week, the bakery can turn out at most 700 items, of which 300 loaves of bread and 100 cakes are required by regular customers. The profit on a loaf of bread is \$2 and on a cake is \$3. How many of each should the bakery produce in order to maximize the profit? What is the maximum profit?

38. Decompose into partial fractions:

$$\frac{-x - 18}{3x^2 - 13x + 4}.$$

39. Solve. $\frac{5}{x} - \frac{4}{y} = 11,$

$\frac{3}{x} + \frac{2}{y} = -11.$

CHAPTER 9 NAME ____________

TEST FORM B CLASS ______ SCORE ________ GRADE ______

Solve.

1. $0.6x + 0.2y = -0.2,$
   $0.1x + 0.01y = 0.06.$

2. An airplane travels 2700 km with a tail wind in 3 hr. It travels 2600 km with a head wind in 4 hr. Find the speed of the plane and the speed of the wind.

3. Solution A is 10% hydrochloric acid. Solution B is 35% hydrochloric acid. How many liters of each should be mixed to get 60 liters of a solution that is 25% acid?

4. In triangle ABC, the measure of angle B is 8° less than the measure of angle A. The sum of the measures of angles A and B is 84° more than the measure of angle C. Find the angle measures.

Solve using matrices. If there is more than one solution, list three of them.

5. $6x + 4y = 2,$
   $3x + 6y = -1$

6. $2x + y + z = 5,$
   $x - 3y + z = -4,$
   $-x + 2y - z = 2$

7. $4x - 3y + z = 2,$
   $-x + y - z = 4,$
   $5x - 4y + 2z = -2$

Classify as consistent or inconsistent, dependent or independent.

8. $3x - 2y = 0,$
   $-3x + 2y = 7$

9. $x + y - z = 4,$
   $2x - y + z = 2,$
   $x - 3y + 2z = -5$

10. Find numbers a, b, and c such that the function $f(x) = ax^2 + bx + c$ fits the data points (0,1), (-1,0), and (2,-9). Then write the equation for the function.

Evaluate.

11. $\begin{vmatrix} \frac{2}{3} & 1 \\ 5 & -6 \end{vmatrix}$

12. $\begin{vmatrix} 3 & -1 & 2 \\ 0 & 5 & 1 \\ 1 & 2 & 3 \end{vmatrix}$

ANSWERS

1. ____________

2. ____________

3. ____________

4. ____________

5. ____________

6. ____________

7. ____________

8. ____________

9. ____________

10. ____________

11. ____________

12. ____________

NAME ____________________

TEST FORM B

ANSWERS

13. ________

14. ________

15. ________

16. ________

17. ________

18. ________

19. ________

20. ________

21. ________

22. ________

23. ________

24. ________

25. ________

26. ________

27. ________

Solve using Cramer's rule.

13. $2x - 3y = -22,$
$-3x + 2y = 23$

14. $3x - y + 2z = 7,$
$5x + 2y - z = -8,$
$x + 3y + z = -7$

For Exercises 15 - 24, let

$$A = \begin{bmatrix} 3 & -2 \\ 4 & 1 \end{bmatrix} \quad B = \begin{bmatrix} 2 & -1 & 3 \\ -3 & 4 & 1 \end{bmatrix} \quad C = \begin{bmatrix} 2 & 0 \\ 3 & 1 \\ -1 & 4 \end{bmatrix}$$

$$D = \begin{bmatrix} 1 & 3 & -2 \\ 3 & -1 & 2 \\ 0 & 5 & 1 \end{bmatrix} \quad E = \begin{bmatrix} -2 & 4 & -1 \\ 3 & 0 & -2 \\ -1 & 4 & 1 \end{bmatrix} \quad F = \begin{bmatrix} 2 & 1 & -3 \end{bmatrix}$$

$$G = \begin{bmatrix} -2 & 4 \\ -1 & 3 \end{bmatrix} \quad O = \begin{bmatrix} 0 & 0 \\ 0 & 0 \end{bmatrix} \quad I = \begin{bmatrix} 1 & 0 \\ 0 & 1 \end{bmatrix}$$

Find each of the following, if possible.

15. GC
16. O + A
17. A + I
18. D + E
19. A - G
20. BC
21. B + C
22. -F
23. 2D - E
24. GI

Find A^{-1}, if it exists.

25. $A = \begin{bmatrix} 2 & -1 \\ 3 & 2 \end{bmatrix}$

26. $A = \begin{bmatrix} 1 & 4 & 0 \\ 0 & -2 & 3 \\ 2 & -1 & 1 \end{bmatrix}$

27. $A = \begin{bmatrix} 3 & 4 & 1 \\ -9 & 5 & -3 \\ 6 & -1 & 2 \end{bmatrix}$

NAME ______________________

TEST FORM B

28. Let $A = \begin{bmatrix} 3 & -2 & 1 \\ 4 & 0 & -1 \\ 1 & -2 & 3 \end{bmatrix}$. Find a_{23}, M_{23}, and A_{23}.

Evaluate.

29. $\begin{vmatrix} 2 & 0 & 5 \\ 1 & -3 & 2 \\ 4 & 1 & 6 \end{vmatrix}$

30. $\begin{vmatrix} 2 & 1 & 3 \\ 1 & 1 & 1 \\ -4 & -2 & 6 \end{vmatrix}$

31. $\begin{vmatrix} 2 & -1 & -2 & 3 \\ 5 & 0 & 0 & 0 \\ 4 & 3 & 0 & 1 \\ -6 & 2 & 0 & -1 \end{vmatrix}$

32. $\begin{vmatrix} 3 & -3 & 2 \\ 1 & 0 & 5 \\ -1 & 2 & 4 \end{vmatrix}$

33. Factor: $\begin{vmatrix} 1 & 1 & 1 \\ x & y & z \\ x^2 & y^2 & z^2 \end{vmatrix}$.

34. Write a matrix equation equivalent to this system of equations and use the inverse of the coefficient matrix to solve the system. Show all your work.

$$3x - y = 9,$$
$$x + 2y = -4$$

35. Graph: $3x - 2y \geq 6$.

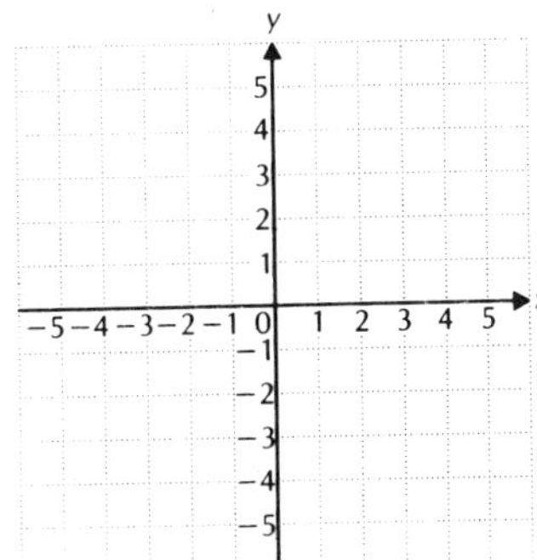

ANSWERS

28. ____________

29. ____________

30. ____________

31. ____________

32. ____________

33. ____________

34. ____________

35. See graph.

ANSWERS

36. ______

37. ______

38. ______

39. ______

36. Maximize and minimize $T = 3x + 8y$ subject to

$$x + 2y \le 6,$$
$$-1 \le x \le 3,$$
$$y \ge 0.$$

37. You are about to take a test which contains questions of Type A worth 10 points each and Type B worth 5 points each. You must complete the test in 60 minutes. Type A questions take 4 minutes and type B questions take 6 minutes. The total number of problems worked must <u>not</u> exceed 10. If you are told that you must work at least 3 questions of Type B, how many of each type of question must you do to maximize your score? What is the maximum score?

38. Decompose into partial fractions:

$$\frac{-5x + 21}{4x^2 + 17x - 15}.$$

39. Two solutions of the equation $y = mx + b$ are $(-4,-5)$ and $(2,-2)$. Find m and b.

CHAPTER 9 *NAME* ________________

TEST FORM C *CLASS* ______ *SCORE* ______ *GRADE* ______

Solve.

1. $0.4x - 0.7y = 1.5,$
   $0.03x + 0.1y = -0.04.$

2. Two cars leave town traveling in opposite directions. One travels at a speed of 65 km/h and the other at 76 km/h. In how many hours will they be 705 km apart?

3. Bill is half as old as his sister Ann. Ten years from now, Bill would be $\frac{3}{4}$ as old as Ann. How old are they now?

4. A person receives $143 per year in simple interest from three investments totaling $1920. Part is invested at 5%, part at 8%, and part at 10%. There is $420 more invested at 8% than at 5%. Find the amount invested at each rate.

Solve using matrices. If there is more than one solution, list three of them.

5. $3x - 8y = -3,$
   $6x + 4y = -1$

6. $4x + 3y - z = 5,$
   $x - 2y + 3z = 4,$
   $5x + y + 2z = 9$

7. $3x - y - z = 0,$
   $4x + 2y + z = 9,$
   $x - 2y + z = -2$

Classify as consistent or inconsistent, dependent or independent.

8. $5x - y = 4,$
   $-10x + 2y = -8$

9. $x - y + z = 4,$
   $x + 2y + z = 7,$
   $-x + y - z = -3$

10. Find numbers a, b, and c such that the function $f(x) = ax^2 + bx + c$ fits the data points (0,1), (2,-3), and (-1,6). Then write the equation for the function.

Evaluate.

11. $\begin{vmatrix} 2 & -\frac{1}{2} \\ 4 & 3 \end{vmatrix}$

12. $\begin{vmatrix} 2 & 1 & 4 \\ 0 & 3 & 2 \\ 1 & 5 & 1 \end{vmatrix}$

ANSWERS

1. ________
2. ________
3. ________
4. ________
5. ________
6. ________
7. ________
8. ________
9. ________
10. ________
11. ________
12. ________

NAME ______________________

TEST FORM C

ANSWERS

13. ______
14. ______
15. ______
16. ______
17. ______
18. ______
19. ______
20. ______
21. ______
22. ______
23. ______
24. ______
25. ______
26. ______
27. ______

Solve using Cramer's rule.

13. $5x - 3y = 4,$
$2x + 4y = -14$

14. $5x + 2y + 3z = -5,$
$x - y + z = -2,$
$2x + 3y + 4z = 3$

For Exercises 15 - 24, let

$$A = \begin{bmatrix} -3 & 1 \\ 0 & 2 \end{bmatrix} \quad B = \begin{bmatrix} -2 & 3 & 0 \\ 1 & 2 & 4 \end{bmatrix} \quad C = \begin{bmatrix} 5 & -2 \\ 2 & 0 \\ 1 & -3 \end{bmatrix}$$

$$D = \begin{bmatrix} 1 & 0 & -2 \\ 4 & -1 & 5 \\ 2 & -3 & 4 \end{bmatrix} \quad E = \begin{bmatrix} 6 & -2 & 1 \\ 4 & -3 & 5 \\ 0 & 2 & 1 \end{bmatrix} \quad F = \begin{bmatrix} 5 & -2 & 7 \end{bmatrix}$$

$$G = \begin{bmatrix} -3 & -1 \\ -2 & -4 \end{bmatrix} \quad O = \begin{bmatrix} 0 & 0 \\ 0 & 0 \end{bmatrix} \quad I = \begin{bmatrix} 1 & 0 \\ 0 & 1 \end{bmatrix}.$$

Find each of the following, if possible.

15. GC
16. O + A
17. A + I
18. D + E
19. A - G
20. BC
21. B + C
22. -F
23. 2D - E
24. GI

Find A^{-1}, if it exists.

25. $A = \begin{bmatrix} -1 & 3 \\ 2 & 4 \end{bmatrix}$

26. $A = \begin{bmatrix} 1 & 0 & -3 \\ 0 & 2 & -4 \\ -2 & 3 & 0 \end{bmatrix}$

27. $A = \begin{bmatrix} -1 & 1 & 2 \\ -2 & 3 & 1 \\ 4 & 0 & -3 \end{bmatrix}$

28. Let $A = \begin{bmatrix} 1 & -3 & 2 \\ 4 & 5 & -2 \\ 2 & 0 & 1 \end{bmatrix}$. Find a_{31}, M_{31}, and A_{31}.

Evaluate.

29. $\begin{vmatrix} 5 & 1 & 3 \\ 2 & -1 & 4 \\ 3 & 0 & 5 \end{vmatrix}$

30. $\begin{vmatrix} 3 & -1 & 2 \\ 1 & 1 & 1 \\ -6 & 2 & -4 \end{vmatrix}$

31. $\begin{vmatrix} -1 & 2 & -2 & 1 \\ 0 & 0 & 0 & 3 \\ 4 & 2 & 1 & -5 \\ 1 & -2 & 3 & 4 \end{vmatrix}$

32. $\begin{vmatrix} 8 & 0 & 1 \\ -3 & 5 & 7 \\ 0 & 2 & 4 \end{vmatrix}$

33. Factor: $\begin{vmatrix} x & y & z \\ x^2 & y^2 & z^2 \\ 1 & 1 & 1 \end{vmatrix}$.

34. Write a matrix equation equivalent to this system of equations and use the inverse of the coefficient matrix to solve the system. Show all your work.

$$2x - 4y = 6,$$
$$3x + 2y = 1$$

35. Graph: $x - 5y \le 5$.

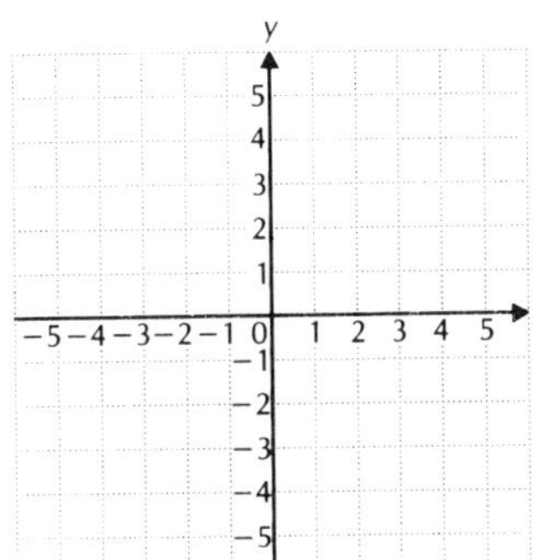

ANSWERS

28. ______________

29. ______________

30. ______________

31. ______________

32. ______________

33. ______________

34. ______________

35. See graph.

NAME ______________________

TEST FORM C

ANSWERS

36. ______________

37. ______________

38. ______________

39. ______________

36. Maximize and minimize $T = 8x + 12y$ subject to

$$x + y \le 6,$$
$$x \ge 0,$$
$$-1 \le y \le 4.$$

37. A tailoring firm can produce either slacks or skirts. In a given week, the firm can turn out at most 200 items, of which 40 pairs of slacks and 50 skirts are required by regular customers. The profit on a pair of slacks is \$12 and on a skirt is \$15. How many of each should the firm produce in order to maximize profit? What is the maximum profit?

38. Decompose into partial fractions:

$$\frac{-x - 23}{2x^2 - 7x - 4}.$$

39. Solve:

$$w + x - y + z = 2,$$
$$w - x + y + z = 6,$$
$$w + 2x - y - z = -6,$$
$$w + x + y - z = -4.$$

CHAPTER 9 NAME ____________________

TEST FORM D CLASS ______ SCORE ________ GRADE ______

Solve.

1. $0.4x + 0.5y = -0.7,$
$0.02x + 0.1y = 0.04.$

2. A train leaves a station and travels north at a speed of 90 km/h. Three hours later, a second train leaves on a parallel track and travels north at 120 km/h. How far from the station will they meet?

3. A collection of 27 coins consists of dimes and nickels. The total value is $1.90. How many dimes and how many nickels are there?

4. When machines A, B, C are all working, they can produce 1040 gadgets in one day. When only A and C are working, 770 gadgets can be produced in one day. When only B and C are working, 690 gadgets can be produced in one day. How many gadgets can be produced in a day by each machine?

Solve using matrices. If there is more than one solution, list three of them.

5. $12x - 4y = -11,$
$8x + 10y = -1$

6. $5x - y - 2z = 0,$
$x - 2y + z = 6,$
$x - y - z = -1$

7. $3x + y + z = 0,$
$x - 2y + z = 4,$
$x + 5y - z = -8$

Classify as consistent or inconsistent, dependent or independent.

8. $3x - 2y = 4,$
$-3x - 2y = 4$

9. $x - 2y + z = 4,$
$2x - y - z = 5,$
$x + y - 2z = 1$

10. Find numbers a, b, and c such that the function $f(x) = ax^2 + bx + c$ fits the data points (0,-3), (1,-2), and (-2,7). Then write the equation for the function.

Evaluate.

11. $\begin{vmatrix} \frac{3}{4} & -2 \\ 3 & 8 \end{vmatrix}$

12. $\begin{vmatrix} 5 & 1 & 3 \\ 0 & 2 & 1 \\ 3 & 1 & 4 \end{vmatrix}$

ANSWERS

1. ____________

2. ____________

3. ____________

4. ____________

5. ____________

6. ____________

7. ____________

8. ____________

9. ____________

10. ____________

11. ____________

12. ____________

NAME ______________________

TEST FORM D

ANSWERS
13. ______
14. ______
15. ______
16. ______
17. ______
18. ______
19. ______
20. ______
21. ______
22. ______
23. ______
24. ______
25. ______
26. ______
27. ______

Solve using Cramer's rule.

13. $2x - 4y = 22,$
    $3x + 7y = 7$

14. $x - 3y + z = 4,$
    $3x - y + 2z = 13,$
    $2x + 2y - z = 11$

For Exercises 15 - 24, let

$$A = \begin{bmatrix} -2 & -3 \\ 1 & 4 \end{bmatrix} \quad B = \begin{bmatrix} 3 & -1 & 2 \\ 0 & 2 & 5 \end{bmatrix} \quad C = \begin{bmatrix} 5 & -1 \\ 1 & 3 \\ 0 & 4 \end{bmatrix}$$

$$D = \begin{bmatrix} 0 & 1 & 3 \\ -2 & 1 & 4 \\ 7 & 0 & 2 \end{bmatrix} \quad E = \begin{bmatrix} -2 & 3 & 1 \\ 5 & 0 & -4 \\ 3 & -1 & 5 \end{bmatrix} \quad F = \begin{bmatrix} -2 & 3 & 5 \end{bmatrix}$$

$$G = \begin{bmatrix} -2 & -3 \\ 4 & 6 \end{bmatrix} \quad O = \begin{bmatrix} 0 & 0 \\ 0 & 0 \end{bmatrix} \quad I = \begin{bmatrix} 1 & 0 \\ 0 & 1 \end{bmatrix}.$$

Find each of the following, if possible.

15. GC
16. O + A
17. A + I
18. D + E
19. A - G
20. BC
21. B + C
22. -F
23. 2D - E
24. GI

Find A^{-1}, if it exists.

25. $A = \begin{bmatrix} -4 & 1 \\ 2 & 3 \end{bmatrix}$

26. $A = \begin{bmatrix} 1 & 1 & -1 \\ -1 & 2 & 3 \\ -2 & -1 & 1 \end{bmatrix}$

27. $A = \begin{bmatrix} 5 & 0 & -1 \\ -3 & 0 & 4 \\ 2 & 0 & 1 \end{bmatrix}$

NAME ______________________

TEST FORM D

28. Let $A = \begin{bmatrix} 4 & 0 & -2 \\ 3 & 1 & 5 \\ 2 & -3 & 6 \end{bmatrix}$. Find a_{23}, M_{23}, and A_{23}.

Evaluate.

29. $\begin{vmatrix} 3 & -1 & 3 \\ 0 & 2 & 1 \\ 5 & 6 & -4 \end{vmatrix}$

30. $\begin{vmatrix} 5 & -1 & 3 \\ 1 & 1 & 1 \\ -10 & 2 & -6 \end{vmatrix}$

31. $\begin{vmatrix} 1 & 3 & 0 & 2 \\ 1 & 0 & -2 & 1 \\ -1 & 0 & 1 & -1 \\ 1 & 0 & 4 & 0 \end{vmatrix}$

32. $\begin{vmatrix} 2 & 6 & 2 \\ 3 & 5 & 4 \\ 0 & -3 & -1 \end{vmatrix}$

33. Factor: $\begin{vmatrix} 1 & a^2 & a \\ 1 & b^2 & b \\ 1 & c^2 & c \end{vmatrix}$.

34. Write a matrix equation equivalent to this system of equations and use the inverse of the coefficient matrix to solve the system. Show all your work.

$$5x - y = 11,$$
$$2x + 4y = 22$$

35. Graph: $2x - 3y \geq -6$.

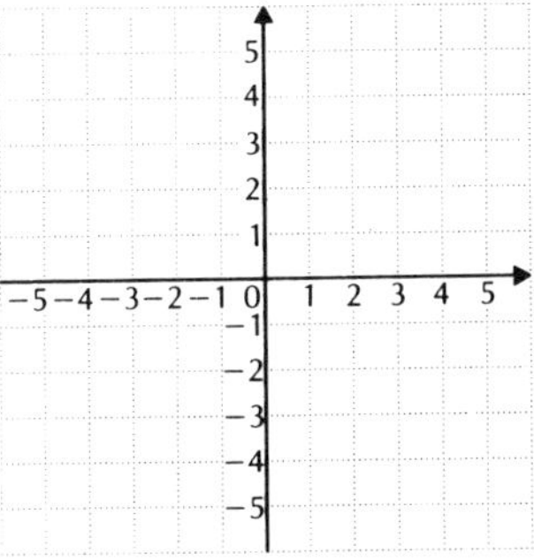

ANSWERS

28. ______________

29. ______________

30. ______________

31. ______________

32. ______________

33. ______________

34. ______________

35. See graph.

NAME ______________________

TEST FORM D

ANSWERS

36. ____________

37. ____________

38. ____________

39. ____________

36. Maximize and minimize $T = 15x + 2y$ subject to

$$3x + 4y \le 12,$$
$$x \ge -2,$$
$$-3 \le y \le 3.$$

37. A college snack bar cooks and sells hamburgers and hot dogs during the lunch hour. To stay in business, it must sell at least 20 hamburgers but cannot sell more than 60. It must also sell at least 15 hot dogs but cannot cook more than 80. It cannot cook more than 100 sandwiches altogether. The profit on a hamburger is \$0.75 and on a hot dog is \$0.50. How many of each kind of sandwich should they sell in order to make the maximum profit? What is the maximum profit?

38. Decompose into partial fractions:

$$\frac{-3x + 34}{5x^2 + 17x - 12}.$$

39. Solve: $\begin{vmatrix} x & 2 \\ -8 & x \end{vmatrix} = 20.$

CHAPTER 9 *NAME* ____________________

TEST FORM E *CLASS* ________ *SCORE* ________ *GRADE* ________

Solve.

1. $7x - 3y = 5,$
   $6x + 4y = 1$

2. $0.4x + 0.1y = 1.4,$
   $0.3x + 0.4y = -0.9$

3. Two investments are made totaling $15,000. For a certain year these investments yield $1095 in simple interest. Part of the $15,000 is invested at 6% and the rest at 9%. Determine the amount invested at each rate.

4. In a factory there are three machines A, B, and C. When all three are running, they produce 915 skateboards per day. If A and C work but B does not, they produce 695 skateboards per day. If A and B work but C does not, they produce 570 skateboards per day. How many skateboards per day can machine A produce?

Solve using matrices. If there is more than one solution, list three of them.

5. $x - 2y = 0,$
   $-2x + 4y = 0$

6. $4x + 3y - z = -3,$
   $x - y + 5z = 8,$
   $3x + 4y - 2z = -7$

7. Classify as consistent or inconsistent, dependent or independent.

$$\begin{aligned} 2x + y - z &= 0, \\ -4x - 3y &= 3, \\ 5x + 4y - z &= -3 \end{aligned}$$

8. Find numbers a, b, and c such that the function $f(x) = ax^2 + bx + c$ fits the data points (0,-4), (-1,-11), and (2,-2). Then write the equation for the function.

ANSWERS

1. ________________

2. ________________

3. ________________

4. ________________

5. ________________

6. ________________

7. ________________

8. 

NAME ______________________

TEST FORM E

ANSWERS

9. ______

10. ______

11. ______

12. ______

13. ______

14. ______

15. ______

16. ______

9. Evaluate: $\begin{vmatrix} -\sqrt{3} & -1 \\ -4 & \sqrt{3} \end{vmatrix}$.

10. Solve using Cramer's Rule. Show all your work.

$$2x - z = 2,$$
$$x - 3y + z = 10,$$
$$-3x + y + 4z = 6$$

Evaluate.

11. $\begin{vmatrix} 6 & 1 & -2 \\ -6 & 3 & -4 \\ 6 & 1 & 11 \end{vmatrix}$

12. $\begin{vmatrix} 3 & 5 & -2 \\ 1 & 0 & -5 \\ 4 & -1 & -1 \end{vmatrix}$

13. $\begin{vmatrix} 2 & 6 & 5 \\ 0 & 0 & 0 \\ -7 & -4 & -1 \end{vmatrix}$

14. $\begin{vmatrix} 5 & 0 & 2 & -3 \\ 0 & 9 & -1 & 4 \\ 0 & 0 & -2 & 8 \\ 0 & 0 & 0 & -3 \end{vmatrix}$

15. Factor: $\begin{vmatrix} x & 1 & yz \\ y & 1 & xz \\ z & 1 & xy \end{vmatrix}$.

16. Write a matrix equation equivalent to this system of equations and use the inverse of the coefficient matrix to solve the system. Show all your work.

$$-3x + 4y = 10,$$
$$2x - 5y = -9$$

NAME ______________________

TEST FORM E

Let

$$A = \begin{bmatrix} -1 & 0 & 2 \\ 0 & 2 & -3 \\ 4 & 0 & -2 \end{bmatrix} \quad B = \begin{bmatrix} 1 & 0 \\ -4 & 2 \\ -2 & -3 \end{bmatrix} \quad C = \begin{bmatrix} 1 & 4 & -3 \\ 0 & -10 & 2 \end{bmatrix}$$

$$D = \begin{bmatrix} 0 & 4 \\ 5 & -2 \end{bmatrix} \quad E = \begin{bmatrix} 1 & 5 & 4 \\ 0 & -2 & 0 \\ -1 & -1 & 1 \end{bmatrix} \quad F = \begin{bmatrix} 3 & -1 \\ 0 & 0 \end{bmatrix}$$

$$G = \begin{bmatrix} 7 & 0 & -2 \end{bmatrix}$$

Matching:
Find each of the following, if possible. Place the letter of the answer in the answer blank. Some letters may be used more than once, and some may not be used.

Questions

17. D + F
18. FB
19. -G
20. DC
21. A^{-1}
22. E - 2A
23. D^{-1}

Answers

a) $\begin{bmatrix} -7 & 0 & 2 \end{bmatrix}$

b) $\begin{bmatrix} 0 & 16 & -12 \\ 0 & -50 & -4 \end{bmatrix}$

c) $\begin{bmatrix} 1 & 0 \\ 4 & -10 \\ -3 & 2 \end{bmatrix}$

d) $\begin{bmatrix} 3 & 5 & 0 \\ 0 & -6 & 6 \\ -9 & -1 & 5 \end{bmatrix}$

e) $\begin{bmatrix} 7 \\ 0 \\ -2 \end{bmatrix}$

f) $\begin{bmatrix} 3 & 3 \\ 5 & -2 \end{bmatrix}$

g) $\begin{bmatrix} \frac{1}{3} & 0 & \frac{1}{3} \\ 1 & \frac{1}{2} & \frac{1}{4} \\ \frac{2}{3} & 0 & \frac{1}{6} \end{bmatrix}$

h) $\begin{bmatrix} \frac{1}{5} & 0 & \frac{1}{5} \\ \frac{3}{5} & \frac{3}{10} & \frac{3}{20} \\ \frac{2}{5} & 0 & \frac{1}{10} \end{bmatrix}$

i) $\begin{bmatrix} 0 & 5 & 6 \\ 0 & 0 & -3 \\ 3 & -1 & -1 \end{bmatrix}$

j) $\begin{bmatrix} 0 & -40 & 8 \\ 5 & 40 & -19 \end{bmatrix}$

k) $\begin{bmatrix} \frac{1}{10} & \frac{1}{5} \\ \frac{1}{4} & 0 \end{bmatrix}$

ℓ) Not possible

m) None of these answers is correct.

ANSWERS

17. ______________
18. ______________
19. ______________
20. ______________
21. ______________
22. ______________
23. ______________

NAME ______________________

TEST FORM E

ANSWERS	
24. See graph.	24. Graph: $-2x \geq 3y - 6$.
25. ______	25. Maximize and minimize $T = 20x + 15y$ subject to $3x + 2y \leq 6$, $x + y \leq 3$, $x \geq 0$, $y \geq 1$
26. ______	26. It takes a tailoring firm 1 hr of cutting and 4 hr of sewing to make a dress. To make a suit it takes 1 hr of cutting and 6 hr of sewing. At most, 5 hr per day are available for cutting, and at most 24 hr per day are available for sewing. The profit on a dress is \$25 and on a suit is \$35. How many of each type of garment should be made to maximize profit? What is the maximum profit?

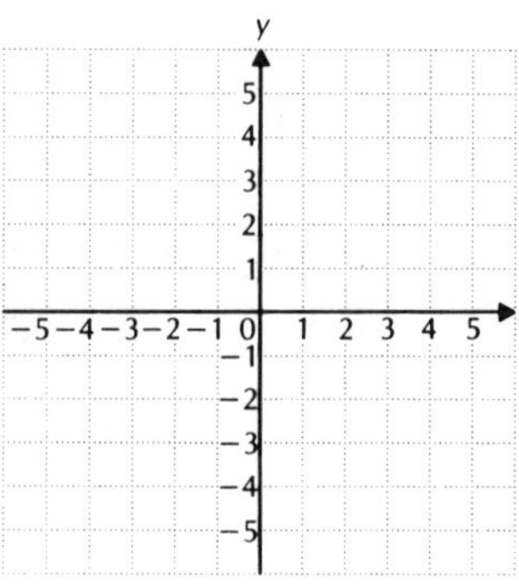

$$3x + 2y \leq 6,$$
$$x + y \leq 3,$$
$$x \geq 0,$$
$$y \geq 1$$

CHAPTER 9 NAME ____________________

TEST FORM F CLASS ________ SCORE __________ GRADE ________

Write the letter of your response on the answer blank.

ANSWERS

1. Solve. $4x + 5y = -2,$
   $3x + 4y = -1$

   a) $\left(-2, \frac{6}{5}\right)$ or $(-3, 2)$ b) $\left(\frac{5}{7}, -\frac{2}{7}\right)$ or $\left(1, -\frac{6}{5}\right)$

   c) $(3, 2)$ or $(-5, 7)$ d) $(6, -6)$ or $(4, -5)$

1. ____________

2. Solve: $\frac{6}{x} + \frac{2}{y} = -2,$
   $\frac{7}{x} - \frac{3}{y} = -13.$

   The solution set

   a) is empty.

   b) contains one ordered pair.

   c) contains four ordered pairs.

   d) contains two ordered pairs.

2. ____________

3. The sum of a certain number and a second number is 15. The second number minus the first number is -119. Find the smaller number.

   a) 67 b) -52 c) 52 d) -67

3. ____________

4. Two investments are made totaling \$19,000. For a certain year these investments yield \$2520 in simple interest. Part of the \$19,000 is invested at 12% and the rest at 14%. How much is invested at 12%?

   a) \$12,000 b) \$8,500 c) \$9,000 d) \$7000

4. ____________

5. Solve using matrices: $2x + 3y - 5z = -10,$
   $-3x + 2y + z = 2,$
   $5x - y + 3z = 13.$

   Find the sum of the x, y, and z values.

   a) 0 b) -4 c) -1 d) 5

5. ____________

NAME ______________________

TEST FORM F

ANSWERS

6. ______

7. ______

8. ______

9. ______

10. ______

6. Classify this system as consistent or inconsistent, dependent or independent.

$$2x - 5y = 4,$$
$$-6x + 15y = 6$$

a) Consistent, dependent

b) Consistent, independent

c) Inconsistent, dependent

d) Inconsistent, independent

7. Classify this system as consistent or inconsistent, dependent or independent.

$$4x + z = 5,$$
$$x + y + z = -2,$$
$$5x + y + z = 3$$

a) Consistent, dependent

b) Consistent, independent

c) Inconsistent, dependent

d) Inconsistent, independent

8. Find numbers a, b, and c such that the function $f(x) = ax^2 + bx + c$ fits the data points (0,1), (-1,9), and (2,3).

a) $a = -1,\ b = 4,\ c = 1$ b) $a = 1,\ b = -2,\ c = -5$

c) $a = 3,\ b = -5,\ c = 1$ d) $a = 3,\ b = 1,\ c = -3$

9. Evaluate: $\begin{vmatrix} -\frac{1}{4} & -\sqrt{2} \\ \sqrt{2} & 8 \end{vmatrix}$.

a) -4 b) -6 c) 2 d) 0

10. Evaluate: $\begin{vmatrix} 4 & 2 & -3 \\ -2 & -3 & 0 \\ 1 & 2 & 2 \end{vmatrix}$.

a) 14 b) 35 c) -13 d) -37

NAME ______________________

TEST FORM F

ANSWERS

11. Solve using Cramer's Rule: $2x - 3y = 13$, $9x + 5y = 9$.

Find the y-coordinate of the solution.

a) $\dfrac{\begin{vmatrix} 2 & 13 \\ 9 & 9 \end{vmatrix}}{\begin{vmatrix} 2 & -3 \\ 9 & 5 \end{vmatrix}}$ b) $\dfrac{\begin{vmatrix} 13 & 13 \\ 9 & 9 \end{vmatrix}}{\begin{vmatrix} 2 & -3 \\ 9 & 5 \end{vmatrix}}$ c) $\dfrac{\begin{vmatrix} 13 & 9 \\ 9 & 5 \end{vmatrix}}{\begin{vmatrix} 2 & -3 \\ 9 & 5 \end{vmatrix}}$ d) $\dfrac{\begin{vmatrix} 2 & -3 \\ 9 & 5 \end{vmatrix}}{\begin{vmatrix} 2 & -3 \\ 9 & 5 \end{vmatrix}}$

11. ______________

12. Solve using Cramer's Rule: $7x - y + z = -3$, $x + 3y + 2z = 2$, $-3x - 2y - z = 2$.

Find the sum of the x, y, and z values.

a) -1 b) 2 c) -4 d) 1

12. ______________

Let

$A = \begin{bmatrix} 4 & 7 & 0 \\ -2 & 0 & 3 \\ 0 & 1 & -2 \end{bmatrix}$ $B = \begin{bmatrix} 2 & -2 & 3 \\ 0 & 1 & -4 \\ -1 & -3 & 2 \end{bmatrix}$ $C = \begin{bmatrix} 0 & 2 \\ -2 & 1 \\ 3 & -1 \end{bmatrix}$

$D = \begin{bmatrix} -1 & 0 & -2 \\ 4 & 0 & -2 \end{bmatrix}$ $E = \begin{bmatrix} -2 & -2 \\ 4 & 4 \end{bmatrix}$ $F = \begin{bmatrix} 0 & 2 \\ -3 & 1 \end{bmatrix}$

$G = \begin{bmatrix} 5 \\ 0 \\ -2 \end{bmatrix}$ $H = \begin{bmatrix} 2 & -8 & 4 \end{bmatrix}$ $I = \begin{bmatrix} 1 & 0 & 0 \\ 0 & 1 & 0 \\ 0 & 0 & 1 \end{bmatrix}$

$O = \begin{bmatrix} 0 & 0 \\ 0 & 0 \end{bmatrix}$

Find each of the following (Questions 13 - 24), if possible.

13. ______________

13. CD

a) $\begin{bmatrix} 0 & 0 & 2 \\ -20 & -1 & 2 \end{bmatrix}$ b) $\begin{bmatrix} 8 & 0 & -4 \\ 6 & 0 & 2 \\ -7 & 0 & -4 \end{bmatrix}$

c) $\begin{bmatrix} -20 & -1 & 8 \end{bmatrix}$ d) $\begin{bmatrix} 20 & 2 & 4 \\ 6 & 1 & -2 \\ -7 & -1 & 4 \end{bmatrix}$

NAME ______________________

TEST FORM F

ANSWERS

14. ____________

15. ____________

16. ____________

17. ____________

18. ____________

14. 3D

a) $\begin{bmatrix} 3 & 0 & 6 \\ 4 & 0 & -2 \end{bmatrix}$ b) $\begin{bmatrix} -3 & 0 & -2 \\ 12 & 0 & -2 \end{bmatrix}$

c) $\begin{bmatrix} -3 & 0 & -6 \\ 12 & 0 & -6 \end{bmatrix}$ d) $\begin{bmatrix} 4 & 3 & 5 \\ -7 & 2 & 1 \end{bmatrix}$

15. G + H

a) $\begin{bmatrix} 3 \\ -8 \\ 2 \end{bmatrix}$ b) $\begin{bmatrix} 7 & -8 & 2 \end{bmatrix}$

c) $\begin{bmatrix} 3 & -8 & 2 \end{bmatrix}$

d) Not possible

16. E - F

a) $\begin{bmatrix} -2 & -4 \\ 7 & 3 \end{bmatrix}$ b) $\begin{bmatrix} -2 & -4 \\ 1 & 3 \end{bmatrix}$ c) $\begin{bmatrix} 6 & -6 \\ -12 & 12 \end{bmatrix}$ d) $\begin{bmatrix} -2 & 0 \\ 1 & 5 \end{bmatrix}$

17. DF

a) $\begin{bmatrix} -12 & -8 \\ 3 & -1 \\ 4 & 2 \end{bmatrix}$ b) $\begin{bmatrix} 0 & 0 & 3 \\ -12 & 6 & 2 \end{bmatrix}$

c) $\begin{bmatrix} 6 & 0 & 12 \\ -60 & -6 & -12 \end{bmatrix}$ d) Not possible

18. -A

a) $\begin{bmatrix} -3 & 14 & 21 \\ -4 & -8 & -12 \\ -2 & -4 & 14 \end{bmatrix}$ b) $\begin{bmatrix} \frac{3}{40} & -\frac{7}{20} & -\frac{21}{40} \\ \frac{1}{10} & \frac{1}{5} & \frac{3}{10} \\ \frac{1}{20} & \frac{1}{10} & -\frac{7}{20} \end{bmatrix}$

c) $\begin{bmatrix} 4 & -2 & 0 \\ 7 & 0 & 1 \\ 0 & 3 & -2 \end{bmatrix}$ d) $\begin{bmatrix} -4 & -7 & 0 \\ 2 & 0 & -3 \\ 0 & -1 & 2 \end{bmatrix}$

ANSWERS

19. $10A + 2B$

a) $\begin{bmatrix} 44 & 66 & 6 \\ -20 & 2 & 22 \\ -2 & 4 & -16 \end{bmatrix}$ b) $\begin{bmatrix} 36 & 66 & 6 \\ -20 & 2 & 22 \\ -2 & 4 & -24 \end{bmatrix}$

c) $\begin{bmatrix} 36 & 74 & -6 \\ -20 & -2 & 38 \\ -2 & 16 & -24 \end{bmatrix}$ d) $\begin{bmatrix} 44 & -20 & 2 \\ 74 & -2 & 16 \\ -6 & 38 & -24 \end{bmatrix}$

19. ____________

20. E^{-1}

a) $\begin{bmatrix} 2 & 2 \\ -4 & -4 \end{bmatrix}$ b) $\begin{bmatrix} -2 & 4 \\ -2 & 4 \end{bmatrix}$ c) $\begin{bmatrix} -\frac{1}{4} & -\frac{1}{8} \\ \frac{1}{4} & \frac{1}{8} \end{bmatrix}$ d) Not possible

20. ____________

21. $F + O$

a) $\begin{bmatrix} 0 & 2 \\ -3 & 1 \end{bmatrix}$ b) $\begin{bmatrix} 0 & 0 \\ 0 & 0 \end{bmatrix}$ c) $\begin{bmatrix} 0 & -3 \\ 2 & 1 \end{bmatrix}$ d) $\begin{bmatrix} 0 & -2 \\ 3 & -1 \end{bmatrix}$

21. ____________

22. $B + I$

a) $\begin{bmatrix} 2 & -2 & 3 \\ 0 & 1 & -4 \\ -1 & -3 & 2 \end{bmatrix}$ b) $\begin{bmatrix} 3 & -2 & 3 \\ 0 & 2 & -4 \\ -1 & -3 & 3 \end{bmatrix}$

c) $\begin{bmatrix} -2 & 2 & -3 \\ 0 & -1 & 4 \\ 1 & 3 & -2 \end{bmatrix}$ d) $\begin{bmatrix} 3 & -1 & 4 \\ 1 & 2 & -3 \\ 0 & -2 & 3 \end{bmatrix}$

22. ____________

NAME ______________________

ANSWERS
23. ________
24. ________
25. ________
26. ________
27. ________

23. A^{-1}

a) $\begin{bmatrix} -3 & -4 & -2 \\ 14 & -8 & -4 \\ 21 & -12 & 14 \end{bmatrix}$

b) $\begin{bmatrix} \frac{3}{40} & -\frac{7}{20} & -\frac{21}{40} \\ \frac{1}{10} & \frac{1}{5} & \frac{3}{10} \\ \frac{1}{20} & \frac{1}{10} & -\frac{7}{20} \end{bmatrix}$

c) $\begin{bmatrix} -4 & -7 & 0 \\ 2 & 0 & -3 \\ 0 & -1 & 2 \end{bmatrix}$

d) $\begin{bmatrix} -3 & 14 & 21 \\ -4 & -8 & -12 \\ -2 & -4 & 14 \end{bmatrix}$

24. A + B

a) $\begin{bmatrix} 8 & -1 & -16 \\ -7 & -5 & 0 \\ 2 & 7 & -8 \end{bmatrix}$

b) $\begin{bmatrix} 6 & 5 & 3 \\ -2 & 1 & -1 \\ -1 & -2 & 0 \end{bmatrix}$

c) $\begin{bmatrix} 6 & 5 & 0 \\ 0 & 0 & -1 \\ 0 & -2 & 0 \end{bmatrix}$

d) $\begin{bmatrix} 8 & -14 & 0 \\ 0 & 0 & -12 \\ 0 & -3 & -4 \end{bmatrix}$

25. Evaluate: $\begin{vmatrix} 3 & 4 & 2 \\ -3 & 1 & -1 \\ -3 & 0 & -2 \end{vmatrix}$.

a) -48 b) 0 c) -24 d) -12

26. Evaluate: $\begin{vmatrix} 0 & 2 & 4 \\ -1 & 1 & -3 \\ 4 & -5 & 12 \end{vmatrix}$.

a) -84 b) -36 c) 4 d) -84

27. Evaluate: $\begin{vmatrix} -2 & 0 & 0 & 0 \\ 4 & 0 & -6 & 1 \\ 1 & 3 & 0 & 2 \\ -2 & 1 & -3 & 0 \end{vmatrix}$.

a) 6 b) 42 c) -42 d) 0

NAME ____________________

TEST FORM F

ANSWERS

28. Factor: $\begin{vmatrix} 1 & c & c^2 \\ 1 & d & d^2 \\ 1 & e & e^2 \end{vmatrix}$.

a) $(c - d)(e - c)(d - e)$
b) $(d - c)(e - c)(d - e)$
c) $(d - c)(e - c)(e - d)$
d) $(d - c)(c - e)(d - e)$

28. ____________

29. Write a matrix equation equivalent to this system of equations.

$$9x - y = 7,$$
$$3x + 5y = -2$$

a) $\begin{bmatrix} 9 & -1 \\ 3 & 5 \end{bmatrix} \begin{bmatrix} x \\ y \end{bmatrix} = \begin{bmatrix} 7 \\ -2 \end{bmatrix}$

b) $\begin{bmatrix} 9 & -1 \\ 3 & 5 \end{bmatrix} \begin{bmatrix} x & y \end{bmatrix} = \begin{bmatrix} 7 \\ -2 \end{bmatrix}$

c) $\begin{bmatrix} 7 & -1 \\ -2 & 5 \end{bmatrix} \begin{bmatrix} x \\ y \end{bmatrix} = \begin{bmatrix} 9 \\ 3 \end{bmatrix}$

d) $\begin{bmatrix} 9 & -1 & 7 \\ 3 & 5 & -2 \end{bmatrix} \begin{bmatrix} x \\ y \end{bmatrix} = \begin{bmatrix} 7 \\ -2 \end{bmatrix}$

29. ____________

30. Graph: $2x < 4y + 8$.

a)

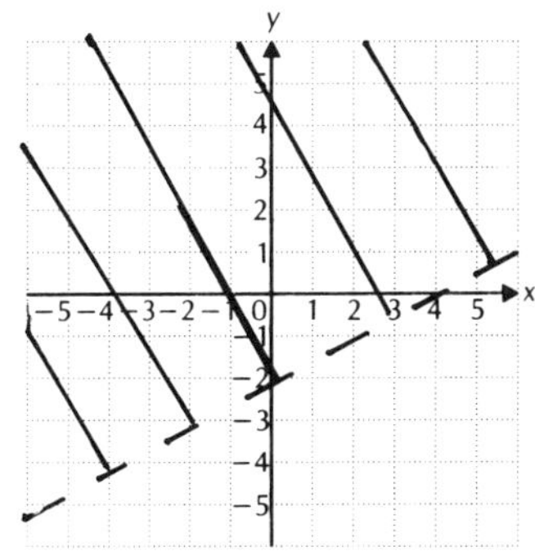

b)

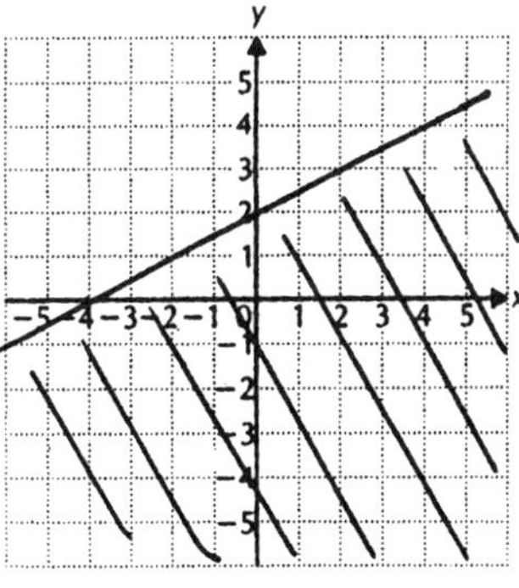

c)

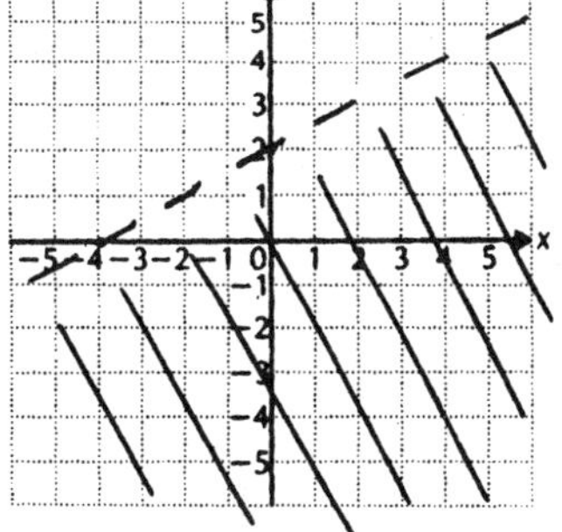

d)

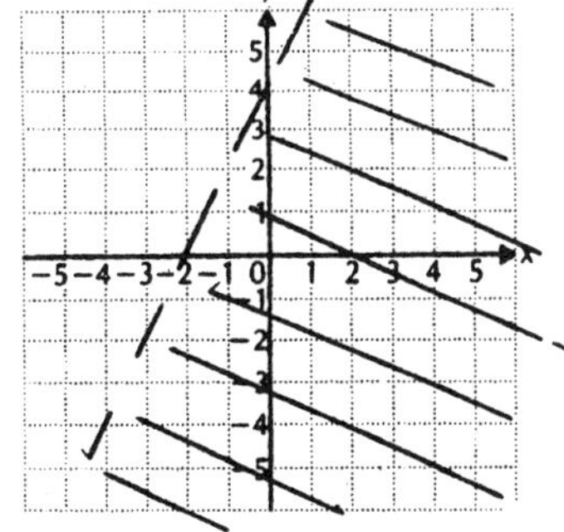

30. ____________

 NAME ____________________

TEST FORM F

ANSWERS	
31. ______	31. Graph this system: $2x + 11y \le 22$, $1 \le x \le 8$, $y \ge 0$. Find the coordinates of any vertices found. a) $(1,0)$, $\left(1,\frac{11}{2}\right)$, $(8,0)$, $\left(8,\frac{38}{11}\right)$ b) $(1,0)$, $\left(1,\frac{20}{11}\right)$, $\left(8,\frac{6}{11}\right)$, $(8,0)$ c) $(0,1)$, $\left(\frac{20}{11},1\right)$, $\left(\frac{6}{11},8\right)$, $(0,8)$ d) $(1,0)$, $(0,2)$, $(8,0)$, $(11,0)$
32. ______	32. Maximize and minimize $P = 11x + 99y$ subject to the system given in Question 31. a) Min. 11 at $(1,0)$, max. 191 at $\left(1,\frac{20}{11}\right)$ b) Min. 11 at $(1,0)$, max. 342 at $\left(1,\frac{30}{11}\right)$ c) Min. 11 at $(1,0)$, max. 792 at $(0,8)$ d) Min. 99 at $(0,1)$, max. 798 at $\left(\frac{6}{11},8\right)$
33. ______	33. You are about to take a test that contains questions of Type A worth 10 points and questions of Type B worth 20 points. You must complete the test in 120 minutes, and the total number of problems worked must <u>not</u> exceed 24. Type A questions take 4 minutes and Type B questions take 12 minutes. If you are told that you must work at least 2 questions of Type B, how many of each type of question must you do to maximize your score and what is the maximum score? a) Type A: 0, Type B: 10, max. Score 200 b) Type A: 12, Type B: 12, max. Score 360 c) Type A: 21, Type B: 3, max. Score 270 d) Type A: 22, Type B: 2, max. Score 260

CHAPTER 10 NAME ____________

TEST FORM A CLASS ______ SCORE ______ GRADE ______

1. Graph: $4x^2 + 7xy - 2y^2 = 0$.

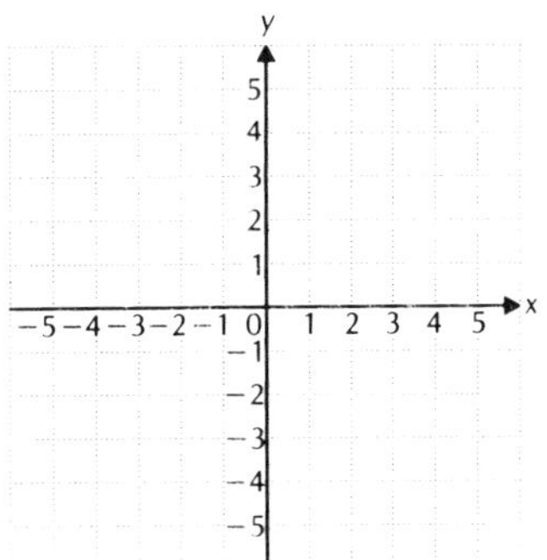

2. Find the center, vertices, and foci of the ellipse $9x^2 + y^2 - 36x - 6y + 36 = 0$, then graph the ellipse.

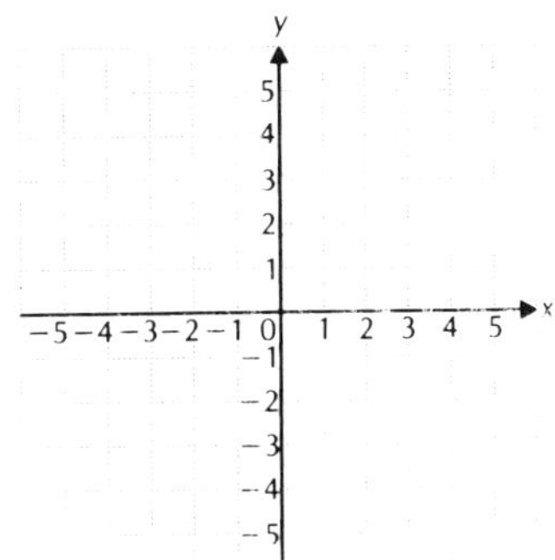

3. Find an equation of the ellipse with vertices (-2,0), (2,0), (0,-4), and (0,4).

4. Find the center, vertices, foci, and asymptotes of the hyperbola $4y^2 - 16x^2 = 64$.

5. Graph: $xy = 2$.

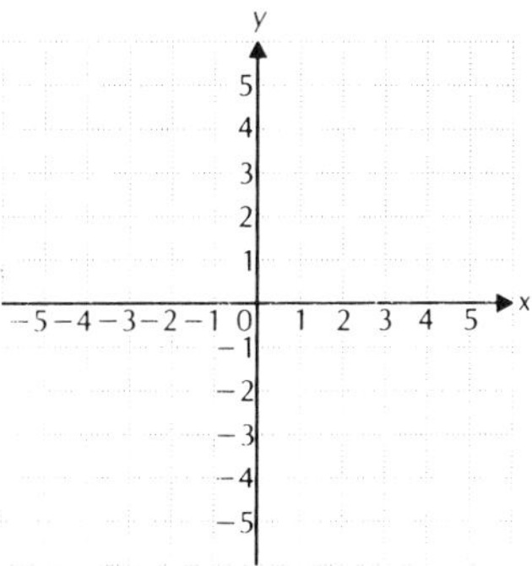

ANSWERS

1. See graph.

2. ____________

See graph.

3. ____________

4. ____________

5. See graph.

NAME ______________________________

TEST FORM A

ANSWERS

6. ________________

7. ________________

8. ________________

9. ________________

10. ________________

6. Find an equation of the parabola with directrix $x = -6$ and focus $(6,0)$.

7. Find the vertex, focus, and directrix of the parabola $x^2 + 2x - 4y + 13 = 0$.

Solve.

8. $x^2 + y^2 = 100,$
   $3x + 4y = 0$

9. $x^2 + y^2 = 2,$
   $xy = 1$

10. The sum of two numbers is 17, and the sum of their squares is 149. What are the numbers?

NAME ____________________

TEST FORM A

Solve.

ANSWERS

11. The diagonal of a rectangle is 3 ft longer than the length of the rectangle and 3 ft shorter than twice the width. Find the dimensions of the rectangle.

11. ____________

12. In a fractional expression, the sum of the values of the numerator and the denominator is 21. The product of their values is 110. Find the values of the numerator and the denominator.

12. ____________

13. ____________

13. The area of a rectangle is 24 ft^2, and the length of a diagonal is $2\sqrt{37}$. Find the dimensions.

14. Find two numbers whose product is 72 if the sum of their squares is 340.

14. ____________

15. Two squares are such that the sum of their areas is 34 ft^2 and the difference of their areas is 16 ft^2. Find the length of a side of each square.

15. ____________

NAME ____________________

ANSWERS

16. ____________
17. ____________
18. ____________
19. ____________
20. ____________
21. ____________
22. ____________
23. ____________
24. ____________

Classify the equation as a circle, an ellipse, a parabola, or a hyperbola.

16. $x = -y^2 + 3y - 2$

17. $3x^2 - xy = 4 + 3x^2$

18. $x^2 + y^2 - 3x + 12y + 29 = 0$

19. $12x^2 + 6y^2 = 6 - 48x$

20. $y = x^2 + 6x - 2 + 23y$

21. $\frac{x^2}{25} - \frac{y^2}{9} = 1$

22. Find an equation of the ellipse with vertices (5,-1), (1,-1), (3,-4), and (3,2).

23. Find an equation of the following parabola: Line of symmetry parallel to the y-axis, vertex (2,-1), and passing through (0,0).

24. The square of a certain number exceeds twice the square of another number by $\frac{1}{18}$. The sum of their squares is $\frac{5}{36}$. Find the numbers.

CHAPTER 10 NAME ____________________

TEST FORM B CLASS ________ SCORE __________ GRADE ______

1. Graph: $2x^2 + 5xy - 3y^2 = 0$.

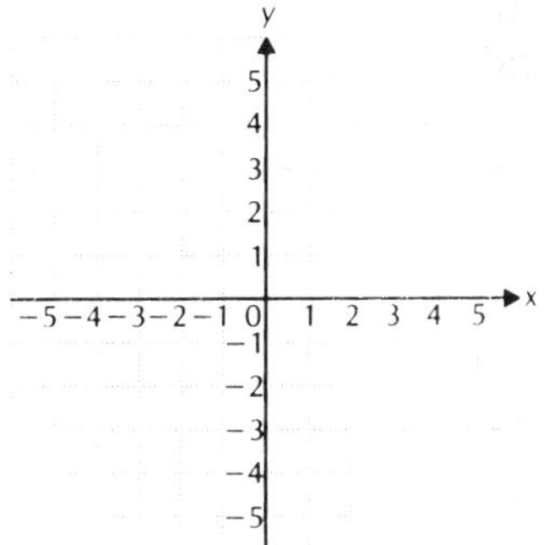

2. Find the center, vertices, and foci of the ellipse $x^2 + 4y^2 - 2x + 8y - 11 = 0$, then graph the ellipse.

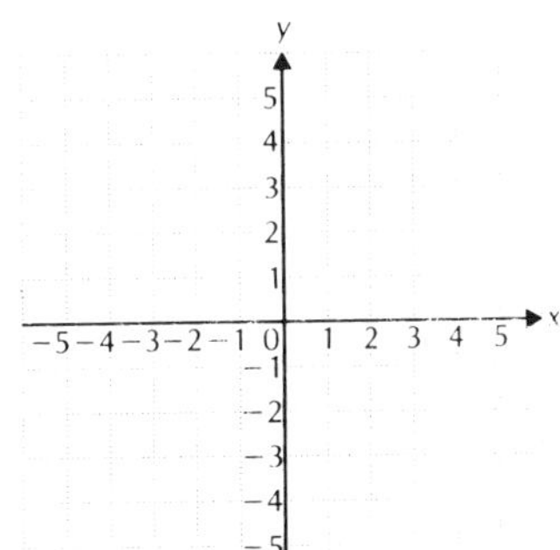

3. Find an equation of the ellipse with vertices (-5,0), (5,0), (0,-1), and (0,1).

4. Find the center, vertices, foci, and asymptotes of the hyperbola $16x^2 - 9y^2 = 144$.

5. Graph: $xy = -4$.

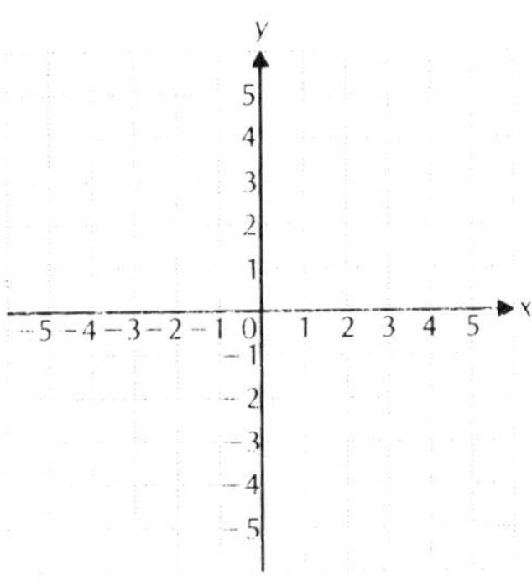

ANSWERS

1. See graph.

2. ____________________

See graph.

3. ____________________

4. ____________________

5. See graph.

ANSWERS

6. ____________

7. ____________

8. ____________

9. ____________

10. ____________

6. Find an equation of the parabola with directrix $y = -8$ and focus $(0,8)$.

7. Find the vertex, focus, and directrix of the parabola $y^2 + 14y + 4x + 33 = 0$.

Solve.

8. $y^2 - x^2 = 16,$
   $x + y = 4$

9. $x^2 + y^2 = 25,$
   $y^2 + x = 5$

10. The sum of two numbers is 25, and the difference of their squares is 125. What are the numbers?

NAME ____________________

TEST FORM B

Solve.

ANSWERS

11. A rectangle has a perimeter of 24 cm and the length of a diagonal is $\sqrt{74}$ cm. Find the dimensions.

11. ______________

12. In a fractional expression, the sum of the values of the numerator and the denominator is 14. The product of their values is 45. Find the values of the numerator and the denominator.

12. ______________

13. A rectangle with diagonal of length $\sqrt{106}$ has an area of 45. Find the dimensions of the rectangle.

13. ______________

14. Find two numbers whose product is 66 if the sum of their squares is 157.

14. ______________

15. A garden contains two square flower beds. Find the length of each bed if the sum of their areas is 113 ft^2 and the difference of their areas is 15 ft^2.

15. ______________

NAME ______________________

TEST FORM B

ANSWERS	
	Classify the equation as a circle, an ellipse, a parabola, or a hyperbola.
16. ________	16. $y^2 + x^2 + 10x - 4y + 28 = 0$
17. ________	17. $25x^2 + 9y^2 + 150x - 18y = -9$
18. ________	18. $y = 2x^2 - 3x + 8$
19. ________	19. $\frac{(y - 3)^2}{4} - \frac{(x - 5)^2}{16} = 1$
20. ________	20. $x = y^2 - 6y + 17 - 3x$
21. ________	21. $xy - 9y^2 = 18 - 9y^2$
22. ________	22. Find an equation of a hyperbola having $y = \frac{3}{4}x$ and $y = -\frac{3}{4}x$ as asymptotes and one vertex (4,0).
23. ________	23. Find two numbers whose product is 10 and the sum of whose reciprocals is $\frac{7}{10}$.
24. ________	24. Find an equation of a circle that passes through the points (1,-1) and (2,-2) and whose center is on the line $3x - y = 5$.

CHAPTER 10 NAME ____________

TEST FORM C CLASS ______ SCORE ________ GRADE ______

1. Graph: $3x^2 + 5xy - 2y^2 = 0$.

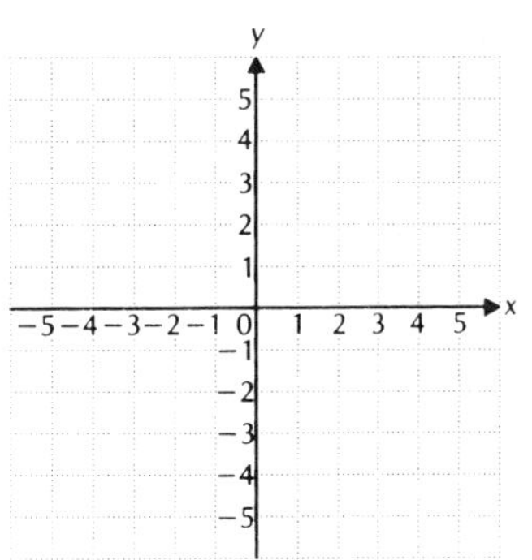

2. Find the center, vertices, and foci of the ellipse $4x^2 + 9y^2 - 16x + 54y + 61 = 0$, then graph the ellipse.

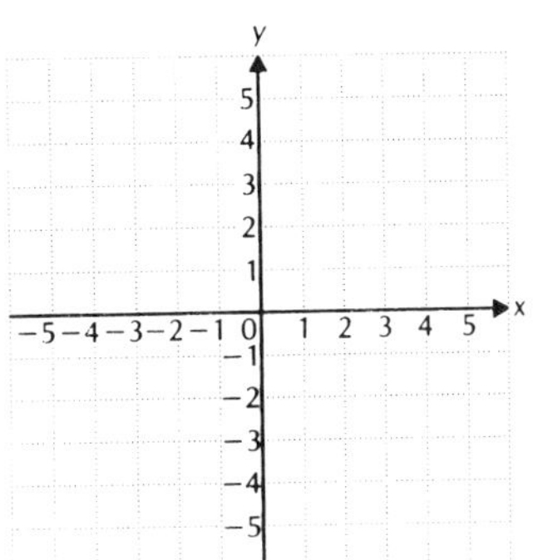

3. Find an equation of the ellipse with vertices (-6,0), (6,0), (0,-5), and (0,5).

4. Find the center, vertices, foci, and asymptotes of the hyperbola $9x^2 - 4y^2 = 36$.

5. Graph: $xy = 3$.

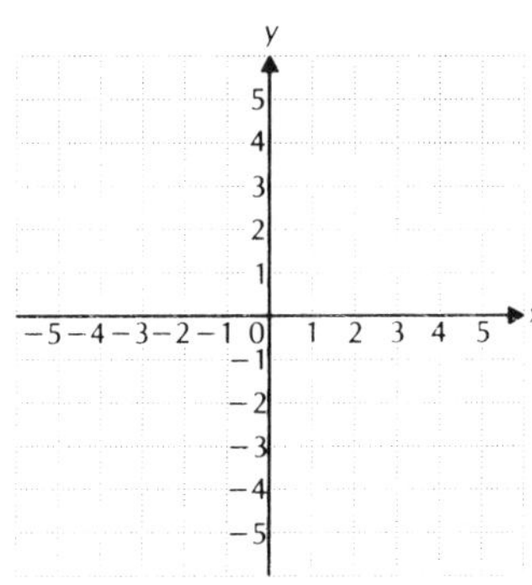

ANSWERS

1. See graph.

2. ____________

See graph.

3. ____________

4. ____________

5. See graph.

NAME ______________________

ANSWERS

6. ____________

7. ____________

8. ____________

9. ____________

10. ____________

6. Find an equation of the parabola with directrix $x = 3$ and focus $(-3,0)$.

7. Find the vertex, focus, and directrix of the parabola $x^2 - 2x + 4y + 21 = 0$.

Solve.

8. $y^2 = x + 5,$
   $3y = x + 7$

9. $2x^2 - y^2 = 4,$
   $x^2 + 2y^2 = 12$

10. The difference of two numbers is 9, and the difference of their squares is 135. What are the numbers?

NAME ____________________

TEST FORM C

	ANSWERS
Solve.	
11. A rectangle has a area of 32 $in.^2$ and a perimeter of of 24 in. Find the dimensions.	11. ________
12. In a fractional expression, the sum of the values of the numerator and the denominator is 21. The product of their values is 104. Find the values of the numerator and the denominator.	12. ________
13. The area of a rectangle is 60 yd^2, and the length of a diagonal is 13 yd. Find the dimensions.	13. ________
14. Find two numbers whose product is 75 if the sum of their squares is 250.	14. ________
15. Two squares are such that the sum of their areas is 53 cm^2 and the difference of their areas is 45 cm^2. Find the length of a side of each square.	15. ________

NAME ____________________

TEST FORM C

ANSWERS

16. ____________

17. ____________

18. ____________

19. ____________

20. ____________

21. ____________

22. ____________

23. ____________

24. ____________

Classify the equation as a circle, an ellipse, a parabola, or a hyperbola.

16. $x^2 + y^2 - 18x + 8y + 33 = 0$

17. $x = 5 + 3y - y^2$

18. $xy + 17 + 9x^2 = 9x^2$

19. $36x^2 + 25y^2 = 900$

20. $x^2 - 25y^2 + 100y - 125 = 0$

21. $x^2 - 46y = 0$

22. Find an equation of the ellipse with vertices (3,3), (-7,3), (-2,7), and (-2,-1).

23. Find two numbers whose product is 12 and the sum of whose reciprocals is $\frac{7}{12}$.

24. Find an equation of a circle that passes through the points (-3,3) and (-2,2) and whose center is on the line $2x + 5y = 4$.

NAME ______________________

CLASS ________ SCORE __________ GRADE ______

ANSWERS

1. Graph: $5x^2 + 3xy - 2y^2 = 0$.

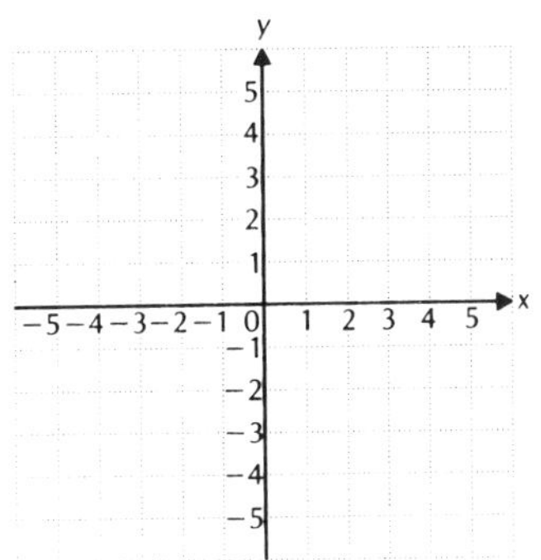

1. See graph.

2. Find the center, vertices, and foci of the ellipse $9x^2 + 16y^2 - 36x - 32y - 92 = 0$, then graph the ellipse.

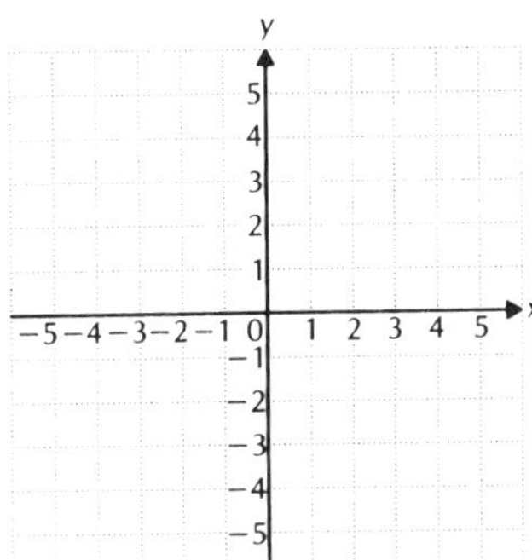

2. ______________

See graph.

3. Find an equation of the ellipse with vertices (-4,0), (4,0), (0,-3), and (0,3).

3. ______________

4. Find the center, vertices, foci, and asymptotes of the hyperbola $25y^2 - 9x^2 = 225$.

4. ______________

5. Graph: $xy = -5$.

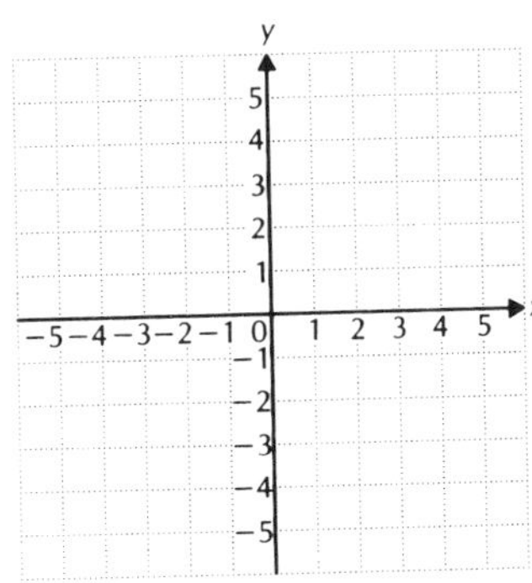

5. See graph.

NAME ____________________

TEST FORM D

ANSWERS

6. ____________________

7. ____________________

8. ____________________

9. ____________________

10. ____________________

6. Find an equation of the parabola with directrix $y = 6$ and focus $(0,-6)$.

7. Find the vertex, focus, and directrix of the parabola $y^2 - 6y - 4x + 17 = 0$.

Solve.

8. $y^2 = x - 4,$
   $3y = x - 2$

9. $x^2 + 9y^2 = 90,$
   $xy = 9$

10. The difference of two numbers is 5, and the sum of their squares is 97. What are the numbers?

NAME ______________________

TEST FORM D

Solve.

ANSWERS

11. It will take 310 yd of fencing to enclose a rectangular field. The area of the field is 4600 yd^2. What are the dimensions?

11. ______________

12. In a fractional expression, the sum of the values of the numerator and the denominator is 23. The product of their values is 112. Find the values of the numerator and the denominator.

12. ______________

13. A rectangle with diagonal of length $\sqrt{241}$ has an area of 60. Find the dimensions of the rectangle.

13. ______________

14. Find two numbers whose product is 96 if the sum of their squares is 292.

14. ______________

15. A garden contains two square flower beds. Find the length of each bed if the sum of their areas is 137 ft^2 and the difference of their areas is 105 ft^2.

15. ______________

NAME ______________________

TEST FORM D

ANSWERS

Classify the equation as a circle, an ellipse, a parabola, or a hyperbola.

16. ______

16. $9 - xy + 10y^2 = 10y^2$

17. ______

17. $x^2 = 15y - 9$

18. ______

18. $\frac{(x - 4)^2}{121} + \frac{(y + 1)^2}{64} = 1$

19. ______

19. $9y^2 - x^2 - 18y - 2x = 1$

20. ______

20. $y = 2x - 15 - 3x^2$

21. ______

21. $x^2 + y^2 + 6x - 2y - 90 = 0$

22. Find an equation of a hyperbola having $y = \frac{2}{3}x$ and $y = -\frac{2}{3}x$ as asymptotes and one vertex (0,-2).

22. ______

23. Find an equation of the following parabola: Line of symmetry parallel to the x-axis, vertex (-3,2), and passing through (1,6).

23. ______

24. The sum of two numbers is 10 and the product is 14. Find the sum of the reciprocals of the numbers.

24. ______

CHAPTER 10 NAME ______________________

TEST FORM E CLASS ________ SCORE ________ GRADE ______

MATCHING:
Place the letter of the answer in the answer blank. Some letters may be used more than once, and some will not be used.

Find:

1. An equation of the ellipse with vertices (-6,-2), (12,-2), (3,1), (3,-5).
2. One of the foci of the ellipse $\frac{x^2}{9} + \frac{y^2}{16} = 1$.
3. One of the vertices of the hyperbola $16x^2 - 9y^2 = 144$.
4. An equation of the hyperbola having asymptotes $y = -\frac{1}{3}x$ and $y = \frac{1}{3}x$ and one vertex (0,-3).
5. The vertex of the parabola $x^2 - 16y + 16 = 0$.
6. An equation of the parabola with directrix $y = -5$ and focus (0,11).

Answers:

a) $(x - 3)^2 + (y + 2)^2 = 1$

b) $\frac{y^2}{9} - \frac{x^2}{81} = 1$

c) $\frac{(x - 3)^2}{81} + \frac{(y + 2)^2}{9} = 1$

d) $(0, \sqrt{7})$

e) (0,5)

f) (0,4)

g) $(x - 3)^2 + (y + 2)^2 = 81$

h) $81y^2 - 9x^2 = 1$

i) $x^2 - 32y + 96 = 0$

j) (0,1)

k) (10,1)

ℓ) $\frac{(x - 3)^2}{81} + \frac{(y + 2)^2}{9} = 1$

m) $(-\sqrt{7}, 0)$

n) (3,0)

o) $y^2 - 32x + 6x - 11 = 0$

Classify each equation as a circle, ellipse, parabola, or hyperbola.

7. $y^2 + 4x - 8 = 0$
8. $25x^2 = 9y^2 - 225$
9. $3x^2 + 3y^2 - 6x + 18y + 3 = 0$
10. $x^2 + 4y^2 + 32y + 48 = 0$

ANSWERS

1. ______
2. ______
3. ______
4. ______
5. ______
6. ______
7. ______
8. ______
9. ______
10. ______

CHAPTER 10

NAME 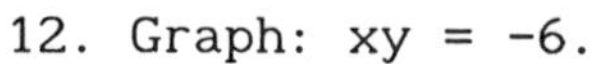

TEST FORM E

ANSWERS

11. See graph.

12. See graph.

13. See graph.

14. See graph.

15. ______

16. ______

17. ______

18. ______

11. Graph:
$10x^2 + 13xy - 3y^2 = 0.$

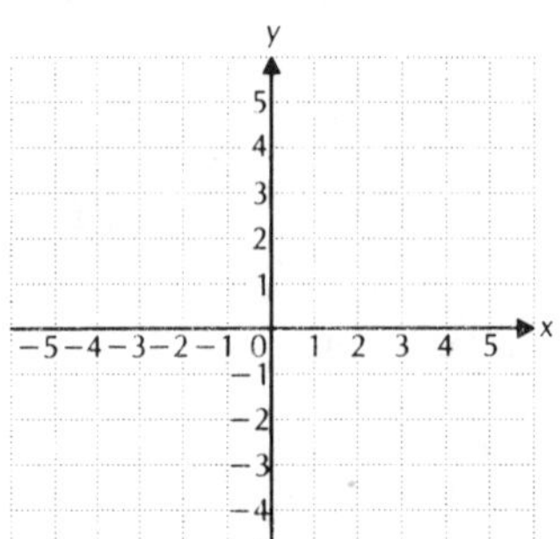

12. Graph: $xy = -6.$

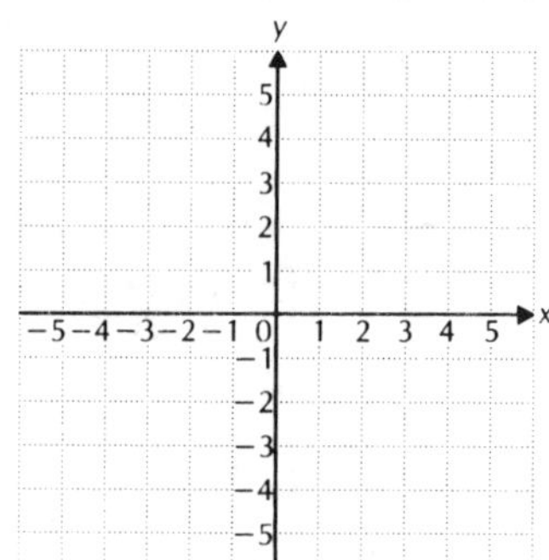

13. Graph:
$x^2 - 4x + 8y + 4 = 0.$

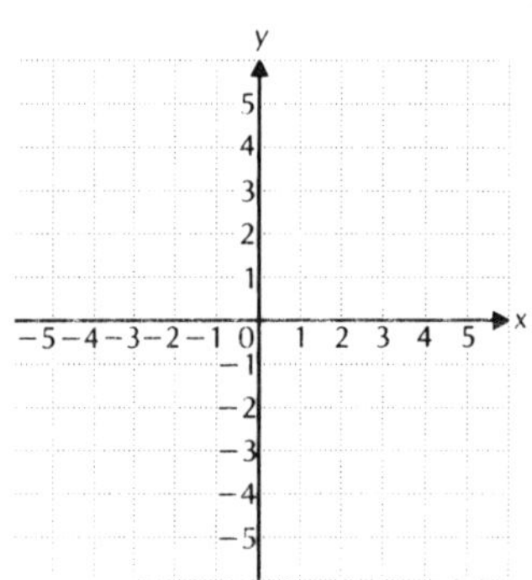

14. Graph:
$4(x - 1)^2 - 9(y - 2)^2 = 36.$

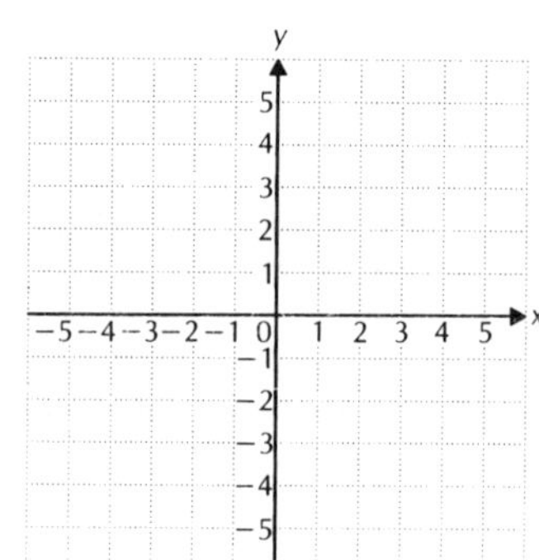

15. Solve: $x^2 + 4y^2 = 16,$
$x + 2y = 4$

16. Solve: $2x^2 - y^2 = 4,$
$x^2 + 2y^2 = 12$

17. Find two numbers whose product is -72 if the sum of their squares is 180.

18. Find an equation of the parabola containing (-2,1) whose line of symmetry is parallel to the x-axis, and whose vertex is (3,-4).

CHAPTER 10 NAME ______________________

TEST FORM F CLASS ________ SCORE __________ GRADE ______

Write the letter of your response on the answer blank.

ANSWERS

1. Graph: $5x^2 - 8xy + 3y^2 = 0$.

a) 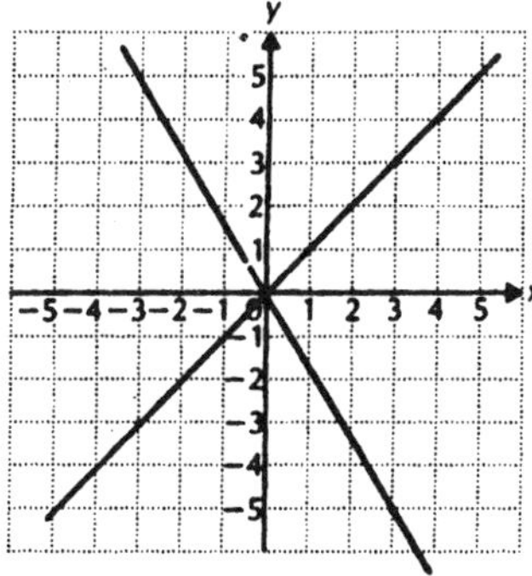

b) 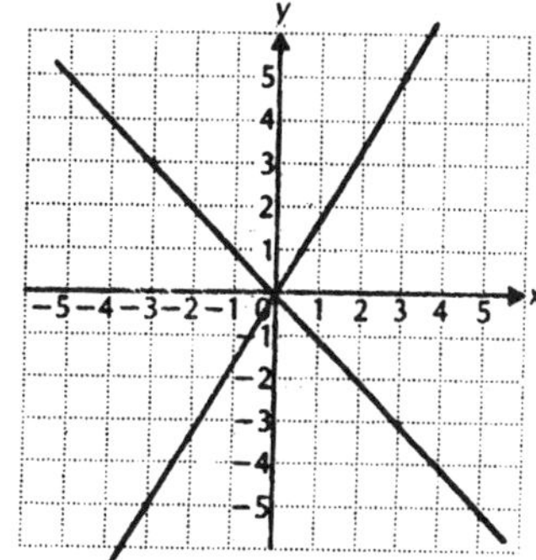

c) 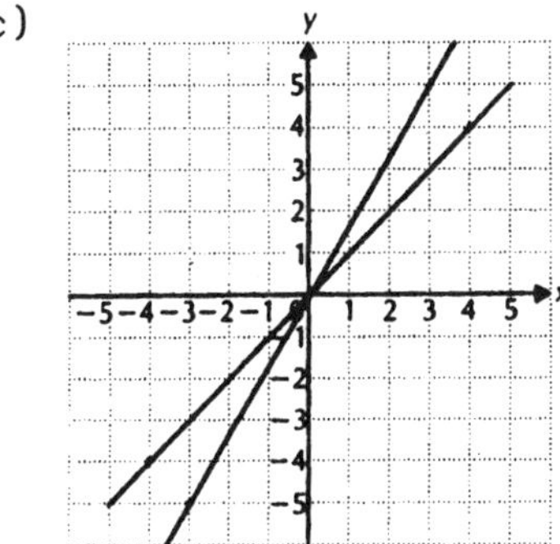

1. ______________

2. ______________

2. Find the center, vertices, and foci of the ellipse $49x^2 + 4y^2 - 294x + 8y + 249 = 0$.

a) C: (3,-1); V: (1,-1), (3,-8), (5,-1), (3,6);
F: $\left(3, -1 + 3\sqrt{5}\right)$, $\left(3, -1 - 3\sqrt{5}\right)$

b) C: (3,-1); V: (4,0), (-4,0), (0,7), (0,-7);
F: $\left(0, 3\sqrt{5}\right)$, $\left(3, -3\sqrt{5}\right)$

c) C: (-3,1); V: (0,1), (-7,1), (-3,6), (-3,-8);
F: $\left(-2, 1 - 3\sqrt{5}\right)$, $\left(-2, 1 + 3\sqrt{5}\right)$

d) C: (3,-1); V: (7,0), (-7,0), (4,0), (-4,0);
F: $\left(3\sqrt{5}, 0\right)$, $\left(-3\sqrt{5}, 0\right)$

CHAPTER 10 NAME ______________________

TEST FORM F

ANSWERS	
3. ______	3. Find an equation of the ellipse with vertices (-6,-2), (12,-2), (3,8), and (3,-12).
4. ______	4. Graph: $x^2 + 4y^2 = 4$.
5. ______	5. Find the vertices, foci, and asymptotes of the hyperbola $9y^2 - 16x^2 = 144$.

3. Find an equation of the ellipse with vertices (-6,-2), (12,-2), (3,8), and (3,-12).

a) $\frac{x^2}{81} - \frac{y^2}{100} = 1$ b) $\frac{(x - 3)^2}{81} + \frac{(y + 2)^2}{100} = 1$

c) $\frac{x^2}{9} - \frac{y^2}{10} = 1$ d) $\frac{(x + 3)^2}{81} - \frac{(y - 2)^2}{100} = 1$

4. Graph: $x^2 + 4y^2 = 4$.

a)

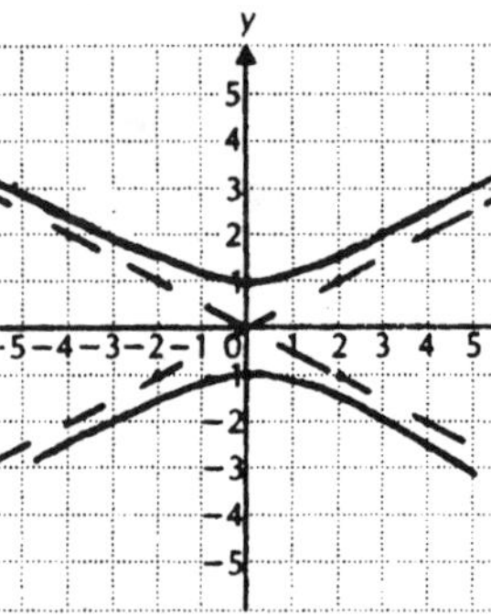

b)

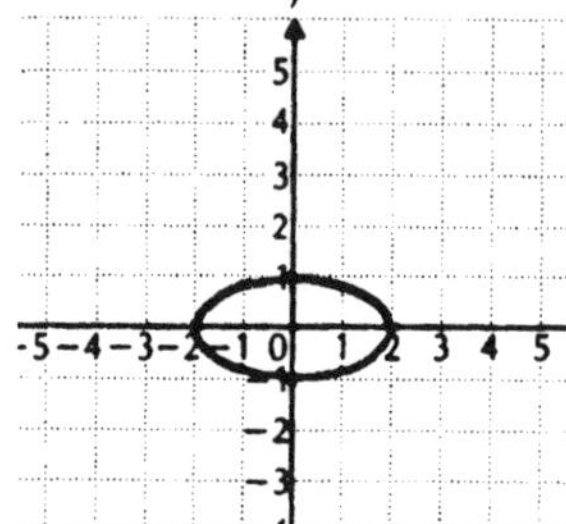

c)

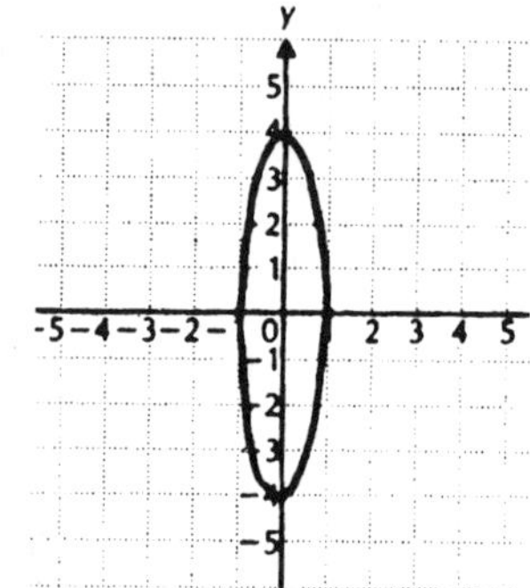

5. Find the vertices, foci, and asymptotes of the hyperbola $9y^2 - 16x^2 = 144$.

a) V: (4,0), (-4,0); F: (8,0),(-8,0);
A: $y = \frac{3}{4}x$, $y = -\frac{3}{4}x$

b) V: (0,3), (0,-3); F: (0,9), (0,-9);
A: $y = 5x$, $y = -5x$

c) V: (0,4), (0,-4); F: (0,8), (0,-8);
A: $y = \frac{1}{5}x$, $y = -\frac{1}{5}x$

d) V: (0,4), (0,-4); F: (0,5), (0.-5);
A: $y = \frac{4}{3}x$, $y = -\frac{4}{3}x$

6. Graph: $xy = 6$.

a) 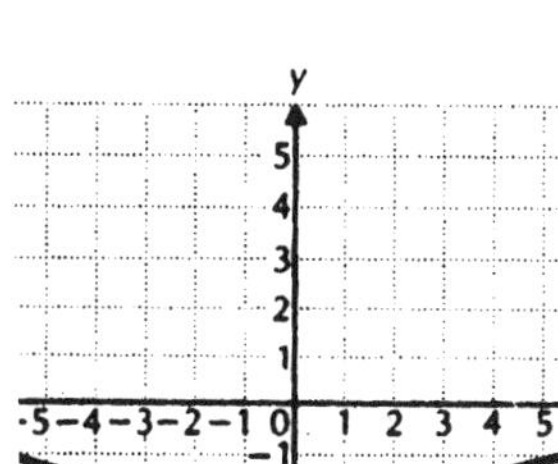

b) 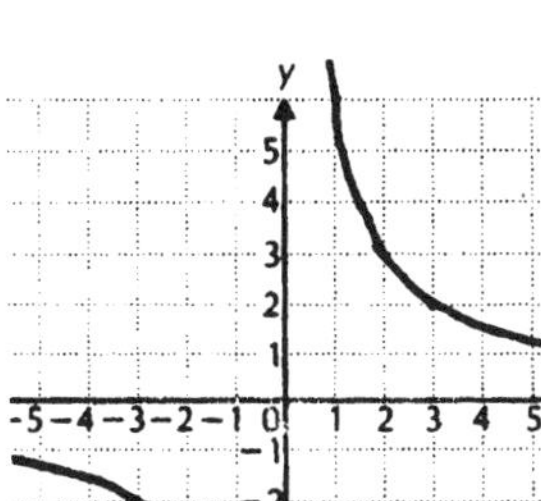

c) 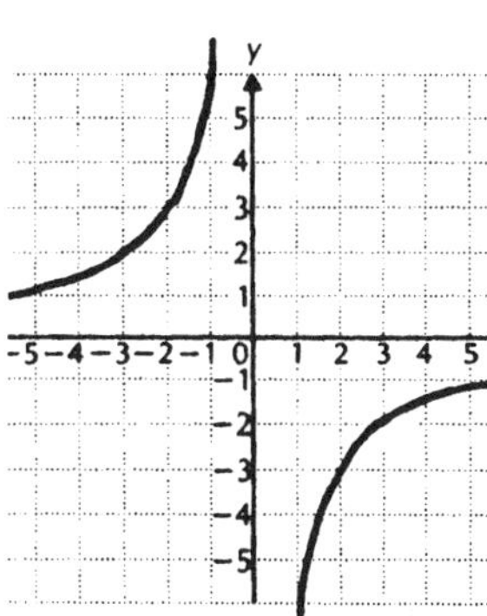

7. Find an equation of the parabola with directrix $x = -2$ and focus $(4,-2)$.

a) $(y + 2)^2 = 8(x + 1)$ b) $(y + 2)^2 = 12(x - 1)$

c) $y^2 = 16x$ d) $(x + 4)^2 = 4(y - 1)$

8. Find the directrix of the parabola $x^2 - 14x - 40y - 31 = 0$.

a) $y = -1$ b) $x = -3$ c) $x = 7$ d) $y = -12$

9. Graph: $x^2 - 8y - 8 = 0$.

a) 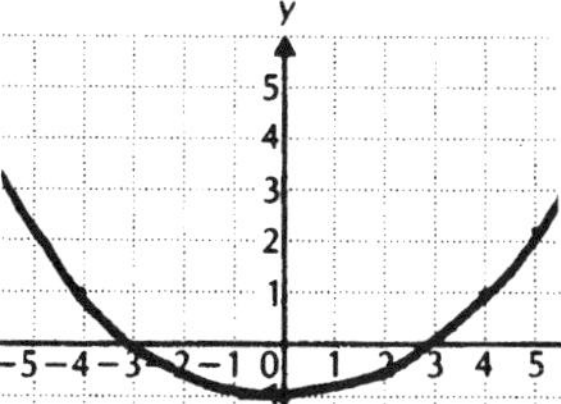

b) c) 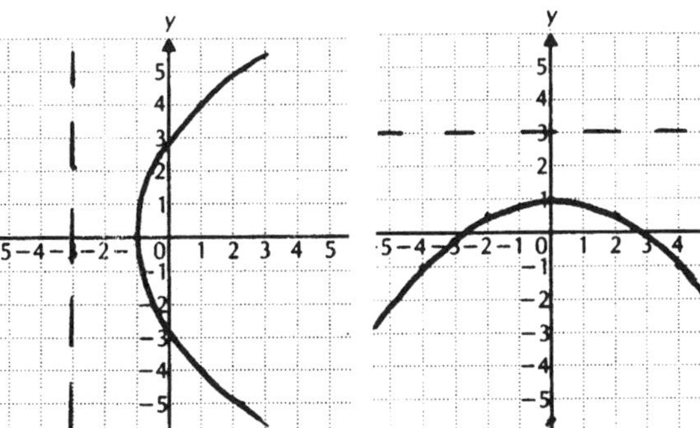

10. Classify the equation: $x^2 - 125 = 25y^2 - 100y$.

a) circle b) ellipse c) hyperbola d) parabola

ANSWERS

6. ______________

7. ______________

8. ______________

9. ______________

10. ______________

 NAME ______________________

ANSWERS	
11. ________	11. Classify the equation: $4x^2 + 2x + 1 = 5y - 4y^2$. a) circle b) ellipse c) hyperbola d) parabola
12. ________	12. Solve: $x + 5y = 6$, $2x^2 + 3y = 5$. The solution set a) is empty. b) consists of two ordered pairs. c) consists of one ordered pair. d) consists of four ordered pairs.
13. ________	13. Solve: $x^2 + y^2 = 4$, $36x^2 + 4y^2 = 144$. The solution set a) is empty. b) consists of two ordered pairs. c) consists of one ordered pair. d) consists of four ordered pairs.
14. ________	14. The sum of two numbers is -3, and the difference of their squares is 57. Find the larger of the two numbers. a) 8 b) 19 c) -11 d) 11
15. ________	15. A rectangle has a perimeter of 40 cm and the length of a diagonal is $4\sqrt{13}$ cm. Find the area of the rectangle. a) 40 cm^2 b) 99 cm^2 c) 96 cm^2 d) 13 cm^2

CHAPTER 11 NAME ______________________

TEST FORM A CLASS ________ SCORE __________ GRADE ______

	ANSWERS
1. Find the 11th term of the arithmetic sequence $\frac{1}{3}, 1, \frac{5}{3}, \frac{7}{3}, \ldots$.	1. ________
2. The 4th term of an arithmetic sequence is 17, and the 10th term is 41. Find the common difference.	2. ________
3. Which term of the arithmetic sequence $\frac{1}{4}, 1, \frac{7}{4}, \frac{5}{2}, \ldots$ is 16?	3. ________
4. Insert three arithmetic means between 5 and 12.	4. ________
5. Find the 10th term of the geometric sequence $\frac{1}{2}, -\frac{1}{4}, \frac{1}{8}, -\frac{1}{16}, \ldots$.	5. ________
6. Evaluate the sum $\sum_{k=1}^{6} 4\left(\frac{1}{2}\right)^{k-1}$.	6. ________
7. Which of the following infinite geometric sequences have sums? a) 4, -20, 100, -500, . . . b) 2, 0.8, 0.32, 0.128, . . . c) $-\frac{1}{3}, \frac{2}{3}, -\frac{4}{3}, 2, \ldots$	7. ________
8. Find the sum of the infinite geometric sequence $36, -6, 1, -\frac{1}{6}, \ldots$.	8. ________
9. If John saves \$15 one week, \$25 the second week, \$35 the third weeek, and continues this pattern, how much money would he save in the 30th week? What is the sum of his savings?	9. ________
10. A rubber ball is dropped from a height of 27 ft and always rebounds $\frac{1}{3}$ of the distance of the previous fall. What distance does it rebound on the 7th time?	10. ________
11. Find fractional notation for $0.177\overline{7}$.	11. ________
12. Use mathematical induction. Prove that for every natural number n, $1 \cdot 2 + 2 \cdot 4 + 3 \cdot 8 + \ldots + n \cdot 2^n = (n-1)2^{n+1} + 2$.	12. See work.

NAME ______________________

TEST FORM A

ANSWERS	
13. ________	13. Find the first 4 terms of this recursively defined sequence. $a_1 = 3,\ a_{k+1} = 1 + a_k^2$
14. a) ________ b) ________ c) ________	14. How many code symbols can be formed using 4 out of 7 of the letters A, B, C, D, E, F, G if the letters: a) can be repeated? b) cannot be repeated? c) cannot be repeated but must begin with B?
15. ________	15. A 5-member team is to be chosen from 11 basketball players. In how many ways can this be done?
16. ________	16. How many committees can be formed from a group of 8 governors and 9 senators if each committee contains 5 governors and 6 senators?
17. ________	17. In how many distinguishable ways can the letters of the word ALGEBRA be arranged?
18. ________	18. Determine the number of subsets of a set of 7 numbers.
19. ________	19. Find the 4th term of $(3a - b)^6$.
20. ________	20. Expand $\left(x + \sqrt{3}\right)^4$.
21. ________	21. What is the probability of getting a total of 5 on a roll of a pair of dice?
22. ________	22. From a deck of 52 cards, 1 card is drawn. What is the probability of drawing a ten or a queen?
23. ________	23. From a bag containing 6 nickels, 9 dimes, and 5 quarters, 8 coins are drawn at random, all at once. What is the probability of getting 3 nickels, 4 dimes and 1 quarter?
24. ________	24. Find decimal notation, rounded to six decimal places, for the first 5 terms of $a_n = \left(\frac{2n + 1}{n}\right)^n$.
25. ________	25. How many games are played in a league with 10 teams if each team plays each other team once?

CHAPTER 11 NAME ____________

TEST FORM B CLASS ______ SCORE ______ GRADE ______

1. Find the 18th term of the arithmetic sequence $\frac{2}{5}, \frac{6}{5}, 2, \frac{14}{5}, \ldots.$

2. The 2nd term of an arithmetic sequence is 9, and the 8th term is 51. Find the common difference.

3. Which term of the arithmetic sequence $\frac{2}{3}, 2, \frac{10}{3}, \frac{14}{3}, \ldots$ is $\frac{38}{3}$?

4. Insert three arithmetic means between 6 and 21.

5. Find the 6th term of the geometric sequence 0.3, 0.6, 1.2, 2.4, ...

6. Evaluate the sum $\sum_{k=1}^{5} \left(\frac{2}{3}\right)^{k-1}$.

7. Which of the following infinite geometric sequences have sums?

   a) $\frac{1}{2}, \frac{3}{2}, \frac{9}{2}, \frac{27}{2}, \ldots$

   b) 0.24, -0.12, 0.06, -0.03, . . .

   c) 46, 4.6, 0.46, 0.046, . . .

8. Find the sum of the infinite geometric sequence $1, \frac{1}{3}, \frac{1}{9}, \frac{1}{27}, \ldots.$

9. A bomb drops from a plane and falls 16 feet the first second, 48 feet the second second, 80 feet the third second, and so on. Find the total number of feet the bomb fell in 18 seconds.

10. A college student borrows $1000 at 10% interest compounded annually. The loan is paid off at the end of 4 years. How much is paid back?

11. Find fractional notation for $0.2424\overline{24}$.

12. Use mathematical induction. Prove that for every natural number n,

$$\frac{1}{1\cdot2\cdot3} + \frac{1}{2\cdot3\cdot4} + \cdots + \frac{1}{n(n+1)(n+2)} = \frac{n(n+3)}{4(n+1)(n+2)}.$$

ANSWERS

1. ____________

2. ____________

3. ____________

4. ____________

5. ____________

6. ____________

7. ____________

8. ____________

9. ____________

10. ____________

11. ____________

12. See work.

NAME ______________________

TEST FORM B

ANSWERS	
13. ____	13. Find the first 4 terms of this recursively defined sequence. $a_1 = -4,\ a_{k+1} = 9 - 2a_k$
14. a) ____ b) ____ c) ____	14. How many code symbols can be formed using 3 out of 8 of the letters B, D, M, P, Q, X, Y, Z if the letters: a) can be repeated? b) cannot be repeated? c) cannot be repeated but must end with XY?
15. ____	15. From a group of 12 gymnasts, 4 are to be chosen to represent their school. In how many ways can the representative be chosen?
16. ____	16. Of the first 8 questions on a test a student must answer 5. Of the next 6 questions the student must answer 3. In how many ways can this be done?
17. ____	17. In how many distinguishable ways can the letters of the word CALCULUS be arranged?
18. ____	18. Determine the number of subsets of a set of 9 numbers.
19. ____	19. Find the 3rd term of $(4a - b)^7$.
20. ____	20. Expand $\left(\sqrt{5} - a\right)^5$.
21. ____	21. What is the probability of getting a total of 9 on a roll of a pair of dice?
22. ____	22. From a deck of 52 cards, 1 card is drawn. What is the probability of drawing a diamond?
23. ____	23. From a group of 9 men and 7 women, a committee of 6 is chosen. What is the probability that 3 men and 3 women will be chosen?
24. ____	24. The value of an office machine is $4000. Its scrap value is 65% of its value the year before. Give a sequence that lists the scrap value of the machine for each year of a 5-year period.
25. ____	25. Solve ${}_nP_8 = 9 \cdot {}_{n-1}P_7$ for n.

CHAPTER 11 *NAME* ______________

TEST FORM C *CLASS* ______ *SCORE* ______ *GRADE* ______

1. Find the 10th term of the arithmetic sequence
$\frac{5}{8}, 1, \frac{11}{8}, \frac{7}{4}, \ldots.$

2. The 3rd term of an arithmetic sequence is 19, and the 9th term is 67. Find the common difference.

3. Which term of the arithmetic sequence
$\frac{3}{8}, \frac{7}{8}, \frac{11}{8}, \frac{15}{8}, \ldots$ is $\frac{51}{8}$?

4. Insert three arithmetic means between 4 and 15.

5. Find the 7th term of the geometric sequence
$6, -3, \frac{3}{2}, -\frac{3}{4}$

6. Evaluate the sum $\sum_{k=1}^{5} (3 - 2^k)$.

7. Which of the following infinite geometric sequences have sums?
a) $8, -2, \frac{1}{2}, -\frac{1}{8}, \ldots$
b) 21, 52.5, 131.25, 328.125, . . .
c) $\frac{7}{8}, \frac{7}{4}, \frac{7}{2}, 7, \ldots$

8. Find the sum of the infinite geometric sequence
7, 0.7, 0.07, 0.007, . . .

9. A hiker walks 3.9 mi the first day, 4.7 mi the second day, 5.5 mi the third day, and so on. What is the total distance hiked in 2 weeks?

10. Grantville has a population of 20,000 now and the population is increasing 8% every year. What will the population be in 5 years?

11. Find fractional notation for $8.555\overline{5}$.

12. Use mathematical induction. Prove that for every natural number n,
$$\frac{1}{3} + \frac{1}{15} + \frac{1}{35} + \ldots + \frac{1}{4n^2 - 1} = \frac{n}{2n + 1}.$$

ANSWERS

1. ______
2. ______
3. ______
4. ______
5. ______
6. ______
7. ______
8. ______
9. ______
10. ______
11. ______
12. See work.

NAME ____________________

TEST FORM C

ANSWERS

13. ______

14. a) ______
b) ______
c) ______

15. ______

16. ______

17. ______

18. ______

19. ______

20. ______

21. ______

22. ______

23. ______

24. ______

25. ______

13. Find the first 4 terms of this recursively defined sequence. $a_1 = 4, \ a_{k+1} = 2a_k - 3$

14. How many code symbols can be formed using 5 out of 6 of the letters C, D, E, F, G, H if the letters:
    a) can be repeated?
    b) cannot be repeated?
    c) cannot be repeated but must begin with EFG?

15. There are 20 students in a fraternity. How many sets of 4 officers may be selected?

16. A committee is to be formed from a group of 12 women and 10 men and is to consist of 5 women and 5 men. How many committees can be formed?

17. In how many distinguishable ways can the letters of the word SCIENCE be arranged?

18. Determine the number of subsets of a set of 5 numbers.

19. Find the 6th term of $(a - 3b)^6$.

20. Expand $\left(\sqrt{7} + a\right)^6$.

21. What is the probability of getting a total of 7 on a roll of a pair of dice?

22. From a deck of 52 cards, 1 card is drawn. What is the probability of drawing a 5 or an ace?

23. If 7 marbles are drawn at random, all at once, from a bag containing 6 green marbles, 8 red marbles, and 3 white marbles, what is the probability that 3 will be green and 4 will be red?

24. Find four numbers in an arithmetic sequence such that the first plus twice the second is 12, and the fourth minus the third is 3.

25. Solve for n: $\binom{n}{n-2} = 45$.

NAME ____________

TEST FORM D

CLASS ______ SCORE ______ GRADE ______

1. Find the 16th term of the arithmetic sequence $\frac{3}{4}, \frac{3}{2}, \frac{9}{4}, 3, \ldots$.

2. The 3rd term of an arithmetic sequence is 16, and the 8th term is 41. Find the common difference.

3. Which term of the arithmetic sequence $4, \frac{11}{2}, 7, \frac{17}{2}, \ldots$ is 43?

4. Insert three arithmetic means between 3 and 18.

5. Find the 9th term of the geometric sequence $36, 12, 4, \frac{4}{3}, \ldots$.

6. Evaluate the sum $\sum_{k=1}^{4} 10\left(\frac{1}{5}\right)^k$.

7. Which of the following infinite geometric sequences have sums?

a) $\frac{1}{4}, \frac{1}{8}, \frac{1}{16}, \frac{1}{32}, \ldots$

b) 1, 1.2, 1.44, 1.728, . . .

c) $-8, -1, -\frac{1}{8}, -\frac{1}{64}, \ldots$

8. Find the sum of the infinite geometric sequence $-8, 2, -\frac{1}{2}, \frac{1}{8}, \ldots$.

9. A theater has 35 seats in the first row, 50 seats in the second row, 65 seats in the third row, and so on, for 20 rows. How many seats are there in the theater?

10. Suppose someone offered you a job for 2 weeks under the following conditions. You will be paid $1.00 for the first day, $2.00 for the second day, $4.00 for the third day, and so on. How much would you earn altogether?

11. Find fractional notation for $5.4545\overline{45}$.

12. Use mathematical induction. Prove that for every natural number n,

$$1^3 + 3^3 + 5^3 + \ldots + (2n - 1)^3 = n^2(2n^2 - 1).$$

ANSWERS

1. ____________

2. ____________

3. ____________

4. ____________

5. ____________

6. ____________

7. ____________

8. ____________

9. ____________

10. ____________

11. ____________

12. See work.

NAME ______________________

TEST FORM D

ANSWERS	
13. ______	13. Find the first 4 terms of this recursively defined sequence. $a_1 = 4,\ a_{k+1} = 5a_k - 2$
14. a) ______ b) ______ c) ______	14. How many code symbols can be formed using 4 out of 9 of the letters A, B, C, D, F, G, M, Y, Z if the letters: a) can be repeated? b) cannot be repeated? c) cannot be repeated but must end with MY?
15. ______	15. On a test, a student is to select 7 out of 12 questions. In how many ways can she do this?
16. ______	16. A restaurant list 8 entrees and 6 vegetables on its menu. If you choose 1 entree and 3 vegetables, how many ways can you make this choice?
17. ______	17. In how many distinguishable ways can the letters of the word ZOOLOGY be arranged?
18. ______	18. Determine the number of subsets of a set of 6 numbers.
19. ______	19. Find the 3rd term of $(2x - 3y)^5$.
20. ______	20. Expand $\left(x - \sqrt{6}\right)^5$.
21. ______	21. What is the probability of getting a total of 10 on a roll of a pair of dice?
22. ______	22. From a deck of 52 cards, 1 card is drawn. What is the probability of drawing a red card?
23. ______	23. From a group of 7 men and 10 women, a committee of 5 is chosen. What is the probability that 2 men and 3 women will be chosen?
24. ______	24. Find three numbers in an arithmetic sequence such that the sum of the first and third is 8 and the product of the first and second is 24.
25. ______	25. Solve ${}_nP_4 = 5 \cdot {}_{n-1}P_3$ for n.

CHAPTER 11 NAME ____________

TEST FORM E CLASS ______ SCORE ______ GRADE ______

ANSWERS

1. Find the 32nd term of the arithmetic sequence 6, 15, 24, ... 1. ________

2. The 3rd term of an arithmetic sequence is 9, and the 9th term is 33. Find the 15th term. 2. ________

3. Which term of the arithmetic sequence $\frac{7}{2}$, 4, $\frac{9}{2}$, . . . is 28? 3. ________

4. Find the sum of the first 10 terms of the arithmetic sequence $q - 3p$, $q - 2p$, $q - p$, ... 4. ________

5. Find the 7th term of the geometric sequence 0.2, 0.6, 1.8, ... 5. ________

6. Find the sum of the first 6 terms of the geometric sequence $\frac{1}{8}$, $-\frac{1}{12}$, $\frac{1}{18}$, ... 6. ________

7. Evaluate the sum $\sum_{k=1}^{5} \frac{4}{2^{k-1}}$. 7. ________

8. Insert five arithmetic means between 6 and 22. 8. ________

9. If you save \$5 the first week, \$7 the second week, \$9 the third week, and so on, how much do you save in 52 weeks? 9. ________

10. A ball dropped from the top of a tower (400 m high) always rebounds 25% of the distance of the previous fall. How far has the ball traveled when it hits the ground for the third time? 10. ________

11. A city has a population of 350,000 now, and the population is only increasing 2% every year. What will the population be in 5 years? 11. ________

12. Find fractional notation for $4.022\overline{2}$. 12. ________

NAME ______________________

TEST FORM E

ANSWERS	
13. See work.	13. Use mathematical induction. Prove for every natural number n. $$1 + 3 + 6 + \ldots + \frac{1}{2}n(n + 1) = \frac{n(n + 1)(n + 2)}{6}$$
14. ______	14. The winner in a bike race gets to select 1 prize from Group A, 2 prizes from Group B, and 1 prize from Group C. If Group A contains 4 trips, Group B contains 6 bicycles, and Group C contains 8 electronic games, how many different selections can be made?
15. ______	15. In how many ways can 7 students be arranged in a straight line? in a circle?
16. ______	16. Suppose the expression $x^4y^3z^2$ is expressed without exponents. In how many ways can this be done?
17. a) ______ b) ______ c) ______	17. How many 4-digit numbers can be named using the digits 1, 2, 3, 4 and 5 if the digits: a) can be repeated? b) are not repeated? c) are not repeated and the numbers must be less than 4000?
18. ______	18. If there are 8 outside doors in a school, in how many ways can a student enter one door and leave by a different door?
19. ______	19. Find the 5th term of $\left(5x - \sqrt{2y}\right)^7$.
20. ______	20. Expand $(2x + y^2)^5$.
21. ______	21. What is the probability of getting a total of 11 on a roll of a pair of dice?
22. ______	22. From a deck of 52 cards, 2 cards are drawn without replacement. What is the probability that both are 5's?
23. ______	23. From a group of 12 senators and 15 representatives, a committee of 5 is chosen. What is the probability that 2 senators and 3 representatives will be chosen?

CHAPTER 11 NAME ____________________

TEST FORM F CLASS ________ SCORE __________ GRADE ______

Write the letter of your response on the answer blank.

ANSWERS

1. Which term of the arithmetic sequence $1, \frac{5}{2}, 4, \frac{11}{2}, \ldots$ is 37?

   a) 25th b) 42nd c) 15th d) 26th

   1. ____________

2. The 5th term of an arithmetic sequence is 17 and the 12th term is 66. Find the common difference.

   a) 3 b) $\frac{21}{6}$ c) 7 d) 21

   2. ____________

3. Find the sum of the first 18 terms of the arithmetic sequence 2, 7, 12, . . .

   a) 783 b) 87 c) 1692 d) 801

   3. ____________

4. Insert three arithmetic means between 10 and 75.

   a) 23, 36, 49 b) $\frac{105}{4}, \frac{85}{2}, \frac{235}{4}$

   c) 30, 45, 60 d) $25, \frac{63}{2}, 60$

   4. ____________

5. Find the 6th term of the geometric sequence 16, -4, 1, ...

   a) $-\frac{1}{64}$ b) $\frac{1}{4}$ c) $-\frac{3}{8}$ d) $-\frac{3}{256}$

   5. ____________

6. Find the sum of the first 6 terms of the geometric sequence $\frac{1}{3}, -\frac{2}{3}, \frac{4}{3}, -\frac{8}{3}, \ldots$

   a) $\frac{11}{3}$ b) -7 c) $\frac{20}{3}$ d) -28

   6. ____________

7. Find the sum of the infinite geometric sequence $3, 1, \frac{1}{3}, \ldots$

   a) 1.5 b) 1 c) 4.5 d) 2

   7. ____________

NAME ______________________

TEST FORM F

ANSWERS

8. ______________

8. Find $\sum_{k=1}^{5} 3k^{k-1}$.

a) 701 b) 3512 c) 1875 d) 2103

9. ______________

9. Find fractional notation for $0.82323\overline{23}$. Simplify the fraction. The sum of the numerator and the denominator is

a) 122 b) 343 c) 361 d) 1822

10. ______________

10. An employee of a company makes deposits in the Credit Union as follows: $15.50 the first month, $17.00 the second month, $18.50 the third month, and so on, for three years. Find the sum of the deposits.

a) $1503 b) $1224 c) $2322 d) $1937.50

11. ______________

11. A ball dropped from the top of a tower (520 ft high) always rebounds $\frac{3}{5}$ of the distance of the previous fall. How high will the fourth bounce be?

a) 187.2 ft b) 67.392 ft c) 52 ft d) 112.32 ft

12. See work.

12. Use mathematical induction. Prove for every natural number n.

$$2 + 6 + 18 + . . . + 2(3^{n-1}) = 3^n - 1$$

13. ______________

13. In how many ways can 6 floats be lined up for a parade?

a) 2^6 b) 1 c) 6! d) 6

NAME ______________________

TEST FORM F

ANSWERS

14. How many different committees consisting of 2 men and 1 woman can be formed from a group of 6 men and 4 women?

a) 2024 b) 120 c) $\frac{3}{10}$ d) 60

14. ____________

15. How many ways can 9 people sit at a round table?

a) ${}_9C_8$ b) ${}_9P_7$ c) 8! d) 9!

15. ____________

16. How many 3-digit numbers can be named using the digits 0, 2, 4, 6, 8 if the digits are not repeated and the numbers must be greater than 500?

a) 10 b) 24 c) 12 d) 60

16. ____________

17. In how many distinquishable ways can the letters of the word TALLAHASSEE be arranged?

a) $\binom{11}{6}$ b) $\frac{6!}{3!2!2!2!}$ c) ${}_{11}P_6$ d) $\frac{11!}{3!2!2!2!}$

17. ____________

18. How many baseball games can be played in a 14-team league if each team plays all other teams once?

a) 91 b) 28 c) 182 d) 46

18. ____________

19. Expand $\left(y - \frac{2}{y}\right)^5$.

a) $y^2 - 5y^3 + 10y - \frac{10}{y} + \frac{5}{y^3} - \frac{1}{y^5}$

b) $y^2 + 5y^3 + 10y + \frac{10}{y} + \frac{5}{y^3} + \frac{1}{y^5}$

c) $y^5 - 10y^3 + 40y - \frac{80}{y} + \frac{80}{y^3} - \frac{32}{y^5}$

d) $y^5 + 10y^3 + 40y + \frac{80}{y} + \frac{80}{y^3} + \frac{32}{y^5}$

19. ____________

ANSWERS

20. ______

21. ______

22. ______

23. ______

24. ______

25. ______

20. Find the 4th term of $(3x - 4y)^8$.

a) $120,960x^5y^4$ b) $-870,912x^5y^3$

c) $96,768x^4y^4$ d) $15,120x^4y^4$

21. Determine the number of subsets of a set of 9 members.

a) ${}_9P_2$ b) 9^2 c) $\binom{9}{2}$ d) 2^9

22. From a deck of 52 cards, 1 card is drawn. What is the probability that it is an ace or a four?

a) $\frac{8}{13}$ b) $\frac{2}{13}$ c) $\frac{1}{26}$ d) $\frac{1}{13}$

23. What is the probability of rolling a 4 on a roll of a pair of dice?

a) 0 b) $\frac{1}{9}$ c) 6 d) $\frac{1}{12}$

24. If 3 marbles are drawn at random, all at once, from a bag containing 5 white marbles and 6 red marbles, what is the probability that 1 will be white and 2 will be red?

a) $\frac{1}{2}$ b) $\frac{3}{11}$ c) $\frac{5}{11}$ d) $\frac{3}{5}$

25. Suppose 4 cards are drawn without replacement from a well-shuffled deck of 52 cards. What is the probability that all four are spades?

a) $\frac{{}_4P_4}{{}_{52}P_4}$ b) $\frac{1}{13}$ c) $\frac{1}{4}$ d) $\frac{\binom{13}{4}}{\binom{52}{4}}$

FINAL EXAMINATION NAME ______________________

TEST FORM A CLASS ________ SCORE __________ GRADE ______

Consider the numbers:

$15,\ -\frac{3}{2},\ -\sqrt[5]{8},\ 0.73,\ 0,\ \sqrt{40},\ -3.7,\ 6.\overline{12},\ -18.$

1. Which are whole numbers?

2. Which are irrational numbers?

3. Which are natural numbers?

4. Which are integers?

Compute. Write scientific notation for each answer.

5. $(4.3 \times 10^{5})(3.1 \times 10^{-11})$

6. $\dfrac{4.8 \times 10^{-15}}{1.5 \times 10^{-10}}$

7. Add and simplify: $\dfrac{x}{x^2 + x - 6} + \dfrac{4}{x^2 - 4}$.

Solve.

8. $3x - 11 \geq 8x + 2$

9. $\dfrac{3}{x^2} - \dfrac{1}{x} = 14$

10. Write a quadratic equation whose solutions are $\sqrt{3}$ and $-2\sqrt{3}$.

11. Find an equation of variation in which y varies jointly as x and z and inversely as the square of w, and $y = 225$ when $x = 3$, $z = 2$, and $w = 8$.

12. Determine the meaningful replacements in the radical expression $\sqrt{18 - 2x}$.

13. Find an equation of the line through (-2,3) with $m = -4$.

14. Find the distance between (3,-2) and (5,1).

ANSWERS

1. ______________

2. ______________

3. ______________

4. ______________

5. ______________

6. ______________

7. ______________

8. ______________

9. ______________

10. ______________

11. ______________

12. ______________

13. ______________

14. ______________

FINAL EXAMINATION NAME ______________________

TEST FORM A

ANSWERS

15. ______________

____

____

____

____

____

16. ______________

____

____

____

____

____

____

____

17. ______________

18. ______________

19. ______________

20. ______________

21. ______________

22. ______________

Consider the functions f and g given by $f(x) = 2x - 4$ and $g(x) = x^2 + 3$ for Questions 15 and 16.

15. Find $(f + g)(x)$, $(f - g)(x)$, $fg(x)$, $(f/g)(x)$, $f \circ g(x)$, and $g \circ f(x)$.

16. Find the domain of f, g, f + g, f - g, fg, f/g, $f \circ g$, and $g \circ f$.

Use the following for Questions 17 and 18.

a) 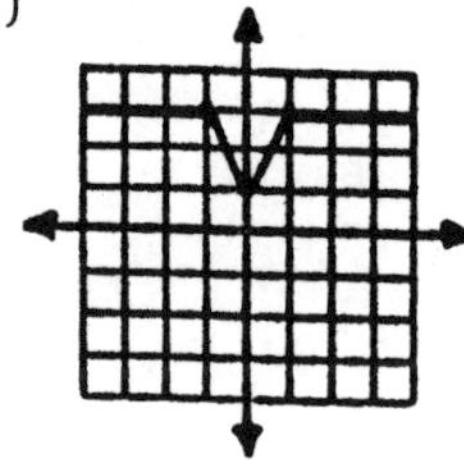 b) c) 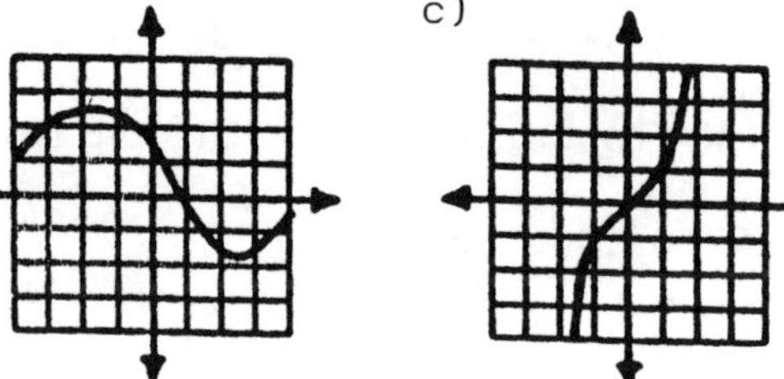

17. Which are increasing?

18. Which are neither increasing nor decreasing?

Solve.

19. $|2x - 5| \leq 7$

20. $6x^2 + 19x - 7 < 0$

21. The sum of the base and height of a triangle is 30 cm. Find the dimensions for which the area is a maximum.

22. Factor the polynomial P(x). Then solve the equation $P(x) = 0$.

$$P(x) = x^3 + x^2 - 7x - 3$$

FINAL EXAMINATION NAME ______________________

TEST FORM A

23. Find a polynomial of lowest degree having roots 2 and -3, and having -1 as a root of multiplicity 2 and 4 as a root of multiplicity 3.

24. Graph: $f(x) = \frac{x^2 - 5x + 5}{x - 2}$.

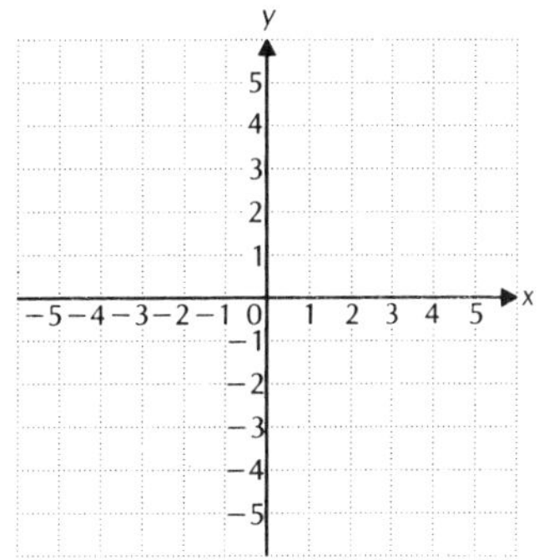

25. Write an equation of the inverse of the relation $y = 2x - 5$.

26. Find $\log_4 13$ using natural logarithms.

27. Solve: $\log_{\sqrt{8}} x = 4$.

28. How many years will it take an investment of $6000 to double if interest is compounded annually at 8%?

Find using a calculator.

29. log 15,750

30. ln 2.35

ANSWERS

23. ______________

24. See graph.

25. ______________

26. ______________

27. ______________

28. ______________

29. ______________

30. ______________

FINAL EXAMINATION NAME ______________________

TEST FORM A

ANSWERS

31. ______________________
____
____

32. See graph.

33. ______________________

34. ______________________

35. ______________________

36. ______________________

37. ______________________

38. ______________________

39. ______________________

40. ______________________

31. The point $\left(\frac{2}{3}, -\frac{\sqrt{5}}{3}\right)$ is on a unit circle. Find the coordinates of its reflection across the origin, the u-axis, and the v-axis.

32. Sketch a graph of y = cos x.

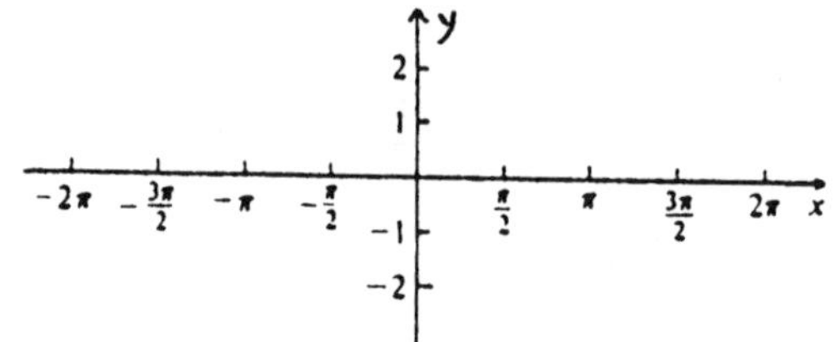

33. What is the range of the cosine function?

34. What is the period of the cosine function?

35. In which quadrants are the signs of the sine and cotangent alike?

Convert to degree measure.

36. $\frac{8\pi}{3}$ radians

37. 2 radians

Use a calculator to find each of the following.

38. sin $15^\circ 12' 13''$

39. sec 1.24

40. A rock on a 3-ft string is rotated at 20 rpm. What is its linear speed, in ft/min?

FINAL EXAMINATION   *NAME* ____________________

TEST FORM A

Complete these identities.

41. $\sin^2 x + \cos^2 x =$ ______

42. $\sec\left(\frac{\pi}{2} - x\right) =$ ______

43. Simplify: $\dfrac{\sin^2 x \cot^2 x - 1}{\sin x \cos^2 x \csc x}$.

44. Use a calculator to find, in degrees, $\tan^{-1}(0.3412)$.

45. Find $\sin\left[\cos^{-1}\left(-\frac{\sqrt{3}}{2}\right)\right]$.

46. Given that $\sin \theta = 0.7986$, $\cos \theta = 0.6018$, $\tan \theta = 1.3270$, $\cot \theta = 0.7536$, $\sec \theta = 1.6617$, $\csc \theta = 1.2522$, find the six function values for $90^\circ - \theta$.

47. An observer stands on level ground, 200 m from the base of a tower, and looks up at an angle of 40° to see the top of the tower. How high is the tower above the observer's eye level?

48. In triangle ABC, $b = 12$, $c = 14$, and $A = 59^\circ$. Find a.

49. Given $\mathbf{u} = \langle 2,-3 \rangle$ and $\mathbf{v} = \langle 5,8 \rangle$, find (a) $\mathbf{u} - \mathbf{v}$ and (b) $|2\mathbf{u} + \mathbf{v}|$.

50. Find cartesian coordinates of the point $(-3, 45^\circ)$.

ANSWERS

41. ______________

42. ______________

43. ______________

44. ______________

45. ______________

46. ______________

____

____

____

____

____

47. ______________

48. ______________

49. a) ______________

b) ______________

50. ______________

FINAL EXAMINATION — NAME ______________________

TEST FORM A

ANSWERS	
	51. Divide 8 cis 39° by 5 cis 15°.
51. ______	
	52. Find polar notation for $-3 + 3i$.
52. ______	
	53. Solve: $0.9x - 1.3y = -0.3$, $0.06x - 0.07y = 0.03$.
53. ______	
	54. Solve using matrices: $x - z = 1$, $3x + y + z = 8$, $x - y - z = 0$.
54. ______	
	55. Find numbers a, b, and c such that the function $f(x) = ax^2 + bx + c$ fits the data points (0,-4), (-1,-9), and (2,-6). Then write the equation for the function.
55. ______	
	56. Evaluate: $\begin{vmatrix} 5 & -1 & 3 \\ 0 & 2 & 4 \\ 1 & -1 & 5 \end{vmatrix}$.
56. ______	
	57. Add: $\begin{bmatrix} 5 & -1 \\ 3 & 7 \end{bmatrix} + \begin{bmatrix} 2 & 0 \\ 1 & 5 \end{bmatrix}$.
57. ______	
	58. Find A^{-1}, if it exists, for $A = \begin{bmatrix} 2 & 3 \\ -1 & 6 \end{bmatrix}$.
58. ______	
	59. Decompose into partial fractions: $\dfrac{3x + 5}{(x + 1)(x + 2)}$.
59. ______	

FINAL EXAMINATION NAME ______________

TEST FORM A

60. Find an equation of the ellipse with vertices (-2,0), (2,0), (0,-3), (0,3).

61. Graph: $xy = -4$.

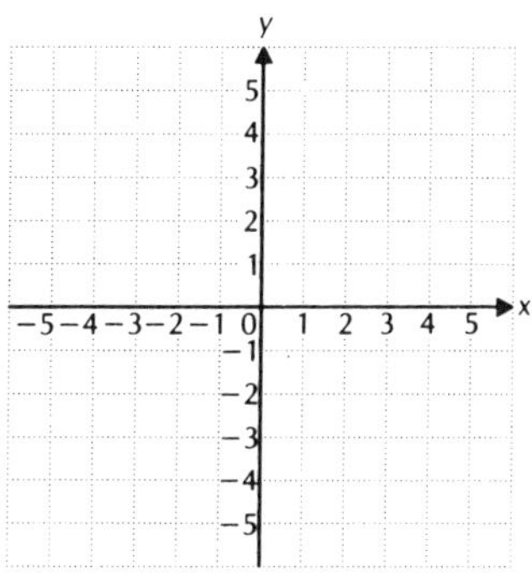

62. The difference of two numbers is 8, and the difference of their squares is 112. What are the numbers?

63. A rectangle has an area of 60 m^2 and a perimeter of 38 m. Find the dimensions.

64. Classify $x^2 + y^2 - 4x + 2y = -2$ as a circle, an ellipse, a parabola, or a hyperbola.

65. A theater has 40 seats in the first row, 65 seats in the second row, 90 seats in the third row, and so on, for 30 rows. How many seats are in the theater?

ANSWERS

60. ______________

61. See graph.

62. ______________

63. ______________

64. ______________

65. ______________

ANSWERS

66. ______

67. a) ______

b) ______

c) ______

68. ______

69. ______

70. ______

66. Evaluate the sum $\sum_{k=1}^{4} 3(5 + 2^k)$.

67. How many code symbols can be formed using 5 out of 9 of the letters A, B, C, D, E, F, G, H, I if the letters:

a) can be repeated?

b) cannot be repeated?

c) can be repeated but must begin with B?

68. Find the first 5 terms of this recursively defined sequence.

$$a_1 = 3$$
$$a_{k+1} = 4a_k - 5$$

69. Expand $\left(x + \sqrt{2}\right)^5$.

70. If 5 marbles are drawn at random, all at once, from a bag containing 8 green marbles, 4 red marbles, and 5 white marbles, what is the probability that 3 will be green and 2 will be white?

FINAL EXAMINATION NAME ______________________

TEST FORM B CLASS _______ SCORE _________ GRADE ______

Compute.

1. $-5.8 - 3.6$

2. $35 + |-35|$

3. $(-17)(-4)$

4. $\frac{56}{-2}$

Factor.

5. $6x^2 - 7x - 20$

6. $a^3 + 343$

7. $6x + 12y - x^2 - 2xy$

8. Find the reciprocal of $2 + 3i$ and express it in the form $a + bi$.

Solve.

9. $12x^2 - x - 63 = 0$

10. $\sqrt{y + 10} = \sqrt{5y - 11} - 3$

11. The length of a rectangle is three times the width. The perimeter is 48 in. Find the dimensions.

12. Determine the nature of the solutions of $x^2 - 6x + 5 = 0$ by evaluating the discriminant.

Consider the relation $\{(1,4), (-2,3), (7,4), (1,8)\}$.

13. Determine whether the relation is a function.

14. Find the range.

15. Find the domain.

ANSWERS

1. ______________

2. ______________

3. ______________

4. ______________

5. ______________

6. ______________

7. ______________

8. ______________

9. ______________

10. ______________

11. ______________

12. ______________

13. ______________

14. ______________

15. ______________

FINAL EXAMINATION *NAME* ____________________

TEST FORM B

ANSWERS

16. ____________

17. See graph.

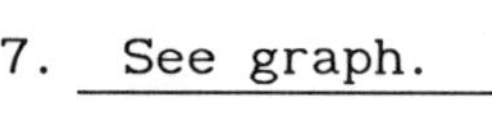

18. ____________

19. See graph.

20. ____________

21. See graph.

22. ____________

23. ____________

16. Find an equation of a circle with center (5,-1) and radius 3.

17. Graph $y = \frac{2}{3}x - 4$ using the slope and the y-intercept.

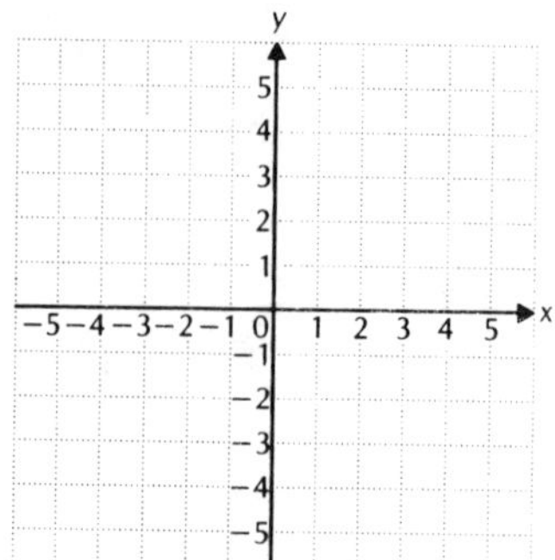

18. Write interval notation for $\{x|x \le -5\}$.

19. Graph: $f(x) = x^3 - x^2 - 9x + 9$.

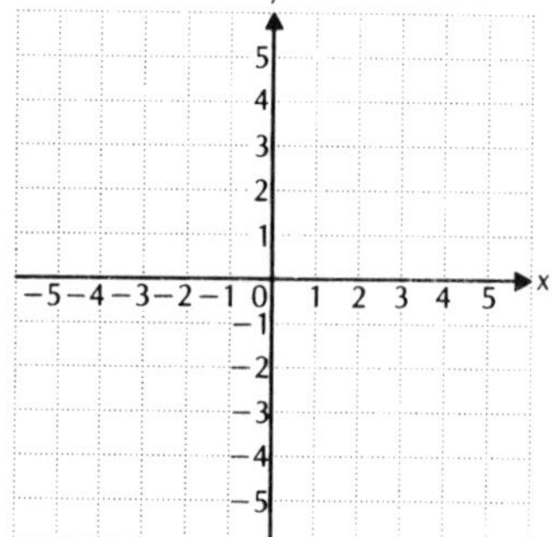

20. Find the x-intercepts of $f(x) = 2x^2 - x - 21$.

21. Graph: $\{x|x < -4\} \cup \{x|x > -2\}$.

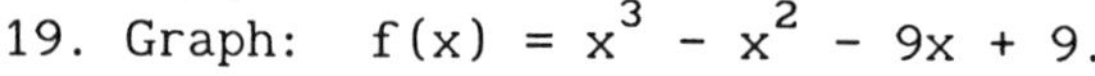

22. Determine whether $x - 5$ is a factor of $x^3 - 4x^2 - 7x + 10$.

23. Find $\{3,5,7,9,13\} \cup \{1,2,5,9,15\}$.

FINAL EXAMINATION NAME ______________________

TEST FORM B

	ANSWERS
24. Solve: $\|2x + 3\| = 11$.	24. ______
25. The population of a certain city was 200,000 in 1990. The exponential growth rate was 3.4% per year. a) Find the exponential growth function. b) Predict the population of the city in the year 2000. c) When will the population be 400,000?	25. a) ______ b) ______ c) ______
26. Graph: $y = \log_3 x$. 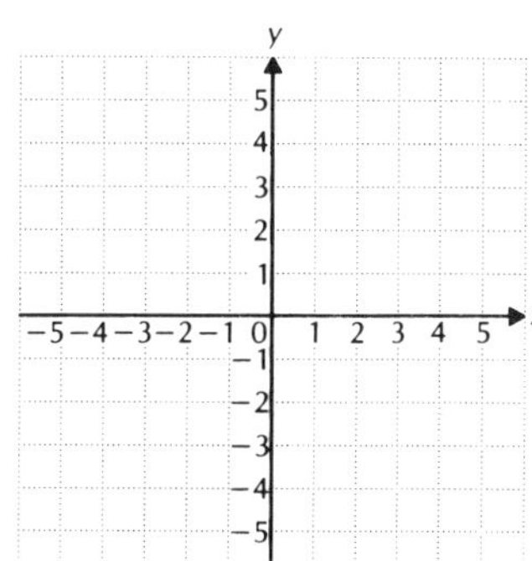	26. See graph.
	27. ______
	28. ______
27. Write an equivalent expression containing a single logarithm: $4 \log_a x - \frac{1}{2} \log_a y - 2 \log_a z$.	
	29. ______
Find using a calculator. 28. ln 0.00027 29. log (-15.21)	30. ______
30. The pH of a substance is 5.4. What is the hydrogen ion concentration?	

FINAL EXAMINATION NAME ____________________

TEST FORM B

ANSWERS

31. See graph.

32. ______

33. ______

34. ______

35. ______

______

______

______

______

______

36. 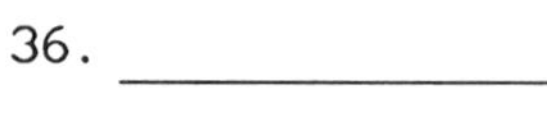

37. ______

38. ______

39. ______

40. ______

31. On a unit circle, mark and label the points determined by $\frac{2\pi}{3}$, $-\frac{5\pi}{4}$, $\frac{13\pi}{6}$.

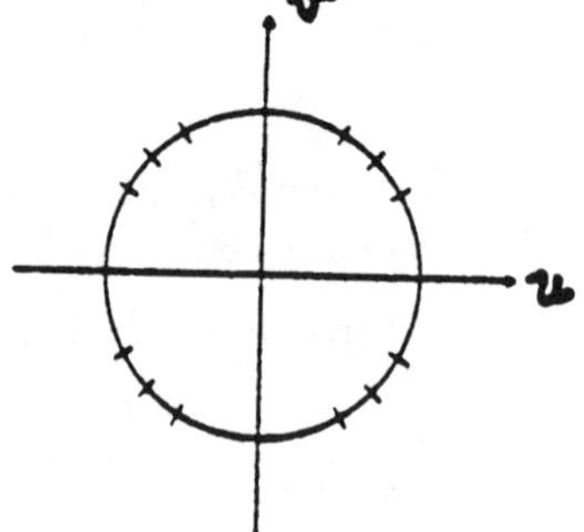

For angles of the following measures, state in which quadrant the terminal side lies, convert to radian measure in terms of π, and convert to radian measure not in terms of π.

32. 75°

33. 134°

34. Find the length of an arc of a circle, given a central angle of $\frac{4\pi}{3}$ and a radius of 6 cm.

35. Find the six trigonometric function values for the angle θ shown.

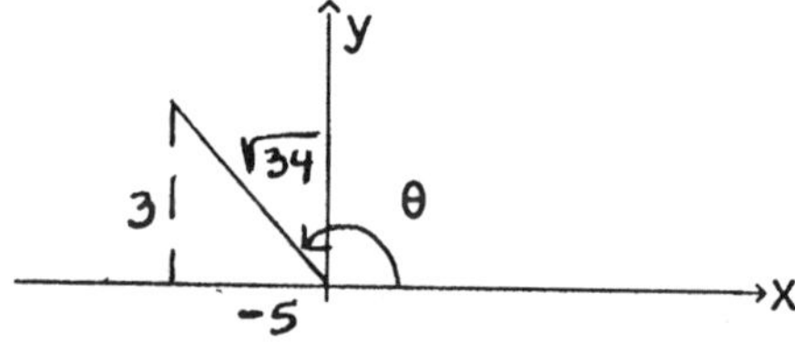

Find the following exactly. Do not use a calculator.

36. $\tan 540^\circ$

37. $\cos(-315^\circ)$

38. Convert $13^\circ 12'$ to degrees and decimal parts of degrees.

Use a calculator to find each of the following.

39. $\tan\left(-\frac{3\pi}{4}\right)$

40. $\csc 29^\circ 15'$

FINAL EXAMINATION NAME ______________________

TEST FORM B

41. Use the sum and difference formulas to write an equivalent expression. You need not simplify.

$$\tan\,(23^\circ + 42^\circ)$$

42. Given that $\cos\alpha = \frac{1}{2}$ and $\sin\beta = \frac{\sqrt{3}}{2}$ and that α and β are between 0 and $\frac{\pi}{2}$, evaluate $\tan\,(\alpha - \beta)$ exactly.

43. Rationalize the denominator: $\sqrt{\frac{\sin x}{\tan x}}$.

44. Find, in radians, $\arccos\left(-\frac{\sqrt{2}}{2}\right)$.

45. Find all solutions of the following equation in $[0,2\pi)$.

$$4\cos^3 x - \cos x = 0$$

46. Express csc x in terms of cot x.

47. Solve the right triangle ABC, given $A = 42.5^\circ$ and $b = 15.2$.

48. Find the area of triangle ABC if $a = 6.2$ cm, $c = 8.5$ cm, and $B = 43.8^\circ$.

49. Find rectangular notation for the vector $(8,150^\circ)$.

50. Convert $r = 8$ to a rectangular equation.

51. Find rectangular notation for $3\text{ cis }225^\circ$.

52. Find the cube root of $-8i$.

ANSWERS

41. ______

42. ______

43. ______

44. ______

45. ______

46. ______

47. ______

48. ______

49. ______

50. ______

51. ______

52. ______

FINAL EXAMINATION NAME ____________________

TEST FORM B

ANSWERS	
53. ______	53. A boat travels 40 km downstream in 2 hr. It travels 50 km upstream in 3 hr. Find the speed of the stream.
54. ______	54. Solve using Cramer's rule: $4x - y = 3,$ $2x + 3y = 7.$
55. See graph.	55. Graph: $2x + 4y \le 8.$ 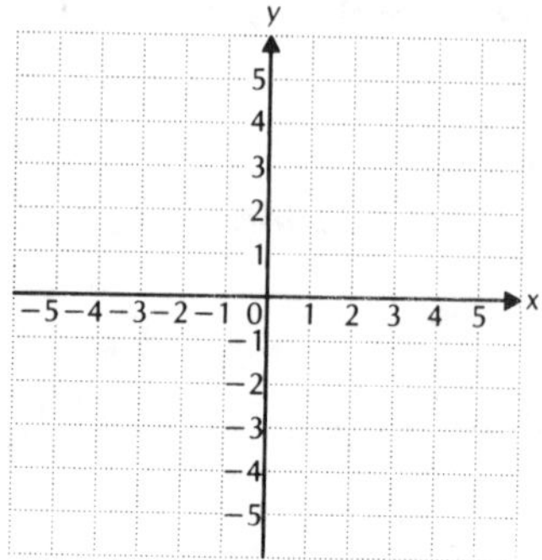
56. ______	56. Let $A = \begin{bmatrix} 5 & 0 & 4 \\ 1 & -1 & 0 \\ 2 & 3 & 7 \end{bmatrix}$. Find a_{21}, M_{21}, and A_{21}.
57. ______	57. Evaluate: $\begin{vmatrix} 0 & 1 & 4 \\ 3 & 2 & 5 \\ 1 & 0 & 6 \end{vmatrix}$.
58. ______	58. Given $A = \begin{bmatrix} 5 & -2 & 3 \\ 1 & 0 & 2 \end{bmatrix}$ and $B = \begin{bmatrix} 5 & 0 \\ -1 & 2 \\ 2 & 3 \end{bmatrix}$, find AB, if possible.
59. ______	59. Classify as consistent or inconsistent, dependent or independent. $5x + 3y = 4,$ $-5x - 3y = 0.$

FINAL EXAMINATION NAME ____________________

TEST FORM B

60. Graph: $2x^2 + 2xy - 12y^2 = 0$.

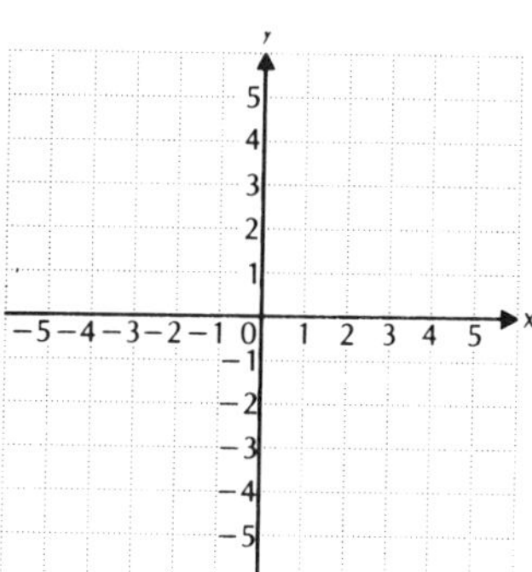

61. Find the center, vertices, foci, and asymptotes of the hyperbola $16x^2 - 9y^2 = 144$.

62. Solve: $y^2 = x + 3$,
$5y = x + 9$.

63. Find two numbers whose product is 132 if the sum of their squares is 265.

64. Classify $x^2 - 4y^2 + 2x + 16y = 19$ as a circle, an ellipse, a parabola, or a hyperbola.

65. Which term of the arithmetic sequence $5, \frac{15}{2}, 10, \frac{25}{2}, \ldots$ is 35?

ANSWERS

60. See graph.

61. ____________

62. ____________

63. ____________

64. ____________

65. ____________

ANSWERS

66. ______________

67. ______________

68. ______________

69. ______________

70. ______________

66. Which of the following infinite geometric sequences have sums?

a) $\frac{1}{3}, \frac{1}{6}, \frac{1}{12}, \frac{1}{24}, \ldots$

b) 3, 3.3, 3.63, 3.993, . . .

c) $-12, -4, -\frac{4}{3}, -\frac{4}{9}, \ldots$

67. What is the probability of getting a total of 7 on a roll of a pair of dice?

68. Determine the number of subsets of a set of 8 members.

69. Find the 5th term of $(3x + 5y)^6$.

70. From a deck of 52 cards, 1 card is drawn. What is the probability of drawing an ace or a three?

FINAL EXAMINATION *NAME* ______________________

TEST FORM C *CLASS* ________ *SCORE* ________ *GRADE* ________

Convert to decimal notation.

1. 6.129×10^{-5}

2. 4.1×10^{4}

3. 2.15 E 3

4. 1.29 E -2

5. Divide and simplify: $\dfrac{x^2 - 4x - 5}{x^2 - x - 6} \div \dfrac{(x + 1)^2}{x^2 - 9}$.

6. Rationalize the denominator: $\dfrac{7 + \sqrt{x}}{7 - \sqrt{x}}$.

7. The diagonal of a square has length $11\sqrt{2}$. Find the length of a side of the square.

Simplify.

8. i^{15}

9. $(2 + 6i) - (-1 - 5i)$

10. $\dfrac{3 + 2i}{1 - 4i}$

11. $(7 - i)(7 + i)$

12. The speed of a boat in still water is 12 mph. It travels 18 mi upstream and 18 mi downstream in a total time of 4 hr. What is the speed of the current?

13. Which of the following are graphs of functions?

a)

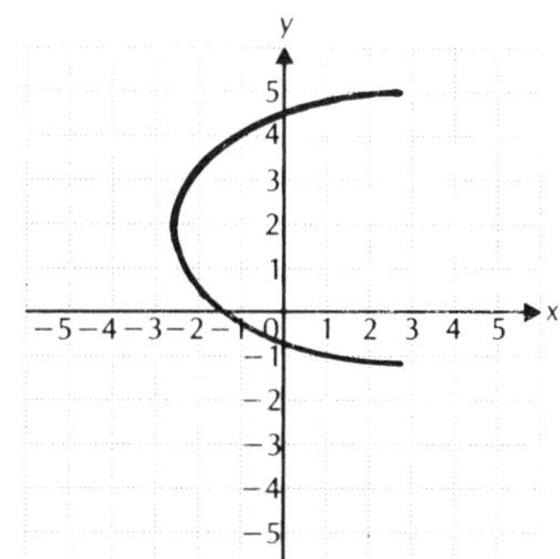

b)

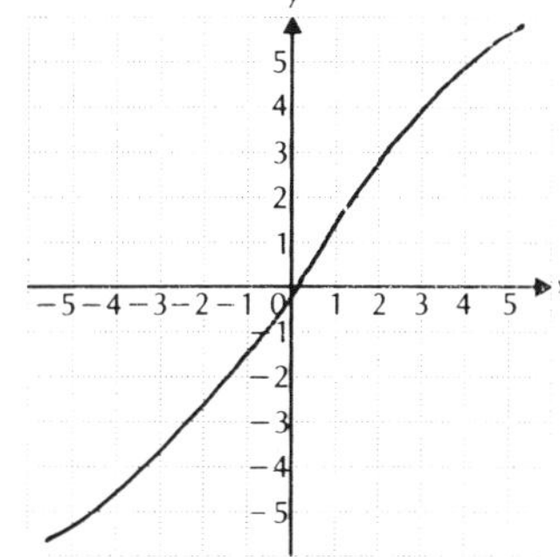

ANSWERS

1. ______________
2. ______________
3. ______________
4. ______________
5. ______________
6. ______________
7. ______________
8. ______________
9. ______________
10. ______________
11. ______________
12. ______________
13. ______________

FINAL EXAMINATION

NAME ______________________

TEST FORM C

ANSWERS

14. ____________

15. ____________

16. ____________

17. ____________

18. ____________

19. See graph.

20. ____________

21. ____________

22. ____________

23. ____________

24. ____________

Consider the following relations for Questions 14 and 15.

a) $x = y^4 - 3$ b) $x^3 - x = y^2$ c) $y^2 - 3 = x$

d) $2x - \frac{3}{y} = 0$ e) $3x = |y|$ f) $x = -3$

14. Which are symmetric with respect to the origin?

15. Which are symmetric with respect to the x-axis?

Use $g(x) = 2x^2 - 5x$ for Questions 16 - 18. Find:

16. $g(-2)$

17. $g(a + 2)$

18. $\dfrac{g(a + h) - g(a)}{h}$

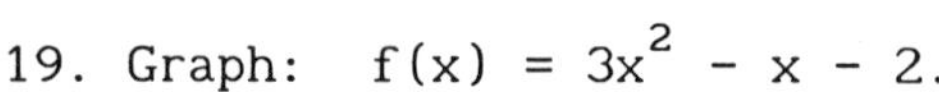

19. Graph: $f(x) = 3x^2 - x - 2.$

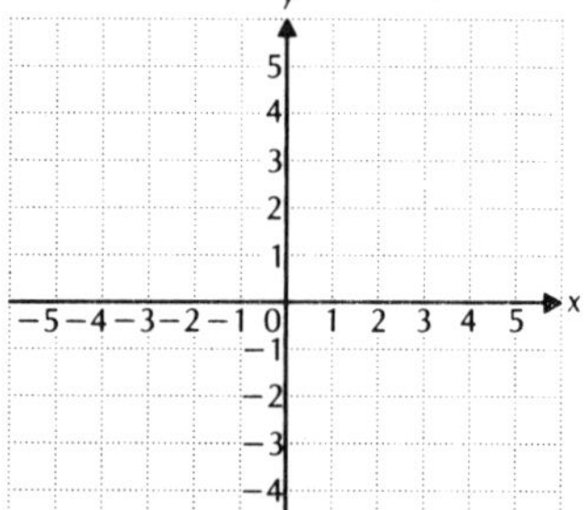

Solve.

20. $|8 - x| > 4$

21. $x^2 + 4x - 21 > 0$

22. $\dfrac{x + 1}{x + 2} < 2$

23. Use synthetic division to find P(-2):

$$7x^4 - 3x^2 + x - 5.$$

24. Factor the polynomial P(x). Then solve the equation for P(x) = 0. $P(x) = x^3 + x^2 - 7x - 3.$

FINAL EXAMINATION NAME ____________________

TEST FORM C

	ANSWERS
25. Which of the following have inverses that are functions?	25. ____

a) 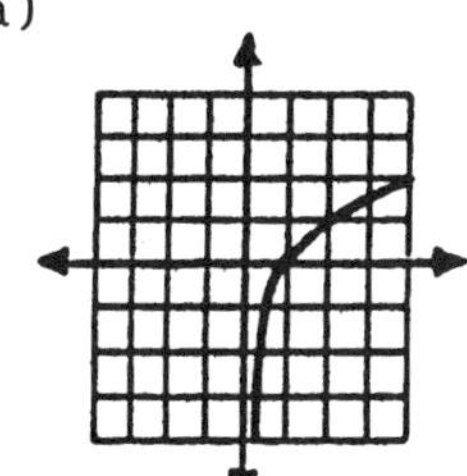

b) 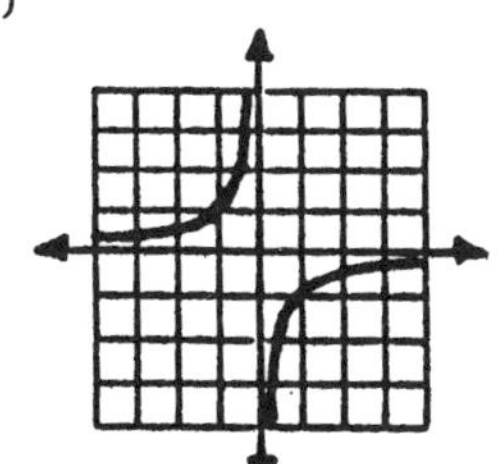

c) 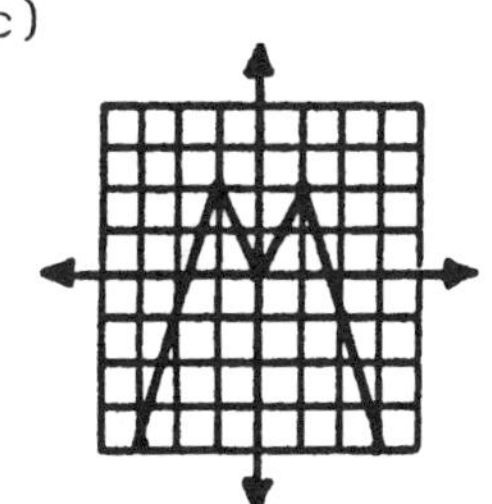

26. ____

27. ____

26. Simplify: $8^{\log_8 3x}$.

Solve.

27. $7^{4x-3} - 17 = 32$

28. ____

28. $\log_2 (x + 1) + \log_2 (x - 1) = 3$

29. How old is an animal bone that has lost 35% of its carbon-14?

29. ____

30. Students in an English class took a final exam. They took equivalent forms of the exam in monthly intervals thereafter. The average score $S(t)$, in percent, after t months was found to be given by
$S(t) = 80 - 12 \log (t + 1)$, $t \geq 0$.

a) What was the average score when they initially took the test, $t = 0$?

b) What was the average score after 6 months?

c) After what time was the average score 60?

30. a) ____

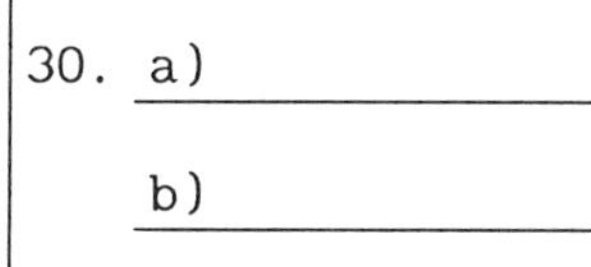

b) ____

c) ____

FINAL EXAMINATION

NAME ______________________________

TEST FORM C

ANSWERS

31. See table.

32. See graph.

33. ____________

34. ____________

35. ____________

36. ____________

37. ____________

38. ____________

39. ____________

40. ____________

31. Complete the following table.

x	$\frac{2\pi}{3}$	$\frac{7\pi}{6}$	$-\frac{\pi}{4}$
sin x			
cos x			

32. Sketch a graph of $y = \tan x$.

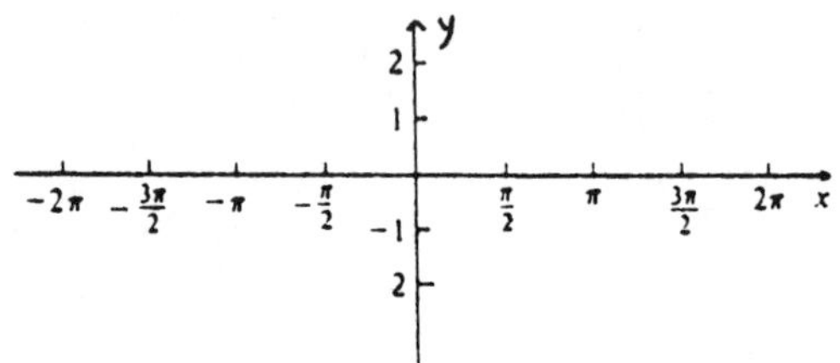

33. What is the period of the tangent function?

34. What is the domain of the tangent function?

35. An arc 15 cm long on a circle of radius 20 cm subtends an angle of how many radians? how many degrees, to the nearest degree?

36. Convert 83.25° to degrees and minutes.

Use a calculator to find θ in the interval indicated.

37. $\sin \theta = -0.8192$ $(180^\circ, 270^\circ)$

38. $\cos \theta = -0.3420$ $\left(\frac{\pi}{2}, \pi\right)$

Use a calculator to find each of the following.

39. cot 3.4937

40. $\cos (28^\circ 12' 14'')$

FINAL EXAMINATION NAME ______________________

TEST FORM C

41. Find $\tan 15^\circ$ exactly.

42. Simplify: $\tan \frac{x}{2} + \tan \frac{x}{2} \cos x$.

43. Find $\sin \frac{5\pi}{12}$ exactly.

44. Find sin [arcsin (0.45)].

45. Find all solutions of the following equation in $[0, 2\pi)$.
$\sin^2 x - 2 \sin x + 1 = 0$

46. Complete this identity: $\sec\left(\frac{\pi}{2} - x\right) =$ ______.

47. Solve the right triangle ABC, given $a = 15$ and $c = 24$.

48. Two forces of 6 newtons and 10 newtons act on an object at right angles. Find the magnitude of the resultant and the angle that it makes with the smaller force.

49. Graph the pair of complex numbers $4 - 2i$ and $-3 + i$ and their sum.

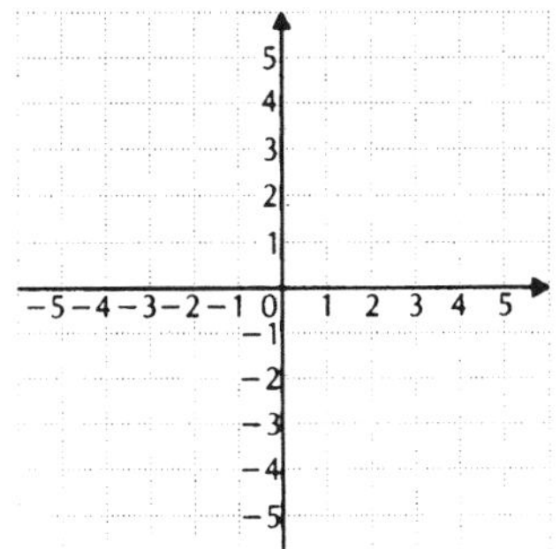

Real

50. Find polar notation for the vector <4,10>.

51. Find the product of $3 \text{ cis } 80^\circ$ and $4 \text{ cis } 15^\circ$.

52. Find $(1 + 2i)^5$. Write polar notation for the answer.

ANSWERS

41. ______________

42. ______________

43. ______________

44. ______________

45. ______________

46. ______________

47. ______________

48. ______________

49. See graph.

50. ______________

51. ______________

52. ______________

FINAL EXAMINATION NAME ______________________

TEST FORM C

ANSWERS	
53. ________	53. A collection of 30 coins consists of nickels and quarters. The total value is \$3.70. How many nickels are there?
54. ________	54. Solve using matrices. If there is more than one solution, list three of them. $$x + y + 2z = 1, \quad 2x + y + z = 2, \quad 3x + 2y + 3z = 3$$
55. ________	55. Evaluate: $\begin{vmatrix} \frac{5}{8} & -3 \\ 2 & 10 \end{vmatrix}$.
56. ________	56. Write a matrix equation equivalent to this system of equations and use the inverse of the coefficient matrix to solve the system. Show all your work. $$5x - y = 11, \quad x + 3y = -1$$
57. ________	57. Maximize and minimize $T = 5x - 3y$ subject to: $$x + y \leq 4, \quad -3 \leq x \leq 0, \quad y \geq 0.$$
58. ________	58. Given $C = \begin{bmatrix} 2 & -1 \\ 4 & 3 \end{bmatrix}$ and $D = \begin{bmatrix} 1 & -2 \\ -1 & 3 \end{bmatrix}$, find $C + 2D$.
59. ________	59. Find A^{-1}, if it exists, for $A = \begin{bmatrix} -2 & 4 \\ -1 & 3 \end{bmatrix}$.

NAME ______________________

60. Graph: $3x^2 + 5xy - 2y^2 = 0$.

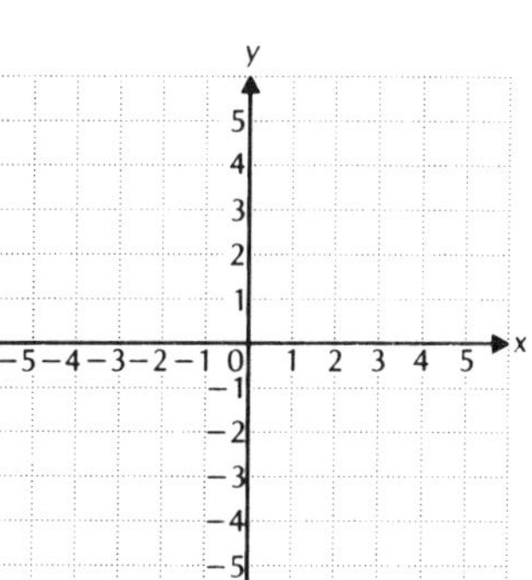

61. Find the vertex, focus, and directrix of the parabola $y^2 - 10y - 10x + 20 = 0$.

62. Solve: $x^2 + 3y^2 = 21$,
$xy = -6$.

63. Two squares are such that the sum of their areas is 106 cm^2 and the difference of their areas is 56 cm^2. Find the length of a side of each square.

64. Classify $\frac{(x - 1)^2}{49} + \frac{(y - 1)^2}{16} = 1$ as a circle, an ellipse, a parabola, or a hyperbola.

65. The third term of an arithmetic sequence is 15, and the ninth term is 45. Find the eighteenth term.

ANSWERS

60. See graph.

61. ______

62. ______

63. ______

64. ______

65. ______

FINAL EXAMINATION *NAME* ______________________

TEST FORM C

ANSWERS	
66. ______	66. Insert three arithmetic means between 5 and 14.
67. ______	67. Find the sum of the infinite geometric sequence $-10, 5, -\frac{5}{2}, \frac{5}{4}, \ldots$.
68. ______	68. Find fractional notation for $8.1515\overline{15}$.
69. ______	69. Of the first 8 questions on a test, a student must answer 5. Of the next 6 questions, the student must answer 3. In how many ways can this be done?
70. ______	70. From a group of 10 men and 6 women, a committee of 5 is chosen. What is the probability that 2 men and 3 women are chosen?

FINAL EXAMINATION NAME ______________________

TEST FORM D CLASS ________ SCORE __________ GRADE ________

Convert to scientific notation.

1. 0.0000034

2. 894.17

Simplify.

3. $(3x^2y^{-5})(-4xy^3)$

4. $\sqrt[3]{-64}$

5. $\left(\sqrt{3} - \sqrt{2}\right)\left(\sqrt{3} + \sqrt{2}\right)$

6. $(5a^3 - 2b^2)^2$

7. Change 80 kg/m to g/cm.

Solve.

8. $(2x - 3)(x - 1)(x + 5) = 0$

9. $15 - 3y < 21$

10. $2(x - 5) = 7 - (x + 6)$

11. $\dfrac{7}{x + 3} = \dfrac{5}{x + 1}$

12. Solve $\dfrac{P_1V_1}{T_1} = \dfrac{P_2V_2}{T_2}$ for V_2.

13. Graph:

$$f(x) = \begin{cases} x - 1, & \text{for } x \le -3 \\ x^2, & \text{for } -3 < x \le 3 \\ x - 5, & \text{for } x > 3 \end{cases}$$

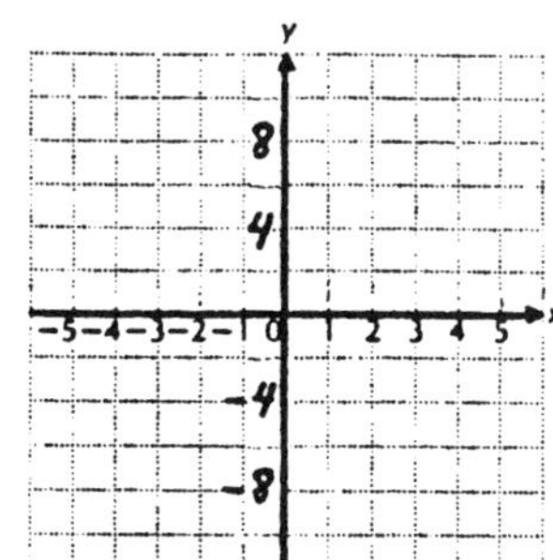

ANSWERS

1. ______________

2. ______________

3. ______________

4. ______________

5. ______________

6. ______________

7. ______________

8. ______________

9. ______________

10. ______________

11. ______________

12. ______________

13. See graph.

FINAL EXAMINATION   *NAME* ______________________

TEST FORM D

ANSWERS

14. ______________

15. ______________

16. ______________

17. ______________

18. ______________

19. ______________

20. ______________

21. ______________

14. Given $R(x) = x^2 + 15x - 340$ and $C(x) = 1.5x^2 + 3x + 20$, find $P(x)$.

15. Find the slope and the y-intercept of the line $15 - 2x = 4y$.

16. Find an equation of the line containing the given point and perpendicular to the given line.

$$(-1,3);\ y - 2 = 4x$$

17. Find the center and the radius of the circle $x^2 + y^2 + 6x - 10y + 18 = 0$.

18. Which of the following functions graphed below are even?

a)

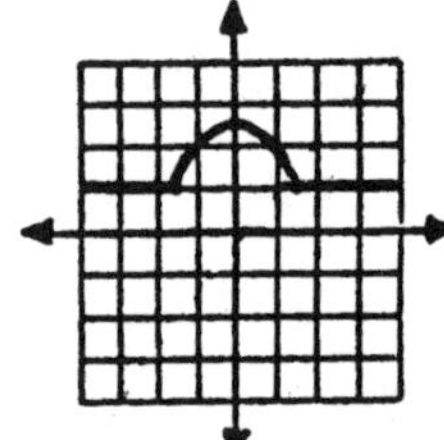

b) c)

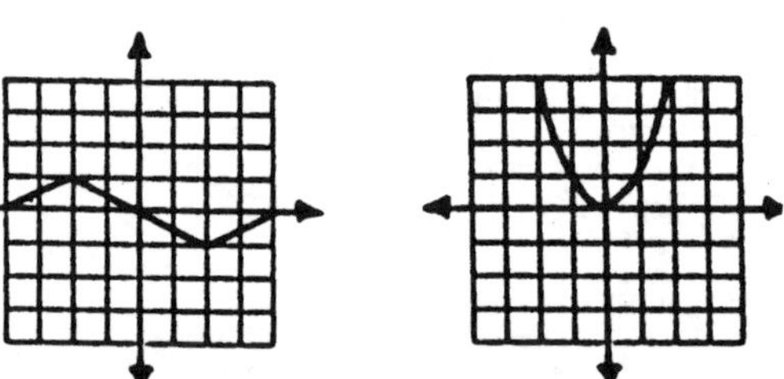

For the functions in Questions 19 and 20:

a) Use completing the square to put each equation into the form $f(x) = a(x - h)^2 + k$;

b) find the vertex; and

c) determine whether there is a maximum or minimum function value and find that value.

19. $f(x) = -x^2 + 5x - 6$ 20. $f(x) = 2x^2 - 4x - 3$

21. Find $\{2,3,5,9,12,15\} \cap \{2,4,8,12,16\}$.

FINAL EXAMINATION *NAME* ______________________

TEST FORM D

ANSWERS

22. Use synthetic division to find the quotient and the remainder. Show all your work.

$$(2x^4 - 6x^3 - x + 7) \div (x - 1).$$

22. ______________

23. Find a polynomial of degree 3 having 2i, -2i, and 3 as roots.

23. ______________

24. What does Descartes' rule of signs tell you about the number of positive real roots and the number of negative real roots of $x^7 - x^4 - x^3 + x - 1$?

24. ______________

25. Find the inverse of the relation H given by H = {(2,-3), (4,-2), (1,5), (-2,-7), (2.1,-1.5)}.

25. ______________

26. Find $h(h^{-1}(-5))$: $h(x) = \dfrac{2x^3 - 5}{7}$.

26. ______________

27. Express in terms of logarithms of a, b, and c:

$$\log \frac{a^2b}{c^5}.$$

27. ______________

28. What is the loudness, in decibels, of a sound whose intensity is $8{,}000 \cdot I_0$?

28. ______________

Given that $\log_{10} 2 = 0.301$, $\log_{10} 3 = 0.477$, and $\log_{10} 10 = 1$, find each of the following.

29. $\log_{10} 5$

30. $\log_{10} 9$

29. ______________

30. ______________

FINAL EXAMINATION NAME ______________________

TEST FORM D

ANSWERS	
31. See graph.	31. Sketch a graph of y = sin x. 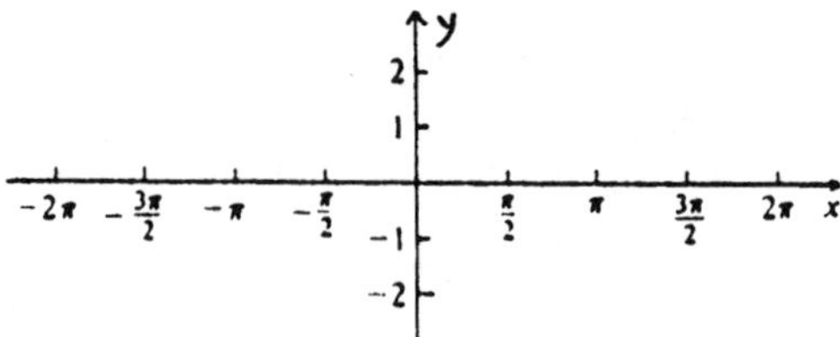
32. ______	32. What is the domain of the sine function?
33. ______	33. What is the amplitude to the sine function?
34. ______	34. A wheel has a 20-ft diameter. The speed of a point on its rim is 10 ft/sec. What is its angular speed, in rad/sec?
35. ______ ______ ______ ______ ______	35. Given that $\sin \theta = \frac{1}{2}$ and that the terminal side is in quadrant II, find the other five function values.
	Use a calculator to find the acute angle θ that is a solution to the equation. Give the answer in degrees and decimal parts of degrees, as well as in radians.
36. ______	36. $\cot \theta = 2.0872$
37. ______	37. $\tan \theta = 0.4663$
	Consider the function $y = -6 \cos\left(3x + \frac{\pi}{3}\right)$ for Questions 38-40.
38. ______	38. What is the amplitude?
39. ______	39. What is the period?
40. ______	40. What is the phase shift?

FINAL EXAMINATION NAME ______________________

TEST FORM D

41. Simplify: $\sin 43^\circ \cos 47^\circ - \cos 43^\circ \sin 47^\circ$.

42. Find an identity for $\sec (x - 630^\circ)$.

Complete these identities.

43. $\sin^2 x + \cos^2 x =$ ______________.

44. $\tan \left(\frac{\pi}{2} - x\right) =$ ______________.

45. Find $\sin 2\theta$, $\cos 2\theta$, $\tan 2\theta$ and the quadrant in which 2θ lies if $\sin \theta = -\frac{4}{5}$ and θ is in quadrant III.

46. Find $\sin^{-1} \frac{1}{2}$.

47. Graph: $r = 1 + 4\cos\theta$.

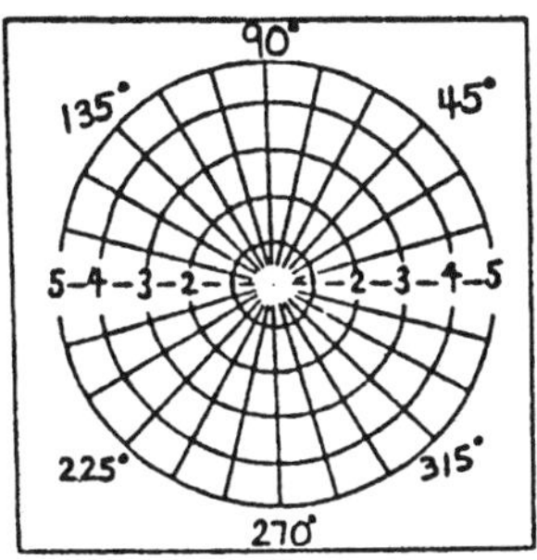

48. Solve the right triangle ABC, given $a = 17$ and $c = 25$.

49. In an isosceles triangle, the base angles each measure 32.5° and the base is 80 cm long. Find the lengths of the other two sides.

50. Find rectangular notation for the vector $(8, 315^\circ)$.

51. Solve: $x^2 = 10i$.

52. Convert $r = 12$ to a rectangular equation.

ANSWERS

41. ______________

42. ______________

43. ______________

44. ______________

45. ______________

46. ______________

47. See graph.

48. ______________

49. ______________

50. ______________

51. ______________

52. ______________

FINAL EXAMINATION NAME ______________________

TEST FORM D

ANSWERS

53. ____________

54. ____________

55. ____________

56. ____________

57. ____________

58. ____________

59. ____________

53. In triangle ABC, the measure of angle B is twice the measure of angle A. The measure of angle C is 20° less than the measure of angle A. Find the measure of angle C.

54. Solve using Cramer's rule:

$$x + 3y - z = 4,$$
$$x + y + 5z = -14,$$
$$5x - 4y + z = 2.$$

For Questions 55 - 58, let

$$A = \begin{bmatrix} 5 & -3 \\ 2 & 0 \end{bmatrix}, \quad I = \begin{bmatrix} 1 & 0 \\ 0 & 1 \end{bmatrix}, \quad F = \begin{bmatrix} 3 & 2 & -4 \end{bmatrix}, \quad 0 = \begin{bmatrix} 0 & 0 \\ 0 & 0 \end{bmatrix}, \quad G = \begin{bmatrix} 1 \\ 5 \\ 0 \end{bmatrix}$$

55. AI

56. A + 0

57. -F

58. FG

59. A small bakery bakes and sells cakes and pies. To stay in business, it must sell at least 8 cakes per day, but cannot bake more than 20. It must also sell at least 16 pies, but cannot bake more than 30. It cannot bake more than 40 cakes and pies altogether. The profit on a cake is \$3.50 and on a pie is \$1.50. How many pies and cakes should be sold in order to maximize profit?

NAME ______________________

ANSWERS

60. Find the center, vertices, and foci of the ellipse $4x^2 + 9y^2 - 16x + 54y + 61 = 0$.

60. ______________________

____

____

61. Find an equation of the parabola with directrix $x = -4$ and focus $(4,0)$.

61. ______________________

62. The area of a rectangle is 96 ft^2, and the length of a diagonal is $4\sqrt{13}$ ft. Find the dimensions.

62. ______________________

63. Solve: $3x^2 - y^2 = 11,$
$x^2 + 2y^2 = 6.$

63. ______________________

64. Classify $x^2 + 2x - 4y + 13 = 0$ as a circle, an ellipse, a parabola, or a hyperbola.

64. ______________________

65. Find the 31st term of the arithmetic sequence $-1, 6, 13, 20, \ldots$.

65. ______________________

FINAL EXAMINATION

NAME ______________________

TEST FORM D

ANSWERS	
66. ______	66. Evaluate the sum $\sum_{k=0}^{3} (1 + 2^k)$.
67. ______	67. Find the 6th term of the geometric sequence 6, 24, 96, ...
68. ______	68. Alex was offered a job paying \$1 the first day, \$2 the second day, and so on, doubling the salary every day thereafter. If he takes the job, how much would he earn in 10 days?
69. ______	69. How many distinguishable ways can the letters of the word BEEKEEPER be arranged?
70. ______	70. Expand: $\left(x - \sqrt{3}\right)^4$.

FINAL EXAMINATION

NAME ______________________

TEST FORM E

CLASS ________ *SCORE* __________ *GRADE* ______

ANSWERS

1. Write an expression containing a single radical.
$\sqrt[3]{p} \cdot \sqrt[4]{q^3} \cdot \sqrt[5]{r^2}$

1. ______________

2. Rationalize the denominator: $\dfrac{2\sqrt{x} + \sqrt{y}}{\sqrt{x} - 3\sqrt{y}}$.

2. ______________

3. Simplify: $\dfrac{x + \frac{27}{x^2}}{1 + \frac{3}{x}}$.

3. ______________

4. Write a quadratic equation whose solutions are 4 and $-\frac{3}{5}$.

4. ______________

5. Solve: $\sqrt{x + 9} = 6 - \sqrt{x - 3}$.

5. ______________

6. Find an equation of variation where y varies jointly as x and the square of z and inversely as w, and $y = 90$ when $x = 18$, $z = 4$, and $w = 12$.

6. ______________

7. If $g(x) = \frac{1}{2}x^2 - x + 4$, find $g(-3)$.

7. ______________

8. Determine whether $y = x^3 - 3$ is symmetric with respect to the x-axis.

8. ______________

9. Find the domain: $f(x) = \dfrac{\sqrt{x - 2}}{x(x^2 - 9)}$.

9. ______________

FINAL EXAMINATION   NAME ______________________

TEST FORM E

ANSWERS	
10. a) ______ b) ______ c) ______	10. For $f(x) = 2x^2 - 6x + 1$, a) find an equivalent equation of the type $f(x) = a(x - h)^2 + k$; b) find the vertex; and c) determine whether there is a maximum or minimum, function value, and find that value.
11. ______	11. Solve: $\lvert 4x - 15\rvert > 9$.
12. ______	12. Solve: $\frac{x - 1}{x - 2} < 2$.
13. a) ______ b) ______	13. Given that $\log_a 2 = 0.301$, $\log_a 6 = 0.778$, and $\log_a 7 = 0.845$, find the following. a) $\log_a \sqrt{14}$ b) $\log_a 9$
14. ______	14. Find log 0.0005431.
15. ______	15. Solve: $\log_3 (\log_4 x) = 0$.
16. ______	16. A surveyor found that the angle of elevation of the top of a flagpole was 32°. After walking 60 ft on horizontal ground toward the base of the pole, the surveyor found its angle of elevation to be 54°. How far is it from the second position to the top of the pole?
17. ______	17. What is the phase shift of $y = -4 \sin \left(2x - \frac{\pi}{2}\right)$?

FINAL EXAMINATION   *NAME* ______________________

TEST FORM E

18. Complete these identities.

a) $1 + \tan^2 x =$ ______   b) $\sin\left(x + \frac{\pi}{2}\right) =$ ______

19. Simplify: $\frac{\sec^2 x - \tan^2 x}{\sin^2 x}$.

20. Convert $-540°$ to radian measure in terms of π.

21. An arc 80 m long on a circle of radius 1200 cm subtends an angle of how many radians?

22. Find sec 2.9473.

23. Simplify: $\sin\left[\arccos\left(-\frac{1}{2}\right)\right]$.

24. Prove the identity: $\cos 2\theta = \frac{\cot^2 \theta - 1}{2 \cot \theta}$.

25. Solve in $[0, 2\pi)$: $\sin 2x \sin x - \cos x = 0$.

ANSWERS

18. a) ______

b) ______

19. ______

20. ______

21. ______

22. ______

23. ______

24. See work.

25. ______

FINAL EXAMINATION NAME ______________________

TEST FORM E

ANSWERS	
26. ________	26. In triangle ABC, $a = 3$, $b = 14$, and $C = 30^\circ 15'$. Find c.
27. ________	27. Convert to a polar equation: $x^2 + y^2 = 121$.
28. ________	28. Find polar notation for $4 - 4i$.
29. ________	29. Simplify: $\frac{3 + 2i}{5 - 4i}$.
30. ________	30. Find the cube roots of -8.
31. ________	31. Solve using Cramer's Rule: $x + y - z = 6$, $2x - y + z = 0$, $x + 2y + z = 1$.
32. ________	32. Maximize $T = 10x + 25y$ subject to: $5x + 4y \le 20$, $x + y \le 6$, $x \ge 0$, $y \ge 1$.
33. ________	33. Find numbers a, b, and c such that the function $f(x) = ax^2 + bx + c$ fits the data points $(0,-4)$, $(3,-4)$, and $(2,-2)$.
34. ________	34. Evaluate: $\begin{vmatrix} 2 & 0 & 3 \\ -1 & 4 & 2 \\ 3 & 1 & 5 \end{vmatrix}$.

FINAL EXAMINATION NAME ______________________

TEST FORM E

	ANSWERS
35. Let $A = \begin{bmatrix} 2 & 4 & 0 \\ -1 & 0 & 1 \\ 0 & -2 & 1 \end{bmatrix}$. Find A^{-1}, if it exists.	35. ______________
36. Let $C = \begin{bmatrix} 1 & -2 \\ 2 & 4 \end{bmatrix}$ and $D = \begin{bmatrix} 0 & 4 \\ 3 & -1 \end{bmatrix}$. Find C(2D - C).	36. ______________
37. Solve: $x^2 + y^2 = 1$, $x^2 + 25y^2 = 25$.	37. ______________
38. Find an equation of the circle having a diameter with endpoints (-3,4) and (3,-4).	38. ______________
39. Find the vertices of the hyperbola: $16y^2 - 25x^2 = 400$.	39. ______________
40. What does Descartes' rule of signs tell you about the number of negative real roots of $x^7 - 5x^3 - 2x + 8$?	40. ______________
41. A polynomial of degree 7 with rational coefficients has 6, -2i, $\sqrt{5}$, and 3 - 2i as roots. Find the other roots.	41. ______________
42. Find the rational roots of $x^4 - 5x^3 + 3x^2 + 7x - 2 = 0$.	42. ______________

FINAL EXAMINATION NAME ______________________

TEST FORM E

ANSWERS	
43. ______	43. Find the sum of the first 20 terms of the arithmetic sequence 0.02, 0.05, 0.08, 0.11, ...
44. ______	44. Find the 9th term of the geometric sequence $\frac{1}{10}, -\frac{1}{5}, \frac{2}{5}, -\frac{4}{5}, \ldots .$
45. See work.	45. Use mathematical induction. Prove for every natural number n. $1 + 6 + 11 + \ldots + (5n - 4) = \frac{n(5n - 3)}{2}$
46. ______	46. How many 4-digit numbers can be named using the digits 3, 4, 5, 6, 7, 8, and 9 if the digits are not repeated and the numbers must be less than 4000?
47. ______	47. Expand: $\left(2x - \sqrt{3y}\right)^4$.
48. ______	48. From a group of 10 senators and 12 representatives, a committee of 8 is chosen. What is the probability that 3 senators and 5 representatives are chosen?

Graph.

49. $f(x) = \begin{cases} x + 2, & \text{for } x < -3 \\ -x, & \text{for } -3 \le x \le 0 \\ \sqrt{x} & \text{for } x > 1 \end{cases}$

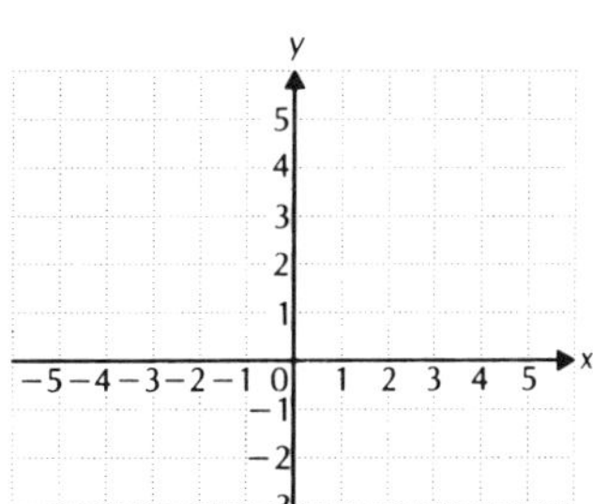

50. $f(x) = \dfrac{x^2 - x - 1}{x - 2}$.

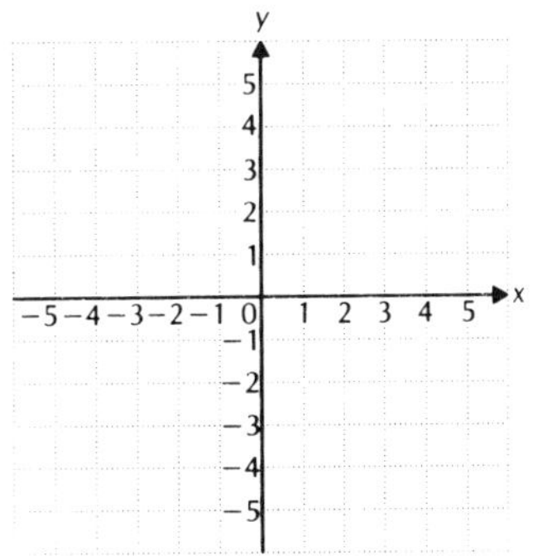

51. $x^2 + 8y = 0$.

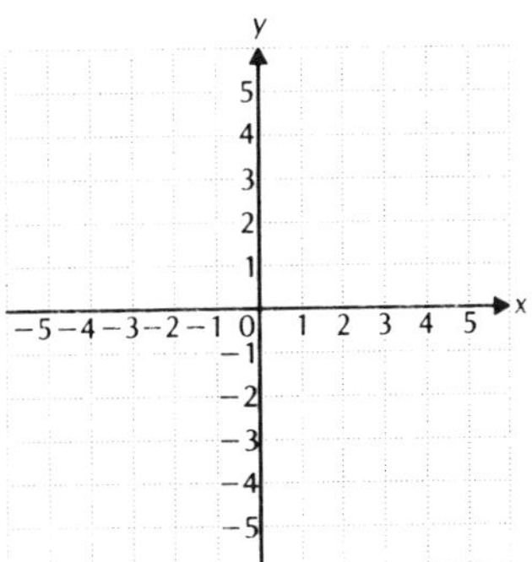

52. $y = \sin x - \cos x$

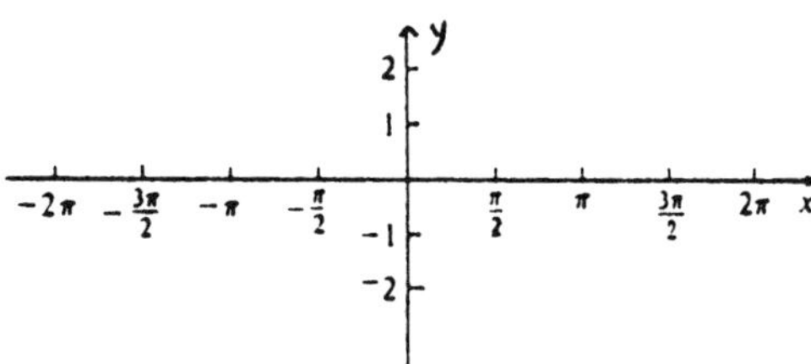

ANSWERS

49. See graph.

50. See graph.

51. See graph.

52. See graph.

FINAL EXAMINATION NAME ______________________

TEST FORM E

ANSWERS	
53. ________	53. How many years will it take an investment of \$8000 to double if interest is compounded annually at 6.5%?
54. ________	54. Find polar notation for the vector <5,-2>.
55. ________	55. The fourth term of an arithmetic sequence is 14, and the seventh term is 26. Find the twentieth term.
56. ________	56. A rock on a 7-ft string is rotated at 30 rpm. What is its linear speed, in ft/sec?
57. ________	57. A rectangle has a diagonal of 26 ft and a perimeter of 68 ft. Find the dimensions.
58. ________	58. Evaluate the sum $\sum_{k=1}^{5} 2(4 - 2^k)$.
59. ________	59. Decompose into partial fractions: $\frac{7x + 5}{3x^2 + 4x + 1}$.
60. ________	60. Convert $29^\circ 15' 24''$ to degrees and decimal parts of degrees.

FINAL EXAMINATION — NAME ____________

TEST FORM F — CLASS ______ SCORE ______ GRADE ______

Write the letter of your response on the answer blank. — **ANSWERS**

1. Subtract and simplify: $\dfrac{3x}{x^2 + 15x + 56} - \dfrac{7}{x^2 + 13x + 42}$.

   Find the numerator.

   a) $3x - 7$ b) $3x^2 + 11x - 56$

   c) $3x + 7$ d) $x - 7$

   1. ____________

2. Convert to scientific notation: 0.0149.

   a) 1.49×10^{-2} b) 0.149×10^{-1}

   c) 1.49×10^{2} d) 149×10^{3}

   2. ____________

3. Rationalize the numerator: $\dfrac{1 - \sqrt{y}}{3 - \sqrt{y}}$.

   Find the denominator of the answer.

   a) $3 + 2\sqrt{y} - y$ b) $3 - y$

   c) $9 - y$ d) $9 - y^2$

   3. ____________

4. Find the center and radius of the circle: $x^2 + y^2 - 4x + 20y + 23 = 0$.

   a) C: (-2,10), R: 3 b) C: (2,-10), R: 9

   c) C: (2,10), R: 9 d) C: (2,-10), R: 3

   4. ____________

5. Solve: $4x^2 - 11x + 3 = 0$. Simplify the answer.

   a) $-\frac{1}{4}, 3$ b) $\dfrac{-11 \pm \sqrt{133}}{4}$

   c) $\dfrac{-11 \pm \sqrt{133}}{8}$ d) $\dfrac{11 \pm \sqrt{73}}{8}$

   5. ____________

6. Solve for t_2: $S = \dfrac{A}{r(t_1 - t_2)}$.

   a) $Srt_1 - A$ b) $\dfrac{Srt_1 - A}{Sr}$

   c) $t_1 - A$ d) $\dfrac{A - Srt_1}{Sr}$

   6. ____________

FINAL EXAMINATION NAME ______________________

TEST FORM F

ANSWERS

7. ____________

8. ____________

9. ____________

10. ____________

11. ____________

12. ____________

7. Insert two arithmetic means between 6 and 10.

a) $7\frac{1}{2}$, 9 b) 8, 9 c) $7\frac{1}{3}$, $8\frac{2}{3}$ d) $7\frac{2}{5}$, $9\frac{1}{5}$

8. Find $f \circ g(x)$: $f(x) = 2x^2 - 3x$, $g(x) = x + 2$.

a) $2x^2 + 5x + 8$ b) $2x^2 - 3x + 2$

c) $2x^2 + 5x + 2$ d) $(2x^2 - 3x)(x + 2)$

9. Determine whether $y = 3x^3 - 2x$ is even, odd, or neither even nor odd.

a) Even b) Odd c) Neither

10. Find the domain: $g(x) = \dfrac{\sqrt{x + 1}}{x^2 - 2x}$.

a) $\{x|x \neq 0, 2\}$ b) $\{x|x \geq -1\}$

c) $\{x|x \neq 2\}$ d) $\{x|x \geq -1, x \neq 0, 2\}$

11. Find an equation of the line containing the given point and parallel to the given line: $\left(\frac{1}{2}, 4\right)$; $y - 4x = 5$.

a) $y = 4x + 2$ b) $y = 4x - 5$

c) $y = 4x - 4$ d) $y = \frac{x}{4} - 2$

12. Solve: $x^2 + 6x - 16 > 0$.

a) $\{x|x < -8\}$ b) $\{x|x \neq -8, 2\}$

c) $\{x|x < -8 \text{ or } x > 2\}$ d) $\{x|x < -2 \text{ or } x > 8\}$

FINAL EXAMINATION NAME ______________________

TEST FORM F

13. Determine whether the second coordinate of the vertex of the function $f(x) = 4x^2 - 20x + 13$ is a maximum or a minimum, and find the maximum or minimum.

a) -12 is a minimum b) -7 is a maximum

c) -7 is a minimum d) 33 is a maximum

14. Write an equivalent expression containing a single logarithm and simplify: $\log_b 3x + \log_b 10 - 2 \log_b x$.

a) $\log_b \frac{30}{x}$ b) $\log_b \frac{5x}{10 + x^2}$

c) 1 d) $\frac{\log_b 5x}{\log_b (10 + x^2)}$

15. If $f(x) = \sqrt{x - 4}$, find a formula for $f^{-1}(x)$.

a) $f^{-1}(x) = x + 4$ b) $f^{-1}(x) = \sqrt{y - 4}$

c) $f^{-1}(x) = x^2 - 4$ d) $f^{-1}(x) = x^2 + 4$

16. Solve: $\log_3 (x + 2) - \log_3 x = 2$.

a) 1 b) $\frac{2}{7}$ c) $\frac{1}{4}$ d) 2

17. In right triangle ABC, b = 8 and A = 50° 15′. Find a. (Standard lettering has been used.)

a) 5.1 b) 6.7 c) 10.4 d) 9.6

18. What is the amplitude of $y = -4 \cos \left(3x - \frac{\pi}{2}\right)$?

a) -4 b) $\frac{\pi}{6}$ c) 4 d) $\frac{\pi}{2}$

ANSWERS

13. ____________

14. ____________

15. ____________

16. ____________

17. ____________

18. ____________

NAME ______________________

ANSWERS

19. ________

19. Complete this identity: $\cos\left(\frac{\pi}{2} + x\right) =$ ______.

a) $\sin \frac{\pi}{2}$ b) $-\sin x$ c) $-\cos x$ d) $\sin x$

20. ________

20. Simplify: $\frac{1 + \cot^3 x}{1 + \cot x}$.

a) $\sec^2 x - \cot x$ b) $\csc^2 x - \cot x$

c) $1 + \cot^2 x$ d) $1 - 2 \cot x$

21. ________

21. Convert 40° to radian measure in terms of π.

a) $\frac{7200}{\pi}$ b) 18π c) $\frac{2\pi}{9}$ d) $\frac{\pi}{18}$

22. ________

22. Find $\cot(225^\circ)$ exactly.

a) 0 b) 1

c) -1 d) Does not exist

23. ________

23. If $\sin \theta = 0.4305$ and $0^\circ < \theta < 90^\circ$, find θ in degrees.

a) 0.4450 b) $64^\circ 30'$ c) 7.5° d) 25.5°

24. ________

24. Find fractional notation for $3.151\overline{5}\overline{15}$.

a) $\frac{63}{20}$ b) $\frac{31}{11}$ c) $\frac{104}{33}$ d) $\frac{34}{11}$

FINAL EXAMINATION NAME ______________________

TEST FORM F

25. Find $\sin 255^\circ$ exactly.

a) $\dfrac{\sqrt{2} - \sqrt{6}}{4}$ b) $\dfrac{\sqrt{2} + \sqrt{6}}{4}$

c) $\dfrac{\sqrt{6} - \sqrt{2}}{4}$ d) $-\dfrac{\sqrt{6} + \sqrt{2}}{4}$

26. Simplify: $\arccos\left[(\sin)\left(-\frac{\pi}{4}\right)\right]$.

a) $\frac{\pi}{4}$ b) $\frac{3\pi}{4}$ c) $\frac{\pi}{3}$ d) $-\frac{\sqrt{2}}{2}$

27. Two ships are on an east-west line from each other. There is a lighthouse on the shore whose position as seen from one ship is N $60^\circ 40'$ W and from the other ship N $55^\circ 10'$ E. The first ship is 80 km from the lighthouse. How far apart are the ships?

a) 42.5 km b) 147 km c) 126 km d) 71.2 km

28. Find polar notation for the vector <4,-2>.

a) $(4.47, 333^\circ)$ b) $(8, 330^\circ)$

c) $(2, 333^\circ)$ d) $(4.47, 225^\circ)$

29. Simplify: $\dfrac{8 + i}{2 - i}$.

a) $\frac{3}{5} + \frac{2}{5}i$ b) $4 - i$ c) $\frac{2}{5} - \frac{1}{5}i$ d) $3 + 2i$

30. Solve: $x^2 = 12i$.

a) $\sqrt{6} + 6i, -\sqrt{6} - 6i$ b) $\sqrt{6} + \sqrt{6}i, -\sqrt{6} - \sqrt{6}i$

c) $6i, -6i$ d) $\sqrt{6}i, -\sqrt{6}i$

ANSWERS

25. ______________

26. ______________

27. ______________

28. ______________

29. ______________

30. ______________

FINAL EXAMINATION NAME ______________________

TEST FORM F

ANSWERS

31. ____________

32. ____________

33. ____________

34. ____________

35. ____________

36. ____________

31. Find rectangular notation for 2 cis 225°.

a) $\frac{\sqrt{2}}{2} - \frac{\sqrt{2}}{2}i$ b) $\sqrt{2} - \sqrt{2}i$

c) $-\sqrt{2} - \sqrt{2}i$ d) $1 - \sqrt{3}i$

32. Solve using matrices:

$$x - y + z = 6,$$
$$2x + y + z = 3,$$
$$3x + 2y - z = -4.$$

Find the sum of the x, y, and z values.

a) 2 b) 0 c) -2 d) -4

33. Classify this system as consistent or inconsistent, dependent or independent.

$$5x - 6y = 4,$$
$$-10x + 12y = -6$$

a) Consistent, dependent b) Consistent, independent

c) Inconsistent, dependent d) Inconsistent, independent

34. The sum of a certain number and a second number is 40. The second number minus the first number is 50. Find the smaller number.

a) 25 b) -5 c) 60 d) 45

35. Let $A = \begin{bmatrix} 4 & -2 \\ 2 & 1 \end{bmatrix}$. Find A^{-1} if it exists.

a) $\begin{bmatrix} 1 & 2 \\ -2 & 4 \end{bmatrix}$ b) $\begin{bmatrix} 4 & 2 \\ -2 & 1 \end{bmatrix}$ c) $\begin{bmatrix} \frac{1}{8} & \frac{1}{4} \\ -\frac{1}{4} & \frac{1}{2} \end{bmatrix}$ d) Does not exist

36. Evaluate: $\begin{vmatrix} 3 & 0 & 4 \\ 1 & 1 & -1 \\ 2 & 0 & 2 \end{vmatrix}$.

a) -2 b) 6 c) 21 d) 14

ANSWERS

37. Let $A = \begin{bmatrix} 1 & -2 & 3 \end{bmatrix}$ and $B = \begin{bmatrix} 3 & 0 & 2 \\ 2 & 4 & -1 \\ 0 & 2 & 1 \end{bmatrix}$. Find AB.

a) $\begin{bmatrix} -1 & -2 & 7 \end{bmatrix}$ b) $\begin{bmatrix} 9 & -9 & -1 \end{bmatrix}$

c) $\begin{bmatrix} 3 & 0 & 6 \\ 2 & -8 & -3 \\ 0 & -4 & 3 \end{bmatrix}$ d) $\begin{bmatrix} -1 \\ -2 \\ 7 \end{bmatrix}$

38. Solve: $x^2 + y^2 = 13,$
$3x^2 - y^2 = 3.$

The solution set

a) is empty.

b) consists of one ordered pair.

c) consists of two order pairs.

d) consists of four ordered pairs.

39. Find the asymptotes of the hyperbola: $x^2 - 49y^2 = 49$.

a) $y = \frac{1}{7}x,\ y = -\frac{1}{7}x$ b) $y = 7x,\ y = -7x$

c) $y = \frac{1}{49}x,\ y = -\frac{1}{49}x$ d) $y = 49x,\ y = -49x$

40. A polynomial of degree 5 with rational coefficients has 3, $1 + \sqrt{5}$, and $1 - 3i$ as roots. Find the other roots.

a) $-3,\ 1 - \sqrt{5},\ 1 + 3i$ b) $-\sqrt{3},\ 3i$

c) $-1 - \sqrt{5},\ -1 + 3i$ d) $1 - \sqrt{5},\ 1 + 3i$

41. Given that $P(x) = 3x^5 - 4x^4 + 2$, find $P(-4)$.

a) -4094 b) -1022 c) -4098 d) 256

42. List two of the roots of $P(x) = x^4 - x^3 - x^2 - x - 2$.

a) $-2,\ i$ b) $1,\ i$ c) $2i,\ -2i$ d) $-1,\ i$

37. ______________

38. ______________

39. ______________

40. ______________

41. ______________

42. ______________

FINAL EXAMINATION NAME ____________________

TEST FORM F

ANSWERS

43. ____________

43. The 3rd term of an arithmetic sequence is 12 and the 10th term is 61. Find the 6th term.

a) 29 b) 7 c) 33 d) 47

44. ____________

44. Find the sum of the first 6 terms of the geometric sequence $\frac{3}{4}, \frac{3}{8}, \frac{3}{16}, \ldots$

a) $\frac{9}{32}$ b) $\frac{189}{128}$ c) $\frac{3}{256}$ d) $\frac{93}{128}$

45. See work.

45. Use mathematical induction. Prove for every natural number n. $1 + 3 + 3^2 + \ldots + 3^{n-1} = \frac{3^n - 1}{2}$

46. ____________

46. How many baseball games can be played in a 14-team league if each team plays each other team twice?

a) 182 b) 14! c) 1001 d) 91

47. ____________

47. Find the 4th term of $(2x + y)^5$.

a) $40x^2y^3$ b) $80x^3y^2$ c) $10xy^4$ d) $20xy^4$

48. ____________

48. Suppose 3 cards are drawn without replacement from a well-shuffled deck of 52 cards. What is the probability that all three are diamonds?

a) $\frac{3}{52}$ b) $\frac{1}{64}$ c) $\frac{1}{7225}$ d) $\frac{11}{850}$

ANSWER KEYS FOR ALTERNATE TEST FORMS

Chapter 1, Test Form A

1. 12, 0 2. $\sqrt[-4]{5}$, $\sqrt{50}$ 3. All 4. All except $\sqrt[-4]{5}$, $\sqrt{50}$ 5. 12

6. 12, 0, -11 7. 0 8. 3.7 9. 45 10. 19 11. -3 12. $\frac{221}{2}$

13. 0.005401 14. 58,000 15. 315,000 16. 0.00617 17. 2.17×10^{-5}

18. 5.9321×10^{2} 19. 3.468×10^{-10} 20. 8.5×10^{-7} 21. $\frac{-10x^7}{y^8}$ 22. $\frac{2a^4b^8}{3c^7}$

23. -5 24. 3 25. 4 26. $\frac{y - x}{(xy)(x + y)}$ 27. $16a^6 + 24a^3b + 9b^2$

28. $7x^3y + 4x^2 - 3xy - y^2 + 8$ 29. $8x^3 - 12x^2 + 6x - 1$ 30. $\sqrt[12]{(a + b)^5}$

31. $\frac{1}{\sqrt[4]{c^3}}$, or $\frac{\sqrt[4]{c}}{c}$ 32. $5(3x + 2)^2$ 33. $(a + 6)(a^2 - 6a + 36)$

34. $b^3(b - 2c)(b + 2c)$ 35. $2(6m^2 + 5)(m^2 - 3)$ 36. $(3 - 2a)(3 + 2a)$
$(81 + 36a^2 + 16a^4)$ 37. $(x + 2y)(5 - x)$ 38. $a^3(a^2 - 3b^2)(a^4 + 3a^2b^2 + 9b^4)$

39. $\frac{x + 3}{(x + 2)^2}$ 40. $\frac{y^2 + 8y + 35}{(y - 3)(y + 3)(y + 7)}$ 41. $\frac{64 + 16\sqrt{x} + x}{64 - x}$

42. $a^{3x} + 9a^{x} + 27a^{-x} + 27a^{-3x}$ 43. $\frac{64 - x}{64 - 16\sqrt{x} + x}$ 44. $\frac{b^{2/3}(a - b)}{a^{1/2}}$

45. 10 46. $\frac{x^2(6x - 5)}{(3x - 1)^{3/2}}$ 47. $\frac{375}{11}\frac{\text{mi}}{\text{hr}}$ 48. $x^{3a} - 3x^{2a}y^{b} + 3x^{a}y^{2b} - y^{3b}$

49. $\frac{(3 + x)\sqrt{2 + x}}{2 + x}$

Chapter 1, Test Form B

1. 13, 0 2. $-\sqrt[3]{9}$, $\sqrt{18}$ 3. All 4. All except $-\sqrt[3]{9}$, $\sqrt{18}$ 5. 13

6. 13, 0, -5 7. $\frac{4}{5}$ 8. -16 9. -2.3 10. 12 11. $-\frac{25}{6}$ 12. $\frac{141}{25}$

13. 613,000 14. 0.04312 15. 79,200 16. 0.0000902 17. 3.2×10^{-6}

18. 4.37452×10^{3} 19. 5.022×10^{-8} 20. 4.5×10^{10} 21. $648x^7$

Chapter 1, Test Form B (continued)

22. $\frac{2x^3}{3yz}$ 23. -10 24. 3 25. 3 26. $\frac{x - y}{2x}$ 27. $25x^4 - 40x^2y^3 + 16y^6$

28. $5x^2y + 4x^3 - xy^2 - 14x - 2y - 6$ 29. $125y^3 - 300y^2 + 240y - 64$

30. $\sqrt[60]{(c + d)^{17}}$ 31. $\sqrt[5]{x^4}$ 32. $6(2x - 5)^2$ 33. $(b - 8)(b^2 + 8b + 64)$

34. $7x^3(2x^2 - z)(2x^2 + z)$ 35. $6(4a^2 - 7)(a^2 + 3)$

36. $(4 + 3a^2)(16 - 12a^2 + 9a^4)$ 37. $(2x - y)(4 - x)$

38. $a^3(a^3 - 3b^2)(a^3 + 3b^2)$ 39. $\frac{(x - 5)(x - 7)}{2x(x - 1)}$ 40. $\frac{3x + 5y}{3x - 5y}$

41. $\frac{16 - 8\sqrt{x} + x}{16 - x}$ 42. $27a^{3x} - 108a^{2x}b^{-x} + 144a^xb^{-2x} - 64b^{-3x}$

43. $\frac{16 - x}{16 + 8\sqrt{x} + x}$ 44. $\frac{a^{3/4}(a + b)}{b^{2/3}}$ 45. 9404.8 ft 46. $\frac{x^2 - 4x - 3}{(x - 3)^{3/2}}$

47. $\frac{27}{25}\frac{\text{ton}}{\text{yd}^3}$ 48. $8a^3 + 12a^2b + 36a^2c + 6ab^2 + 36abc + 54ac^2 + b^3 +$
$9b^2c + 27bc^2 + 27c^3$ 49. $\frac{a^3 - 3a^2b + 3ab^2}{b}$

Chapter 1, Test Form C

1. 415, 0 2. $-\sqrt[3]{4}, \sqrt{13}, -\sqrt{39}$ 3. All 4. All except $-\sqrt[3]{4}, \sqrt{13}, -\sqrt{39}$

5. 415 6. 415, -19, 0 7. -59.2 8. 5 9. 22 10. -26.39 11. $\frac{1}{40}$

12. $\frac{23}{16}$ 13. 937,000 14. 0.000612 15. 40,900,000 16. 0.0853

17. 5.37×10^{-4} 18. 8.2915×10^2 19. 5.428×10^{-7} 20. 2.5×10^{-9}

21. $625x^{12}$ 22. $\frac{ab^8}{4c^5}$ 23. -6 24. 4 25. $11 - 4\sqrt{7}$ 26. $\frac{1}{a}$

27. $6x^3 - 11x^2 + 11x - 12$ 28. $9x^4 - 3x^3 + 2x^3y - 9xy^2 + 2x - 15$

29. $64a^3 - 144a^2 + 108a - 27$ 30. $\sqrt[20]{(a - b)^{11}}$ 31. $\frac{1}{\sqrt[3]{a^2}}$, or $\frac{\sqrt[3]{a}}{a}$

32. $3(5x + 1)^2$ 33. $(a + 9)(a^2 - 9a + 81)$ 34. $6x^2(y^2 - z)(y^2 + z)$

35. $3(6m^2 - 1)(m^2 + 3)$ 36. $8(2 + 3b^2)(4 - 6b^2 + 9b^4)$ 37. $(x - 5y)(4 + x)$

38. $a^2(a^3 + 2b^3)(a^6 - 2a^3b^3 + 4b^6)$ 39. $\frac{(x + 7)(x + 5)}{(x + 3)(x + 4)}$

40. $\frac{3x^2 + 14x - 7}{(5x - 2)(x - 5)(x + 4)}$ 41. $\frac{25 + 10\sqrt{x} + x}{25 - x}$

42. $8a^{3x} - 48a^x + 96a^{-x} - 64a^{-3x}$ 43. $\frac{25 - x}{25 - 10\sqrt{x} + x}$ 44. $\frac{a^2 - b}{a^{2/5}b^{1/2}}$

45. 91.2 ft 46. $\frac{x(-2x^2 + 15)}{(x - 5)^{3/2}}$ 47. $240\ \frac{\text{cg}}{\text{mL}}$ 48. a^{12x}

49. $(x^4 - 5)(x^4 + 5)(x^8 + 25)$

Chapter 1, Test Form D

1. 0 2. $\sqrt{80}$, $-\sqrt[4]{52}$ 3. All 4. All except $\sqrt{80}$, $-\sqrt[4]{52}$ 5. None

6. -19, 0 7. -14.9 8. 85 9. $\frac{23}{9}$ 10. -406 11. 5 12. $\frac{113}{35}$ 13. 0.02175

14. 9100 15. 5,090,000 16. 0.000132 17. 1.32×10^{-3} 18. 5.749×10^2

19. 3.869×10^9 20. 5.5×10^3 21. $\frac{-12}{xy}$ 22. $\frac{c^4}{9d^6}$ 23. -4 24. 2

25. $6 + 2\sqrt{5}$ 26. $\frac{y - x}{xy(2x + y)}$ 27. $x^4 - a^2b^2$ 28. $7pq^2 + 7pq - 14q^2 - 11p +$
$19q - 9$ 29. $27x^3 - 54x^2 + 36x - 8$ 30. $\sqrt[12]{(x + y)^7}$ 31. $\sqrt[8]{y^3}$

32. $2(3x - 4)^2$ 33. $(c - 10)(c^2 + 10c + 100)$ 34. $a^3(a - 4b)(a + 4b)$

35. $2(8p^2 - 3)(p^2 + 2)$ 36. $(8 - 3a^3)(64 + 24a^3 + 9a^6)$ 37. $(x + 4y)(6 - x)$

38. $b^2(2a^2 - b^2)(4a^4 + 2a^2b^2 + b^4)$ 39. $\frac{(x - 6)(x - 2)}{(x - 1)(x + 5)^2}$

40. $\frac{y(2y + 13)}{(y - 6)(y - 1)(y + 2)}$ 41. $\frac{36 - 12\sqrt{x} + x}{36 - x}$ 42. $125x^{3t} + 75x^t + 15x^{-t} + x^{-3t}$

43. $\frac{36 - x}{36 + 12\sqrt{x} + x}$ 44. $\frac{b^{5/8}(b^2 - a)}{a^{1/3}}$ 45. 18.4 m 46. $\frac{-4x^4 + 2x^3 + 5x^2}{(2x + 1)^{3/2}}$

Chapter 1, Test Form D (continued)

47. $400 \frac{g}{cm}$ 48. $\frac{140}{33}$ 49. $\left(y - \frac{5}{8}\right)\left(y + \frac{3}{8}\right)$

Chapter 1, Test Form E

1. $\sqrt{50}, \pi, \sqrt[3]{15}$, 17.12112111211112 . . . 2. 8 3. Associative (x) 4. 22.1

5. -11 6. 140 7. -16 8. 17 9. $-\frac{2}{9}$ 10. 0.0097 11. 5.073×10^6

12. $\frac{16x^3}{y^5}$ 13. $25a - 2$ 14. 7 15. $\frac{4s^7t^9}{5r^4}$ 16. $\frac{a^2 + b^2}{ab}$ 17. $4x^6 - 4x^3y^2 + y^4$

18. $8a^3 - 84a^2b + 294ab^2 - 343b^3$ 19. -1 20. $3x^5 - 3x^2y + 5xy - 6y^2 + 11$

21. 8.7×10^{-12} 22. $3(4x^2 + 1)(2x - 1)(2x + 1)$

23. $(y + 0.3)(y^2 - 0.3y + 0.09)$ 24. $(3x + 2)(x - 9)$

25. $(12a - 4b + 7)(6a - 2b - 3)$ 26. $\sqrt[20]{(b + 3)^7}$ 27. $\frac{1}{\sqrt[11]{z^9}}$, or $\frac{\sqrt[11]{z^2}}{z}$

28. $\frac{x^3y^4}{z^5}$ 29. $\frac{(x - 9)(x - 7)}{x(x - 2)}$ 30. $\frac{7}{a + b}$ 31. $\frac{4a - 9\sqrt{ab} + 5b}{a - b}$ 32. $\frac{3y + 5x}{x^{1/4}y^{1/3}}$

33. 4.5 m 34. $\frac{800}{3} \frac{¢}{hr}$ 35. $y^{2b} - z^{2c}$

Chapter 1, Test Form F

1. b 2. d 3. c 4. c 5. a 6. b 7. b 8. d 9. a 10. a 11. d

12. b 13. d 14. a 15. a 16. c 17. d 18. b 19. b 20. d 21. a

22. b 23. a 24. c 25. a 26. d 27. b 28. c 29. c

Chapter 2, Test Form A

1. i 2. $-\sqrt{35}$ 3. $13 + 18i$ 4. $3 + 10i$ 5. $-\frac{7}{13} + \frac{22}{13}i$ 6. 26 7. $\frac{2}{5} + \frac{1}{5}i$

8. x = 4, y = 7 9. $\left\{\frac{5}{3}, -7, 1\right\}$ 10. $\{\pm i, \pm 2\}$ 11. $\left\{\frac{4}{3}, 3\right\}$ 12. $\left\{\frac{1 \pm \sqrt{29}}{4}\right\}$

13. $\{2\}$ 14. $\left\{\frac{1 \pm i\sqrt{79}}{8}\right\}$ 15. $\{7\}$ 16. $\{-1\}$ 17. $\{-2, 9\}$ 18. $\{y \mid y > -2\}$

Chapter 2, Test Form A (continued)

19. $\{-2, 2, -1\}$ 20. $\{0\}$ 21. 7 m x 21 m 22. 5 mph 23. $\{3 \pm 2\sqrt{3}\}$ 24. One real 25. $x^2 + 2\sqrt{2}x - 6 = 0$ 26. $m_1 = \dfrac{Fd^2}{km_2}$ 27. 6 28. $y = \dfrac{328xz}{w^2}$

29. 50 ft 30. No 31. $\{x|x \le 5\}$ 32. $\left\{\dfrac{-3 \pm \sqrt{17}}{2}\right\}$ 33. $\left\{8 \pm 4\sqrt{3}\right\}$

Chapter 2, Test Form B

1. $-i$ 2. $-\sqrt{30}$ 3. $7 - i$ 4. $11 + 2i$ 5. $1 - 2i$ 6. 50 7. $\dfrac{5}{29} - \dfrac{2}{29}i$

8. $x = 5$, $y = -2$ 9. $\left\{9, -3, \dfrac{1}{3}\right\}$ 10. $\left\{\dfrac{1 \pm \sqrt{13}}{2}, -1, 2\right\}$ 11. $\left\{-\dfrac{5}{2}, 4\right\}$

12. $\left\{\dfrac{1 \pm \sqrt{21}}{5}\right\}$ 13. $\{1\}$ 14. $\left\{\dfrac{5 \pm i\sqrt{47}}{4}\right\}$ 15. $\{15\}$ 16. $\{2\}$ 17. $\{-5, 3\}$

18. $\{x|x < -2\}$ 19. $\{-3, 3, 2\}$ 20. $\{0\}$ 21. \$2203.99 22. $2\frac{1}{7}$ hr

23. $\left\{2 \pm \sqrt{6}\right\}$ 24. Two real 25. $x^2 - 4 = 0$ 26. $\pi = \dfrac{C}{2r}$ 27. 60 ft

28. $y = \dfrac{250xz}{w^2}$ 29. 50 cm 30. Yes 31. $\{x|x \ge 4\}$ 32. 56 mph 33. $\left\{-3, \dfrac{2}{k}\right\}$

Chapter 2, Test Form C

1. $-i$ 2. $\sqrt{15}$ 3. $22 - 3i$ 4. $14 + 7i$ 5. $\dfrac{21}{25} - \dfrac{22}{25}i$ 6. 10 7. $\dfrac{6}{61} + \dfrac{5}{61}i$

8. $x = 5$, $y = 2$ 9. $\left\{\dfrac{2}{5}, -4, 5\right\}$ 10. $\{16\}$ 11. $\left\{-\dfrac{2}{5}, 3\right\}$ 12. $\left\{\dfrac{2 \pm \sqrt{10}}{3}\right\}$

13. $\{10\}$ 14. $\left\{\dfrac{-1 \pm i\sqrt{31}}{4}\right\}$ 15. $\{10\}$ 16. $\{5\}$ 17. $\{-3, 6\}$ 18. $\{x|x > 4\}$

19. $\{-1, 1, -3\}$ 20. $\{0\}$ 21. $A = 100^\circ$, $B = 50^\circ$, $C = 30^\circ$ 22. 36.5 km/h

23. $\left\{2 \pm \sqrt{6}\right\}$ 24. Two nonreal 25. $2x^2 - 5x - 12 = 0$ 26. $\ell = \dfrac{P - 2w}{2}$

27. 7 cm 28. $y = \frac{9}{2}xz$ 29. 3 sec 30. No 31. $\left\{x \middle| x \ge \dfrac{1}{5}\right\}$ 32. $\{-2, 1\}$

33. $\left\{6 \pm 3\sqrt{3}\right\}$

Chapter 2, Test Form D

1. i 2. $\sqrt{30}$ 3. $15 + 16i$ 4. $2 - 5i$ 5. $\frac{23}{25} + \frac{14}{25}i$ 6. 17 7. $\frac{7}{58} - \frac{3}{58}i$

8. $x = 1,\ y = -1$ 9. $\left\{\frac{3}{2}, 4, 1\right\}$ 10. $\{27, -1\}$ 11. $\left\{\frac{2}{7}, -4\right\}$ 12. $\left\{\frac{5 \pm \sqrt{57}}{4}\right\}$

13. $\{2\}$ 14. $\left\{\frac{1 \pm 2i\sqrt{2}}{3}\right\}$ 15. $\{27\}$ 16. $\left\{-\frac{5}{2}, 5\right\}$ 17. $\{-6, 4\}$ 18. $\{x \mid x < 2\}$

19. $\{-4, 4, -1\}$ 20. $\{0\}$ 21. \$1325 22. Bill: 3 hr; Sam: 6 hr

23. $\left\{2 \pm \sqrt{7}\right\}$ 24. Two real 25. $x^2 + 9 = 0$ 26. $V_1 = \frac{P_2V_2T_1}{P_1T_2}$ 27. 15 ft, 36 ft 28. $y = \frac{225xz}{64wp}$ 29. 8.25 hr 30. Yes 31. $\left\{x \mid x \geq \frac{4}{3}\right\}$ 32. 8.26%

33. $\{86\}$

Chapter 2, Test Form E

1. h 2. p 3. a 4. e 5. i 6. n 7. r 8. w 9. z 10. t 11. s

12. o 13. c 14. q 15. v 16. $x = 7,\ y = 10$ 17. 8 hr 18. 8 km/h

19. \$3500 20. $\frac{2 \pm \sqrt{3}}{2}$ 21. Two real 22. Sum: 7, Product: 3

23. $x^2 - 2x + 2 = 0$ 24. $b = \frac{a^2 + c}{2}$ 25. $c = \frac{Ab}{b - A}$ 26. $6\sqrt{2}$ 27. $y = \frac{0.2x^3}{s + w}$

28. $4\frac{2}{3}$ amps 29. $\left\{x \mid x \geq -\frac{5}{12}\right\}$ 30. No

Chapter 2, Test Form F

1. d 2. b 3. d 4. a 5. d 6. c 7. c 8. d 9. b 10. c 11. a

12. a 13. d 14. d 15. b 16. a 17. c 18. a 19. b 20. d 21. c

22. b 23. a 24. c

Chapter 3, Test Form A

1. Yes 2. {-5,1,7,8} 3. {-1,2,5,8} 4.

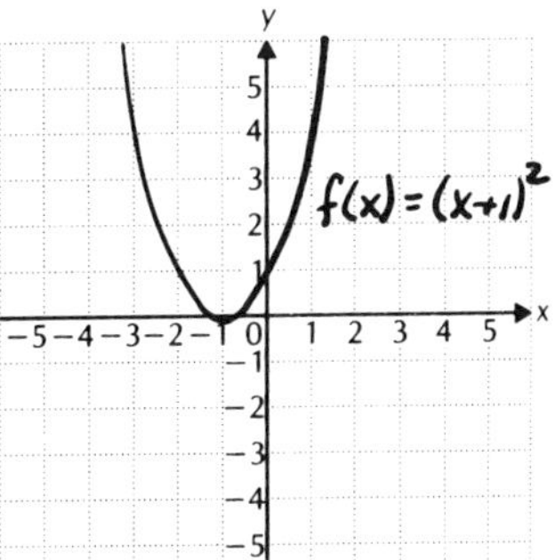

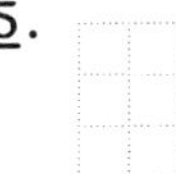

5.

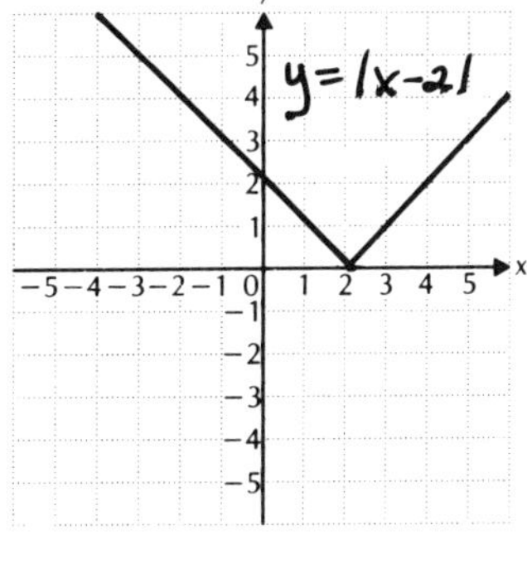

6.

7. a, d 8. c, d 9. $\{x \mid x \geq 2\}$

10. $(f + g)(x) = 2x^2 + 3x - 5$, $(f - g)(x) = 2x^2 - 3x + 5$, $fg(x) = 6x^3 - 10x^2$, $f/g(x) = \frac{2x^2}{3x - 5}$, $f \circ g(x) = 18x^2 - 60x + 50$, $g \circ f(x) = 6x^2 - 5$

11. All reals, all reals, all reals, all reals,

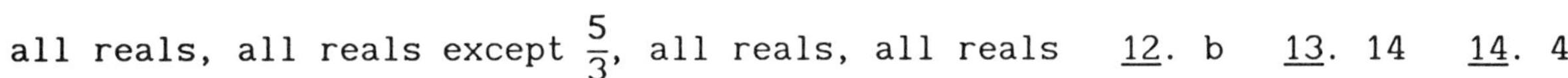

all reals, all reals except $\frac{5}{3}$, all reals, all reals 12. b 13. 14 14. 4

15. $a^2 - 7a + 14$ 16. $2a - 3 + h$ 17. $P(x) = 1.1x^2 - 32x - 30$ 18. $m = \frac{7}{2}$, y-intercept is $-\frac{15}{2}$ 19. $y = 2x + 9$ 20. $y = \frac{3}{2}x - 7$ 21. $\sqrt{20} \approx 4.472$

22. $\left(\frac{9}{2}, -\frac{1}{2}\right)$ 23. Parallel 24.

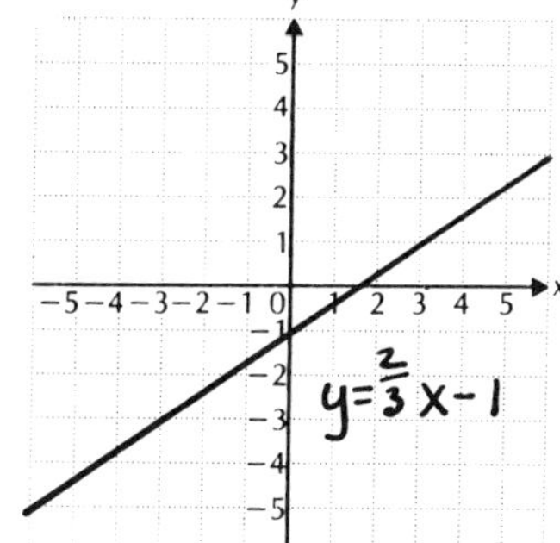

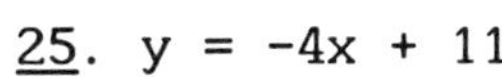

25. $y = -4x + 11$

26. $(x - 3)^2 + (y + 4)^2 = 4$

27. Center: (4,-1);
radius: 4

28. a)

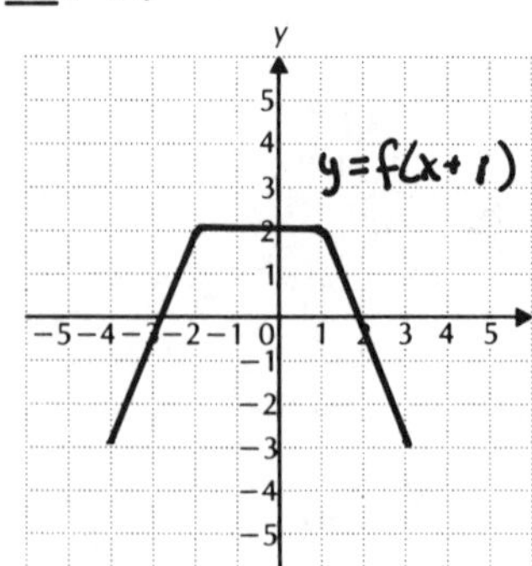

b)

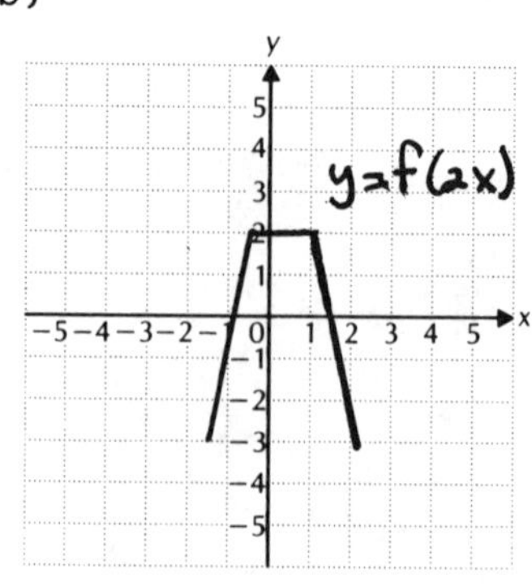

c)

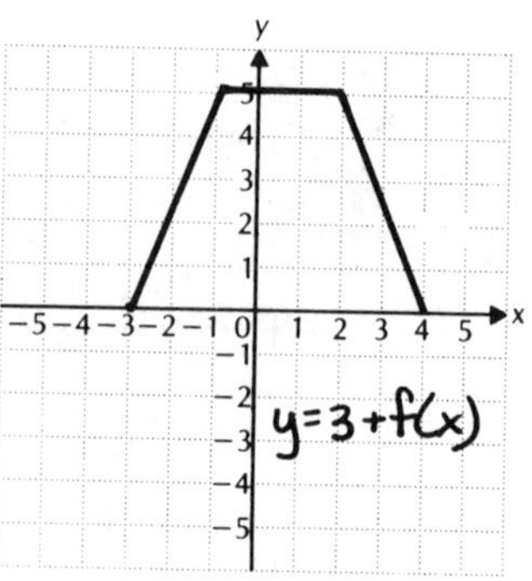

29. b 30. f 31. a, c, d, e 32. (-1,3) 33. [-7,∞) 34. b 35. a

36.

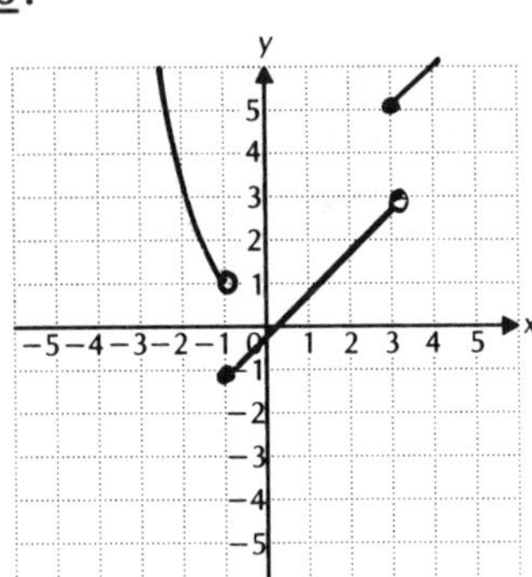

37. $A(x) = x\sqrt{144 - x^2}$ 38. (a) $f(x) = 5|x|$, $g(x) = 3x - 2$; (b) $f(x) = 18 - 2x^2$, $g(x) = 2x - 3$; Answers may vary. 39. $x \neq 0$, $x \neq \pm 2$

Chapter 3, Test Form B

1. No 2. {-3,1,4,5} 3. {-1,2,7} 4.

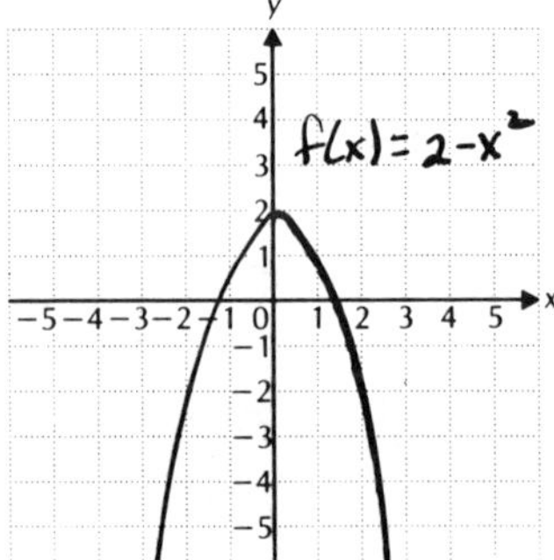

5.

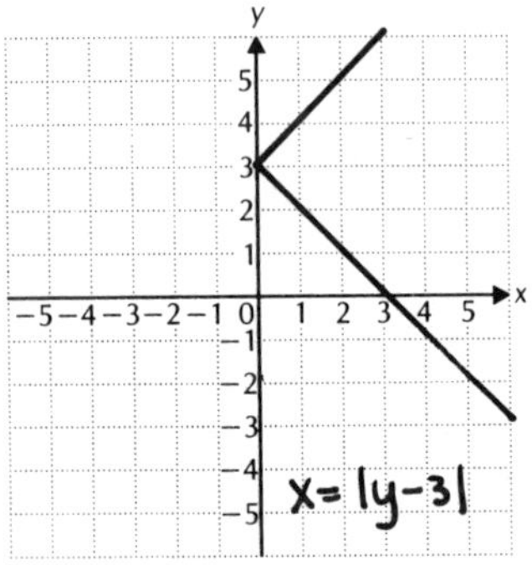

x= |y-3|

6. 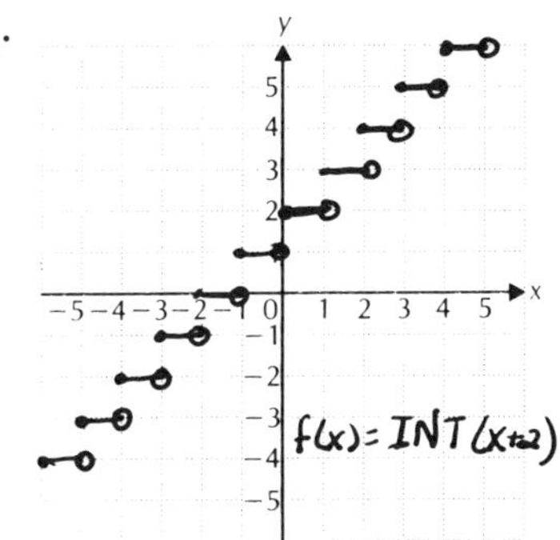

7. d 8. b, e, f 9. $\{x | x \neq 2\}$

10. $(f + g)(x) = x^2 + 5x + 1$, $(f - g)(x) = -x^2 + 5x + 3$, $fg(x) = 5x^3 + 2x^2 - 5x - 2$, $(f/g)(x) = \dfrac{5x + 2}{x^2 - 1}$, $f \circ g(x) = 5x^2 - 3$, $g \circ f(x) = 25x^2 + 20x + 3$

11. All reals, all reals, all reals, all reals, all reals, all reals except ± 1, all reals, all reals 12. a 13. -5 14. 0

15. $3a^2 + 14a + 11$ 16. $6a + 8 + 3h$ 17. $P(x) = 2.3x^2 + 37x + 290$

18. $m = -\frac{3}{2}$, y-intercept is 10 19. $y = -2x + 2$ 20. $y = \frac{1}{6}x - \frac{8}{3}$

21. $\sqrt{17} \approx 4.123$ 22. $\left(\frac{9}{2}, -3\right)$ 23. Perpendicular

24. 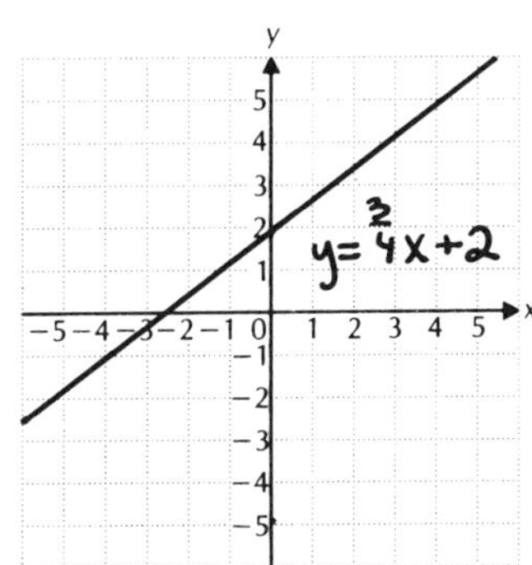

25. $y = 2x + 10$ 26. $(x + 1)^2 + (y + 3)^2 = 25$

27. Center: (2,-5); radius: 5

28. a)

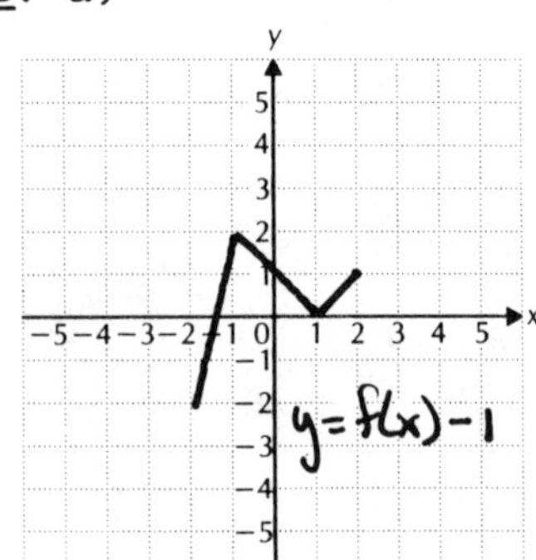

b)

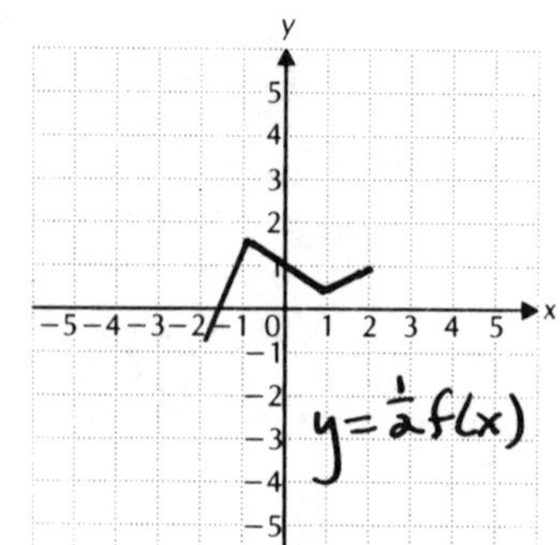

c)

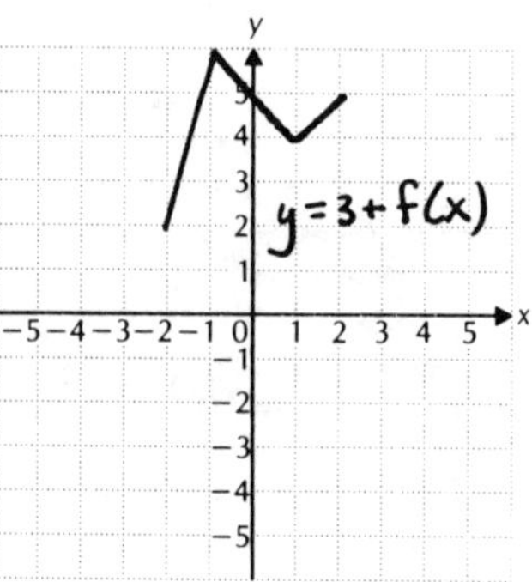

29. b, d, e   30. a, c   31. f   32. $(-\infty,-1]$ 33. $(-9,4]$ 34. b   35. c

36.

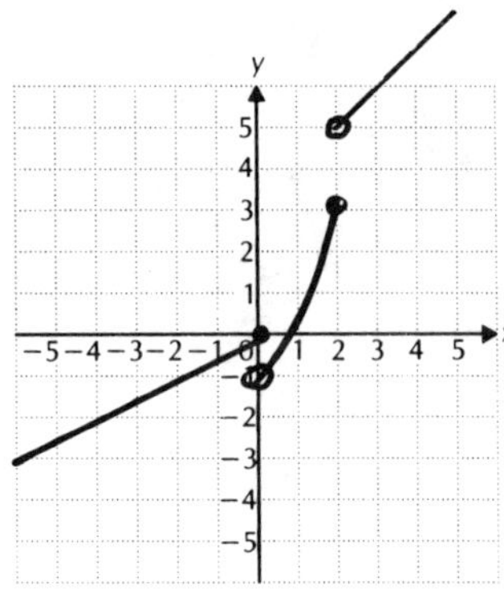

37. $A(h) = \frac{1}{2}h(2h - 4)$ 38. (a) $f(x) = \frac{1}{x^2}$, $g(x) = x + 3$; (b) $f(x) = 5x^3 + 2$, $g(x) = x - 3$; Answers may vary.   39. $k = -4$

Chapter 3, Test Form C

1. Yes   2. $\{-6,-1,3,5\}$ 3. $\{-1,0,3,5\}$ 4.

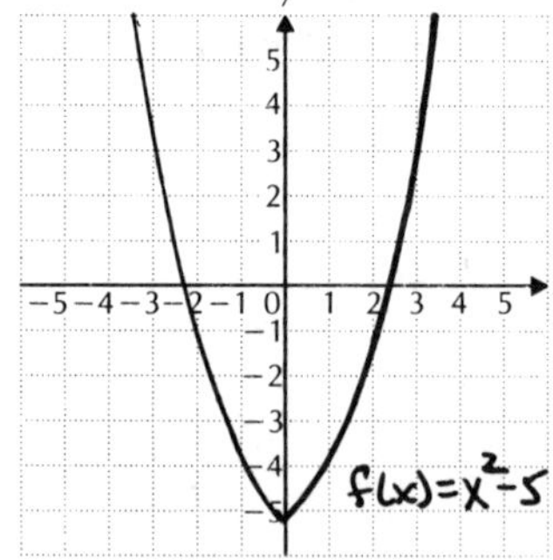

5.

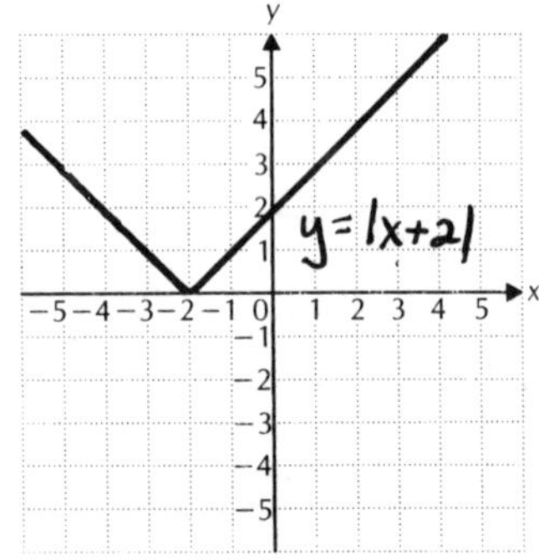

6. 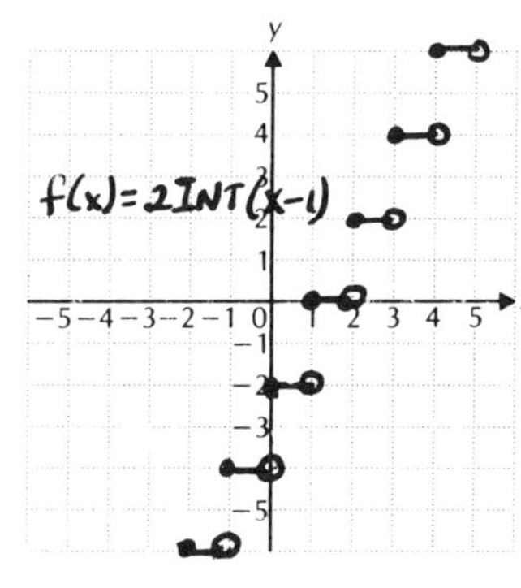

7. d,f 8. f 9. $\{x \mid x \neq 0\}$

10. $(f + g)(x) = x^2 - 4x + 12$, $(f - g)(x) = 6 - 4x - x^2$, $fg(x) = -4x^3 + 9x^2 - 12x + 27$, $(f/g)(x) = \dfrac{9 - 4x}{x^2 + 3}$, $f \circ g(x) = -4x^2 - 3$, $g \circ f(x) = 16x^2 - 72x + 84$

11. All reals, all reals, all reals, all reals, all reals, all reals, all reals, all reals 12. (b) 13. -7 14. -1 15. $b^2 + 9b + 13$

16. $2a + 5 + h$ 17. $P(x) = 0.5x^2 + 64x - 25$ 18. $m = \frac{2}{5}$, y-intercept is $-\frac{9}{5}$

19. $y = 2x - 7$ 20. $y = -\frac{7}{5}x + \frac{18}{5}$ 21. $\sqrt{85} \approx 9.220$ 22. $\left(\frac{1}{2}, -\frac{5}{2}\right)$

23. Neither 24. 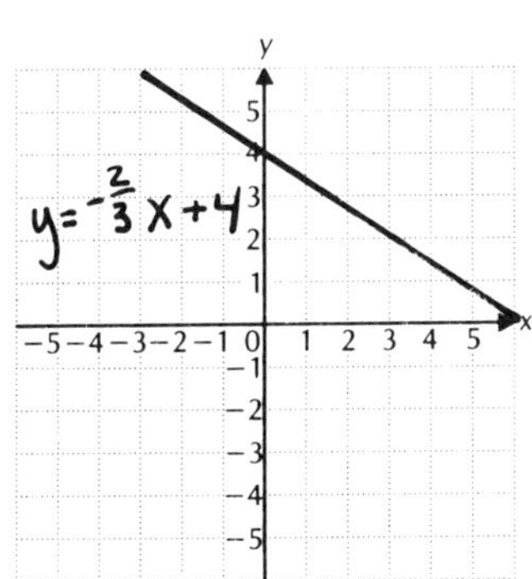

25. $y = -2x - 5$

26. $(x + 1)^2 + (y - 5)^2 = 9$ 27. Center: (3,-2); radius: $\sqrt{5}$

28. a)

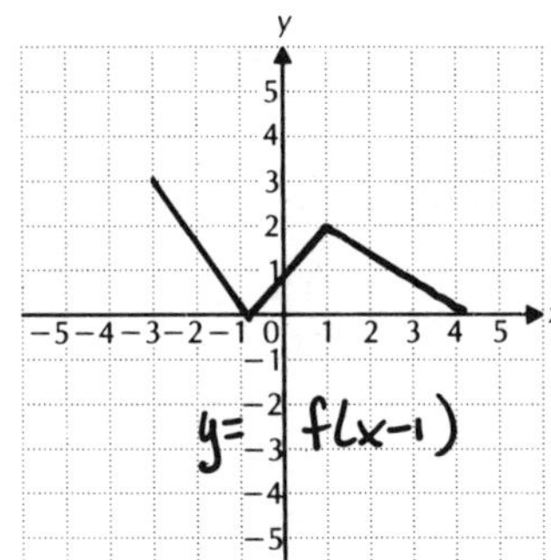

b)

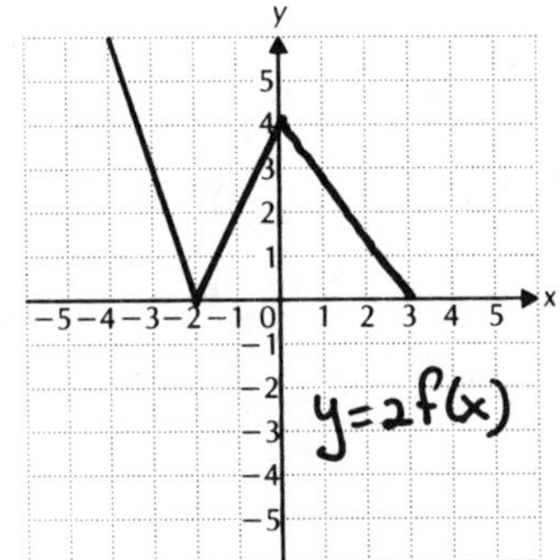

c)

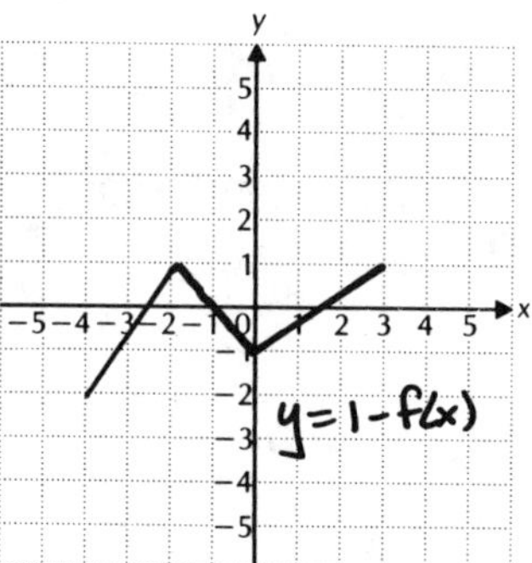

29. b, f 30. c, e 31. a, d 32. (-2,5] 33. $(-\infty,-4)$ 34. (a) 35. (b) and (c) 36.

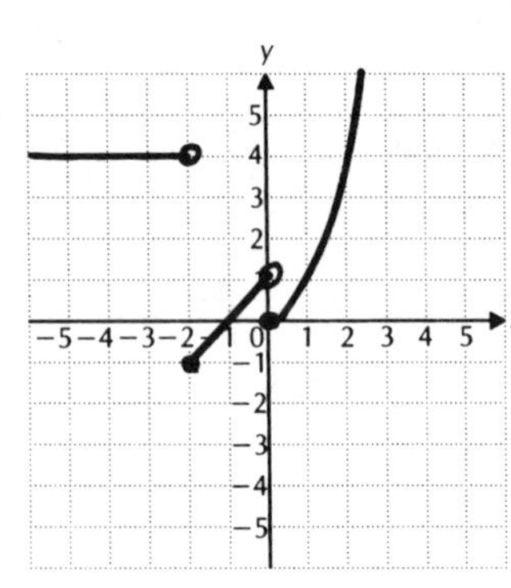

37. $A(x) = x(10 - x)$

38. (a) $f(x) = \sqrt[3]{x}$, $g(x) = 3x - 4$; (b) $f(x) = 2x^2 + x + 7$, $g(x) = x - 4$; Answers may vary. 39. (3,0)

Chapter 3, Test Form D

1. No 2. {-1,3,4,5} 3. {-1,3,5} 4.

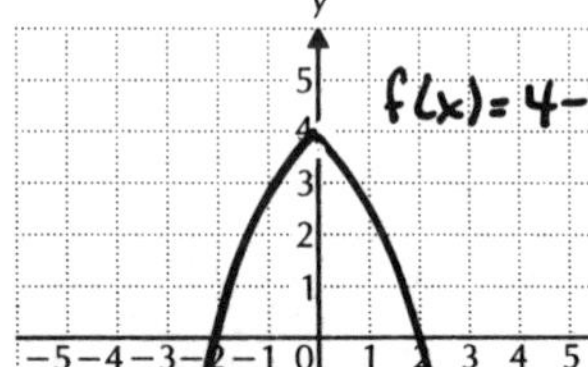

5.

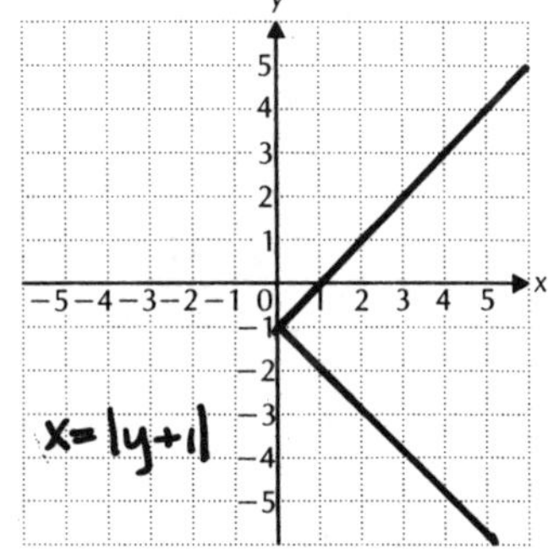

6.

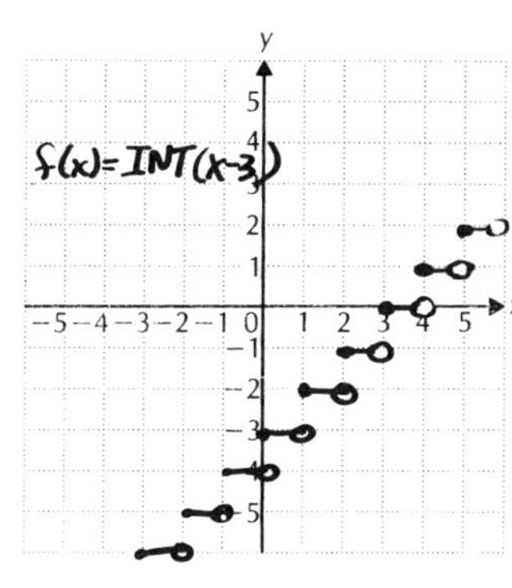

7. d, f   8. none   9. $\{x \mid x \geq -\frac{3}{2}\}$

10. $(f + g)(x) = 3x^2 + 2x + 3$, $(f - g)(x) = 3x^2 - 2x - 5$, $fg(x) = 6x^3 + 12x^2 - 2x - 4$

$(f/g)(x) = \frac{3x^2 - 1}{2x + 4}$, $f \circ g(x) = 12x^2 + 48x + 47$,

$g \circ f(x) = 6x^2 + 2$

11. All reals, all reals, all reals, all reals, all reals, all reals except -2, all reals, all reals   12. a   13. 52   14. 0   15. $2a^2 - 9a + 7$

16. $4a - 5 + 2h$ 17. $P(x) = 1.1x^2 + 18x - 29$ 18. $m = -\frac{2}{5}$, y-intercept is 3

19. $y = -3x + 1$ 20. $y = \frac{4}{7}x + \frac{23}{7}$ 21. $\sqrt{89} \approx 9.434$ 22. $\left(4, -\frac{1}{2}\right)$

23. Perpendicular   24.

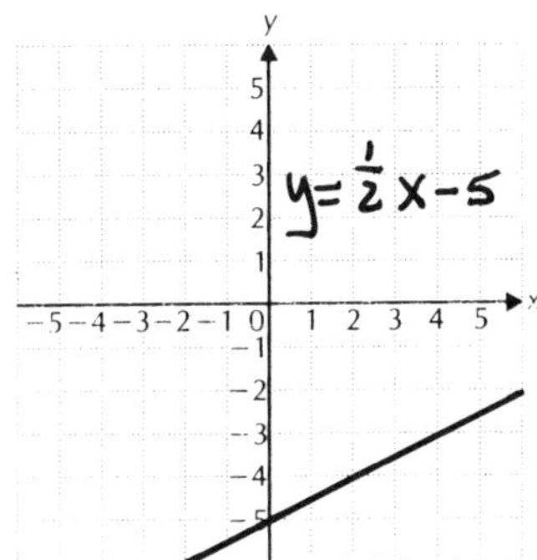

25. $y = \frac{1}{2}x - \frac{7}{2}$ 26. $(x - 2)^2 + (y + 6)^2 = 16$ 27. Center: (1,-6); radius: 6

Chapter 3, Test Form D (continued)

28. a)

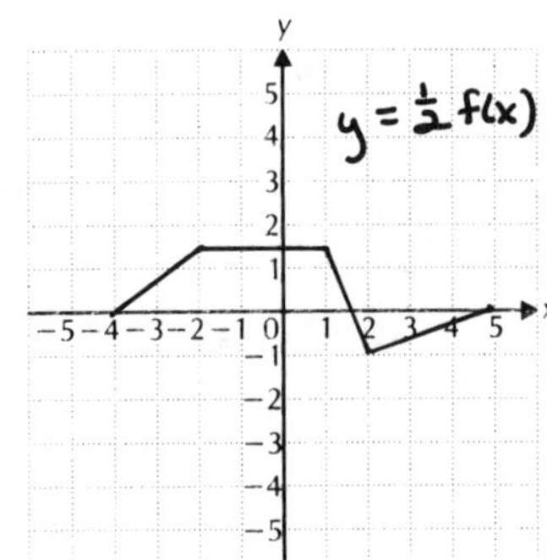

b)

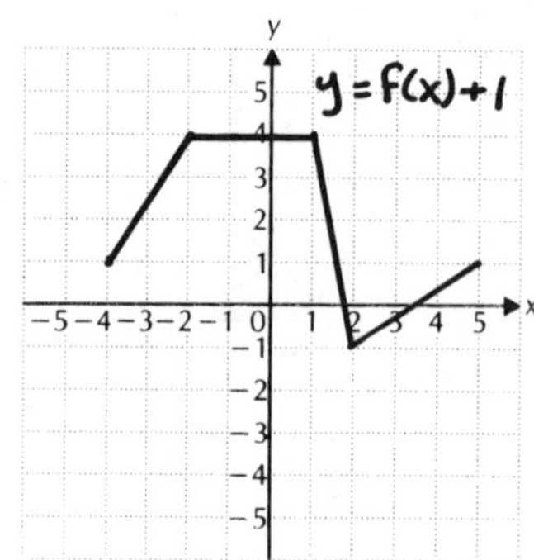

c)

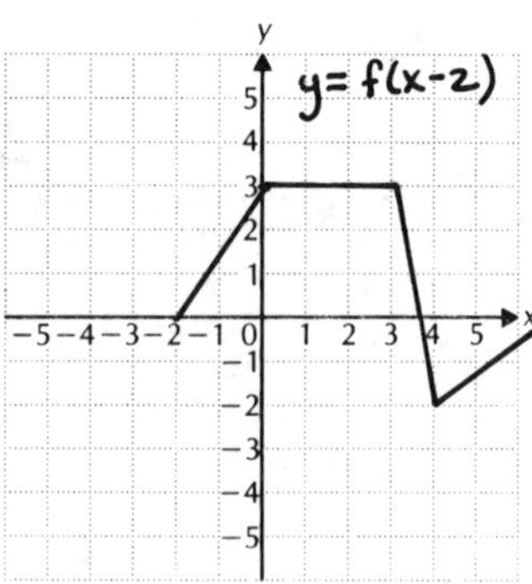

29. b, f 30. a, c 31. d, e 32. $(-5,2]$ 33. $(-\infty,6]$ 34. c 35. None

36.

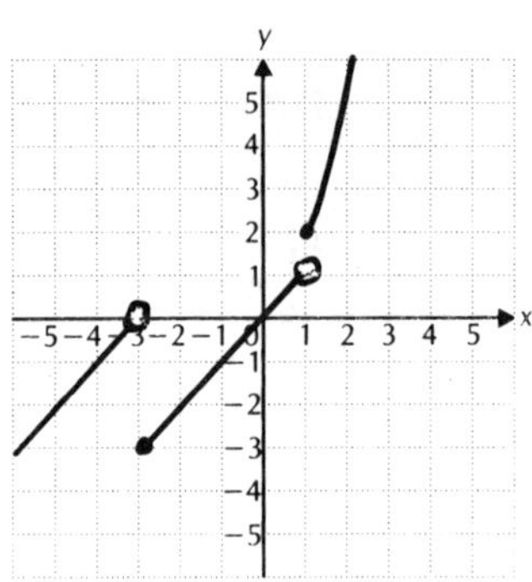

37. $A(x) = x\sqrt{400 - x^2}$

38. (a) $f(x) = \frac{1}{\sqrt{x}}$, $g(x) = 2x - 1$

(b) $f(x) = 6x^2 + 5$, $g(x) = x + 2$ 39. $k = -13$;

Answers may vary.

Chapter 3, Test Form E

1. Yes 2. $\{-5,-2,0,3,5\}$ 3. $\{-1,2,4,5\}$ 4.

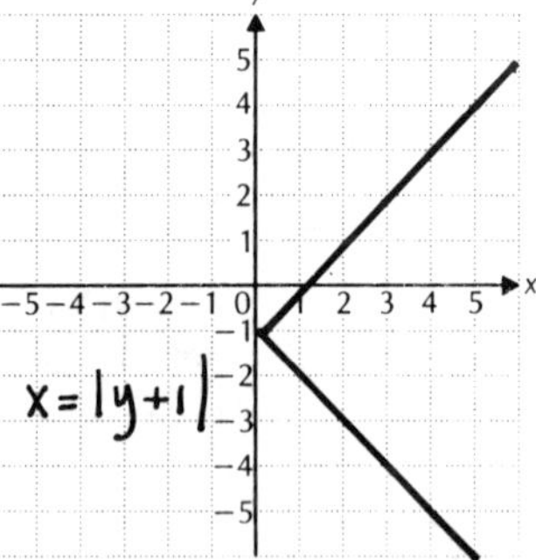

5.

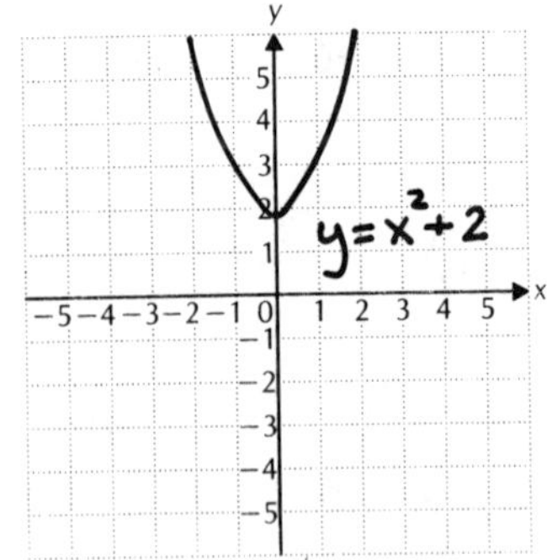

6. 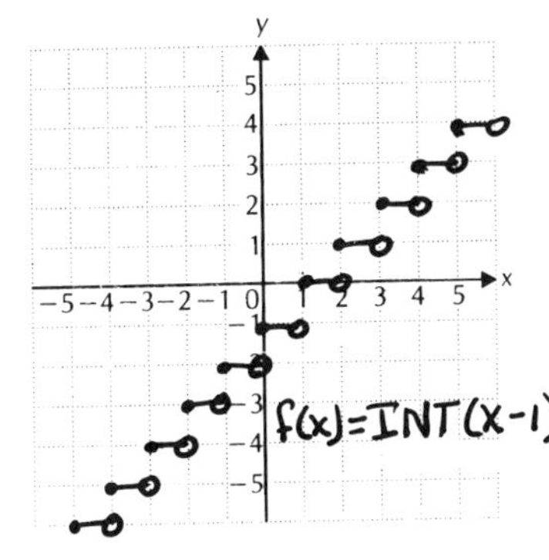

7. f 8. a, d, f 9. c, f 10. $\{x|x \neq -1 \text{ or } x \neq 4\}$

11. $(f + g)(x) = 4x^2 + 2x - 4$, $(f - g)(x) = -4x^2 + 2x - 2$, $fg(x) = 8x^3 - 12x^2 - 2x + 3$, $(f/g)(x) = \dfrac{2x - 3}{4x^2 - 1}$, $f \circ g(x) = 8x^2 - 5$, $g \circ f(x) = 16x^2 - 48x + 35$ 12. All reals, all reals, all reals, all reals, all reals, all reals except $-\frac{1}{2}, \frac{1}{2}$, all reals, all reals 13. b, c, d 14. 49 15. 35 16. $3a^2 - 13a + 19$

17. $6a - 1 + 3h$ 18. $P(x) = -0.2x^2 + 27x + 2$

19. (a)

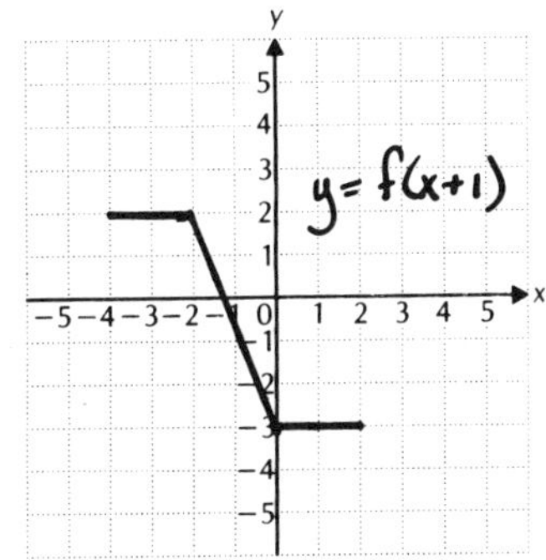

(b)

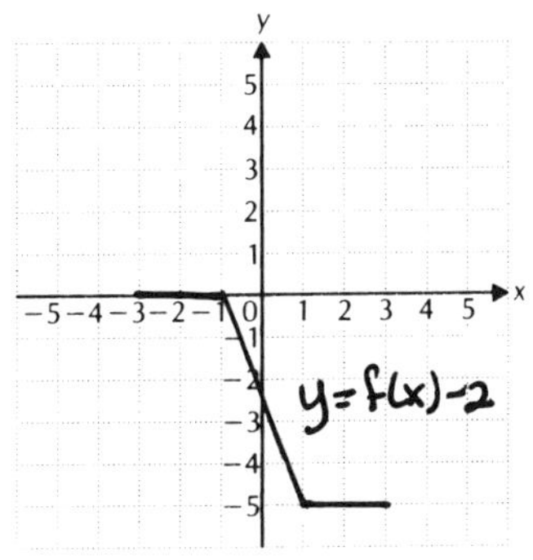

(c)

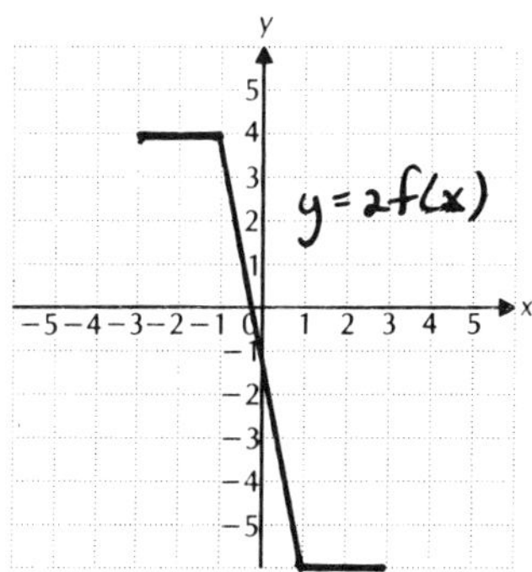

20. c, e 21. a, b, f 22. d 23. [-2,5) 24. $(-3,\infty)$ 25. a

26. b 27. m = -6, y-intercept is 10 28. y = 5x - 7 29. $\sqrt{10} \approx 3.162$

Chapter 3, Test Form E (continued)

30. 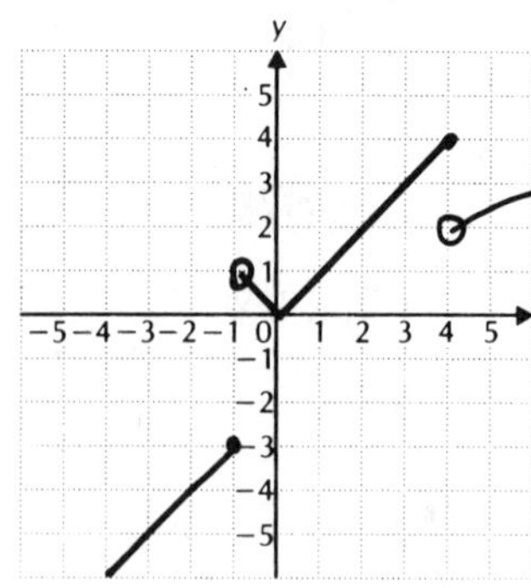

31. $A = t\sqrt{256 - t^2}$

32. $y = -\frac{2}{3}x + \frac{10}{3}$

33. $f(x) = \frac{5\sqrt{x}}{2}$, $g(x) = 3x - 8$

34. $(x + 1)^2 + (y - 5)^2 = 144$

Chapter 3, Test Form F

1. b 2. d 3. b 4. c 5. b 6. c 7. a 8. a 9. d 10. c 11. d

12. b 13. b 14. c 15. d 16. b 17. b 18. b

Chapter 4, Test Form A

1. (a) $f(x) = 2(x - 2)^2 - 7$; (b) $(2,-7)$; (c) minimum: -7

2. (a) $f(x) = -3\left(x - \frac{1}{3}\right)^2 - \frac{2}{3}$; (b) $\left(\frac{1}{3}, -\frac{2}{3}\right)$; (c) maximum: $-\frac{2}{3}$

3. 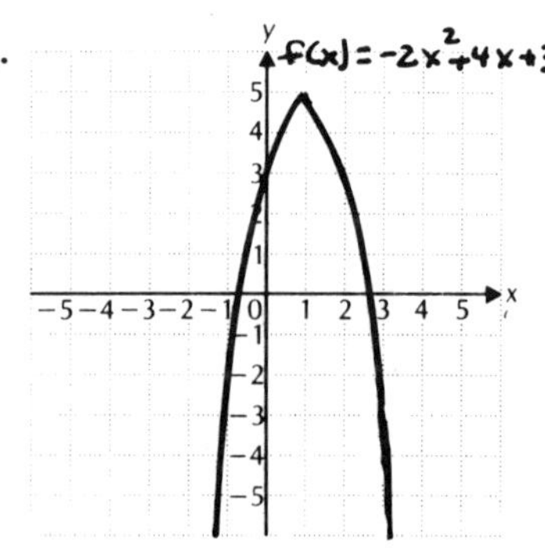

4. $\left(\frac{3 - \sqrt{41}}{8}, 0\right)$, $\left(\frac{3 + \sqrt{41}}{8}, 0\right)$

5. $\{1,2,5,7,9,11\}$

6. 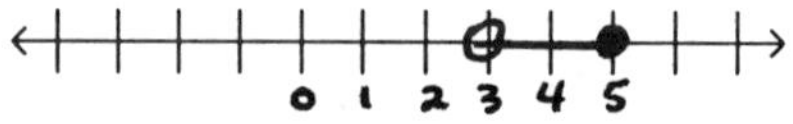 (number line: open circle at 3, closed circle at 5, segment between; labels 0 1 2 3 4 5)

7. $(-8,-1]$

8. $(-\infty,-1] \cup [6,\infty)$

9. $(4,10)$

10. $\{-2, \frac{10}{3}\}$

11. $(\infty,1) \cup (9,\infty)$

12. $\left(-1, \frac{2}{3}\right)$

13. $(-\infty,-11) \cup (-4,\infty)$

14. -8, -8

15. 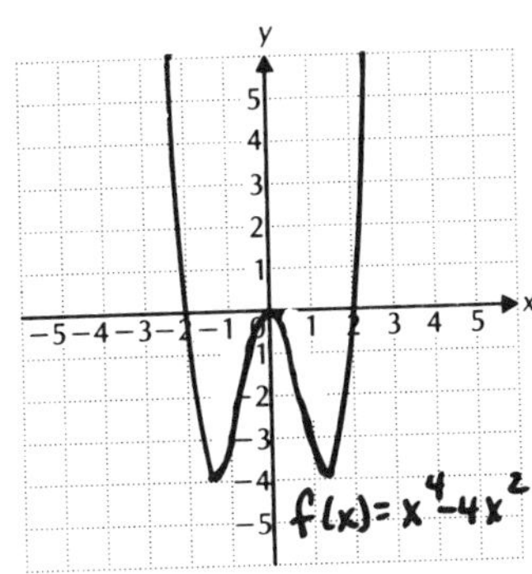

16. 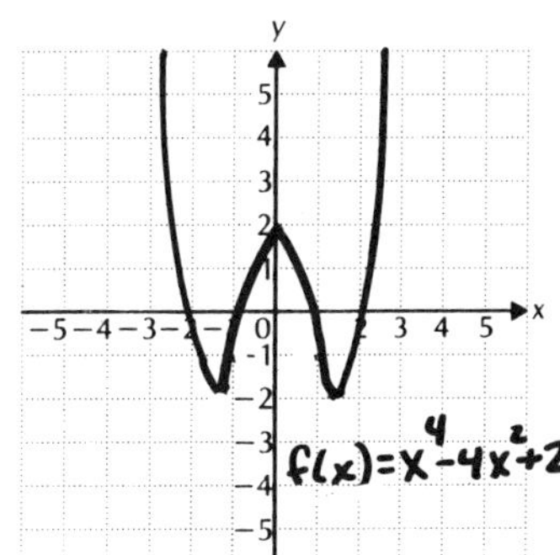

17. 499 18. Yes 19. The quotient is $4x^2 + x + 3$. The remainder is 1.

20. 100 21. $(x - 4)(x - 1)(x + 3)$; 4, 1, -3 22. $x^4 - 2x^3 - 7x^2 + 18x - 18$

23. $(x - 2)(x + 3)(x - 1)^2(x - 5)^3$ 24. $x^3 - 11x^2 + 41x - 51$

25. $\pm \left\{\frac{1}{4}, \frac{1}{2}, \frac{3}{4}, 1, \frac{3}{2}, 2, 3, 6\right\}$ 26. Positive: 3 or 1; Negative: 1

27. $[-1,2]$ 28. $\pm\ 2.3$ 29. 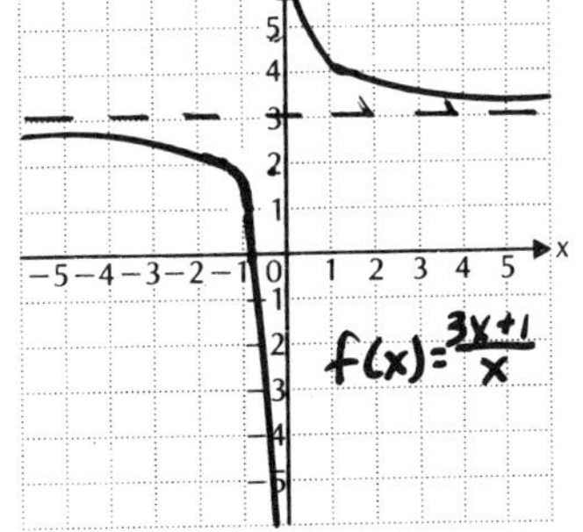

30. $k = \frac{5}{6}$

31. $\left(-\infty, -\frac{3}{2}\right]$

32. $-2 < x < 3$ or $x > 4$

Chapter 4, Test Form B

1. (a) $f(x) = -3(x - 2)^2 + 8$; (b) $(2,8)$; (c) maximum: 8

2. (a) $f(x) = 5\left(x + \frac{1}{5}\right)^2 + \frac{4}{5}$; (b) $\left(-\frac{1}{5}, \frac{4}{5}\right)$; (c) minimum: $\frac{4}{5}$

3. 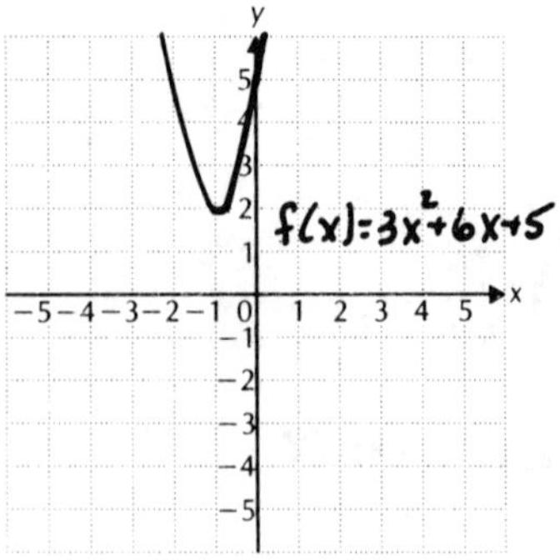

4. $\left(\frac{-5 - \sqrt{97}}{4}, 0\right)$, $\left(\frac{-5 + \sqrt{97}}{4}, 0\right)$ 5. {5,15}

6. 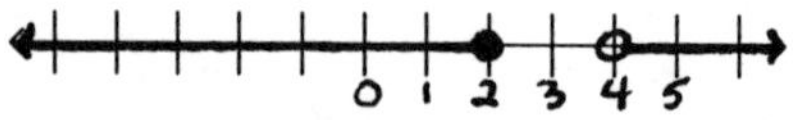

7. ∅

8. $(-\infty,-1) \cup \left(\frac{5}{3}, \infty\right)$ 9. [-2,3] 10. $\{-4, \frac{22}{5}\}$

11. $(-\infty,-4) \cup (2,\infty)$ 12. $\left(-\frac{1}{2}, \frac{1}{5}\right)$ 13. $\left(-\frac{3}{2}, -\frac{2}{3}\right)$

14. -9,9

15. 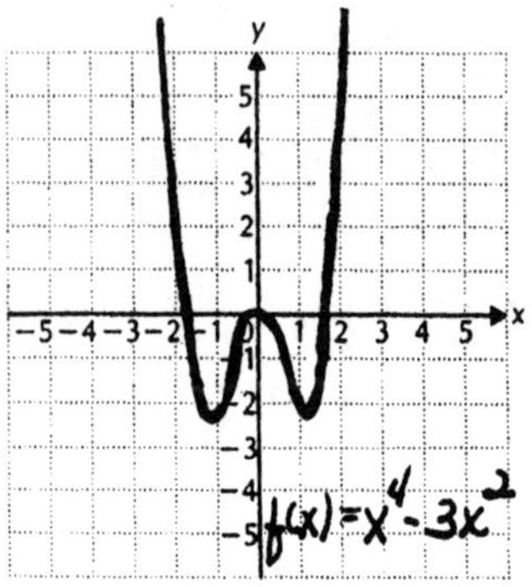

16. 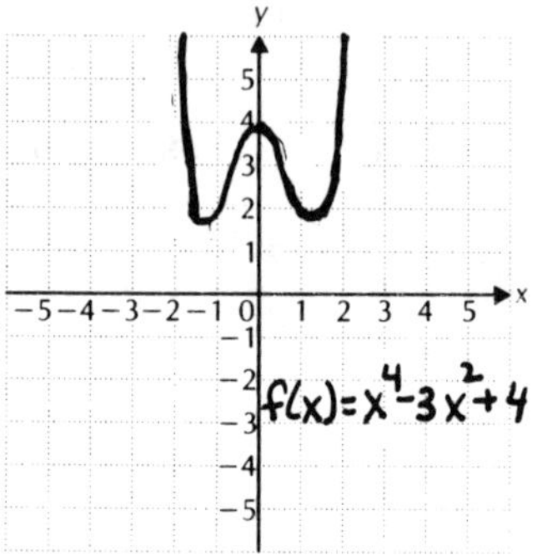

17. 4 18. Yes 19. The quotient is $3x^3 - 6x^2 + 10x - 17$.

The remainder is 33. 20. 323 21. $(x - 5)(x + 1)(x + 2)$; 5, -1, -2

22. $x^3 - 5x^2 - 3x + 15$ 23. $x(x - 5)(x - 3)^2(x + 1)^4$

24. $x^3 - 13x^2 + 49x - 49$ 25. $\pm\left(\frac{1}{2}, 1, 2, 4, 8\right)$ 26. Positive: 3 or 1;

Negative: 3 or 1 27. [-2,1] 28. 1.8

29.

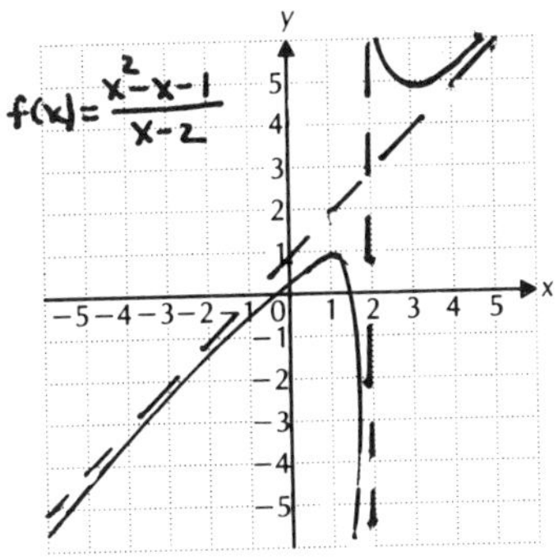

30. 36 31. $\left[-\sqrt{5}, \sqrt{5}\right]$ 32. k = -13

Chapter 4, Test Form C

1. (a) $f(x) = 6(x + 1)^2 - 11$; (b) $(-1,-11)$; (c) minimum: -11

2. (a) $f(x) = -2\left(x - \frac{1}{4}\right)^2 + \frac{25}{8}$; (b) $\left(\frac{1}{4}, \frac{25}{8}\right)$; (c) maximum: $\frac{25}{8}$

3.

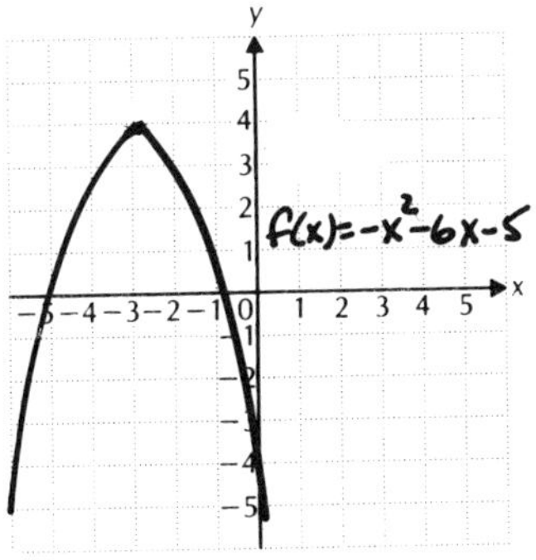

4. $\left(\frac{7 - \sqrt{61}}{6}, 0\right)$, $\left(\frac{7 + \sqrt{61}}{6}, 0\right)$

5. {1,4,7,8,9,11,16,18,20}

6. 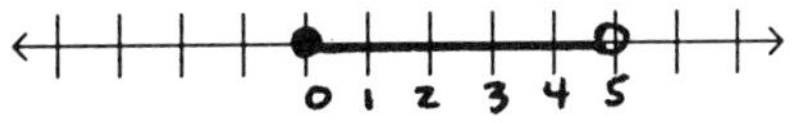

7. [-1,1)

8. $\left(-\infty, -\frac{4}{5}\right] \cup [2,\infty)$ 9. $(-16,-8)$ 10. $\left\{-3, \frac{15}{4}\right\}$

11. $(-\infty,-6) \cup (2,\infty)$ 12. $\left(-\frac{1}{4}, \frac{2}{3}\right)$ 13. $(-4,-1)$

14. Base: 14 cm; height: 14 cm

15. 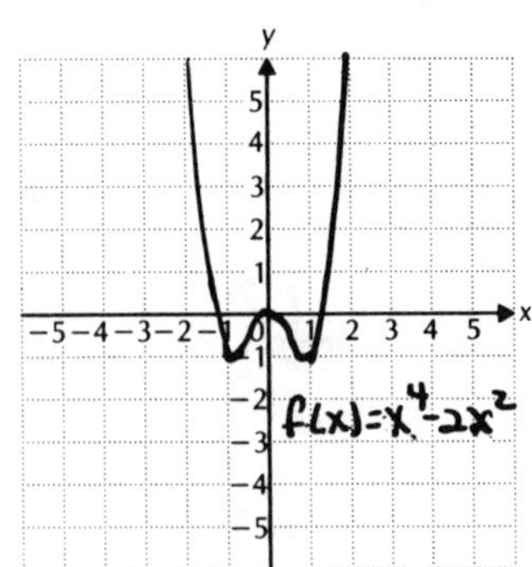

16. 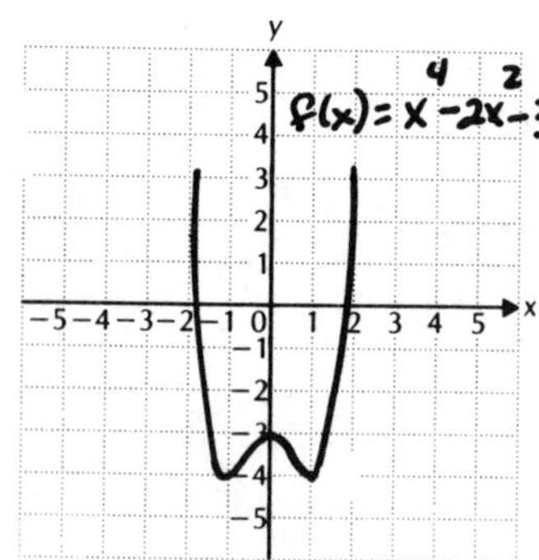

17. 57 18. No 19. The quotient is $5x^2 + 13x + 55$. The remainder is 216. 20. 421 21. $(x - 4)(x - 2)(x + 7)$; 4, 2, -7 22. $x^4 - 7x^2 - 144$

23. $(x + 4)(x + 3)(x - 5)^2(x + 2)^3$ 24. $x^3 + 8x^2 + 46x + 68$

25. $\pm\left(\frac{1}{6}, \frac{1}{3}, \frac{1}{2}, \frac{2}{3}, 1, \frac{4}{3}, 2, 4\right)$ 26. Positive: 2 or 0; Negative: 3 or 1

27. [-2,1] 28. -1.2, 1.4 29.

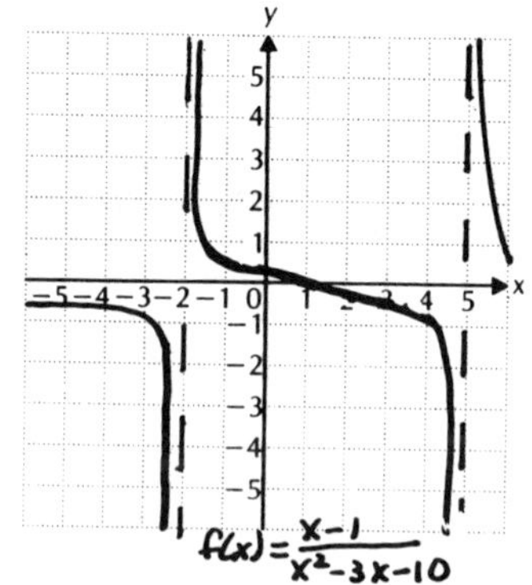

30. $\{x|x > -4\}$, or $(-4,\infty)$

31. $\{h|h > -3 + \sqrt{39}$ cm$\}$

32. $k = -\frac{3}{2}$

Chapter 4, Test Form D

1. (a) $f(x) = -2(x + 2)^2 + 11$; (b) $(-2,11)$; (c) maximum: 11

2. (a) $f(x) = 4\left(x + \frac{5}{8}\right)^2 - \frac{57}{16}$; (b) $\left(-\frac{5}{8}, -\frac{57}{16}\right)$; (c) minimum: $-\frac{57}{16}$

3. 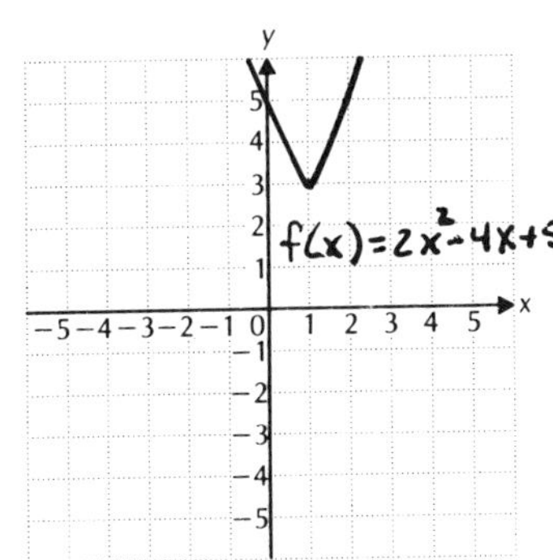

4. $\left(\frac{-5-\sqrt{109}}{6},0\right)$, $\left(\frac{-5+\sqrt{109}}{6},0\right)$ 5. {1,7}

6.
-3 -2 -1 0

7. (-2,3)

8. $(-\infty,-2) \cup (16,\infty)$ 9. $\left[-\frac{7}{3},1\right]$ 10. $\{-4,\frac{3}{2}\}$

11. $(-\infty,-4) \cup (6,\infty)$ 12. $\left(-\frac{1}{7},4\right)$ 13. (-7,8]

14. -11, 11

15. 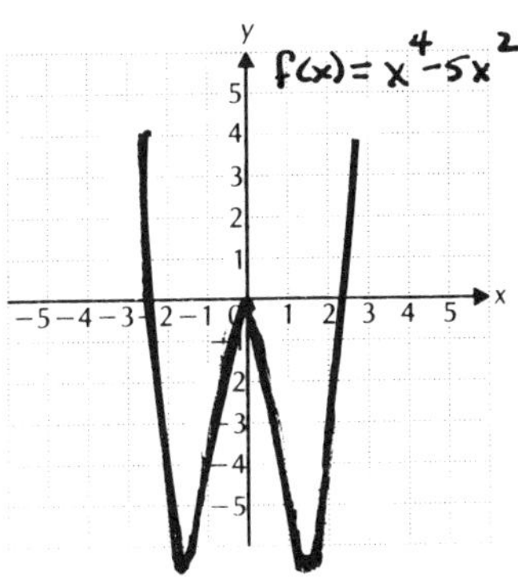

16. 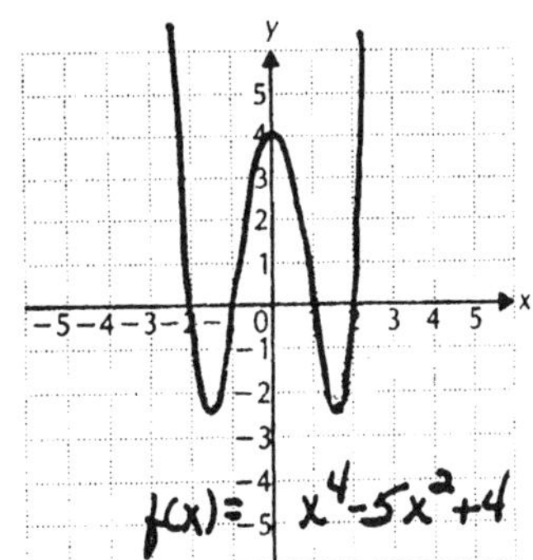

17. 225 18. No 19. The quotient is $x^3 - x^2 + 3x - 10$. The remainder is 34. 20. 2246 21. $(x - 5)(x - 3)(x + 1)$; 5, 3, -1 22. $x^3 - 11x^2 + 27x + 7$

23. $(x - 1)(x + 2)(x + 1)^2(x - 6)^3$ 24. $x^3 - 6x^2 + 4x + 16$

25. $\pm\left(\frac{1}{8}, \frac{1}{4}, \frac{1}{2}, 1, 2, 4\right)$ 26. Positive: 4, 2, or 0; Negative: 0 27. [-1,2]

28. 0.9

Chapter 4, Test Form D (continued)

29.

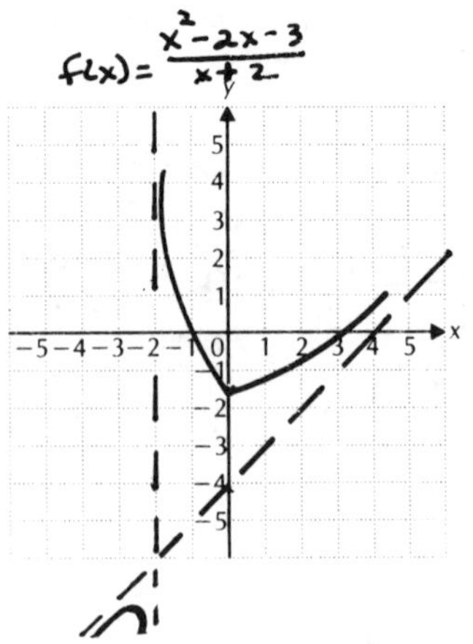

30. $\{x|x > 5\}$, or $(5,\infty)$ 31. (a) 10, 22;

(b) $\{x|10 < x < 22\}$, or $(10,22)$ 32. $k = -7$

Chapter 4, Test Form E

1. (a) $f(x) = 2\left(x - \frac{5}{2}\right)^2 - \frac{35}{2}$; (b) $\left(\frac{5}{2}, -\frac{35}{2}\right)$; (c) minimum: $-\frac{35}{2}$

2.

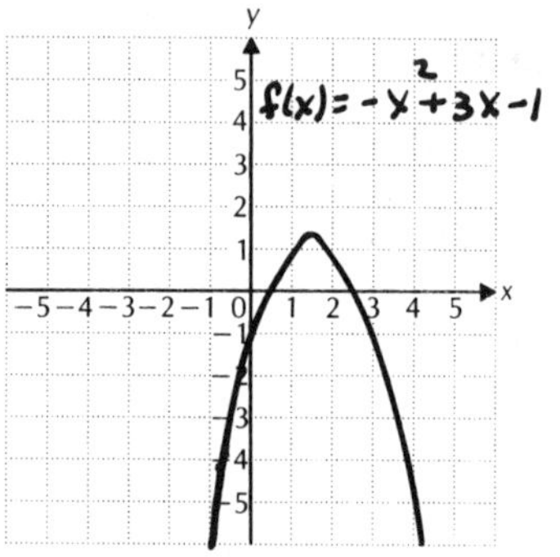

The x-intercepts are $\left(\frac{3-\sqrt{5}}{2},0\right)$ and $\left(\frac{3+\sqrt{5}}{2},0\right)$

3. $\ell = 27$, $w = 27$ 4. $\{b,d,f,h,j,\ell,p\}$

5. 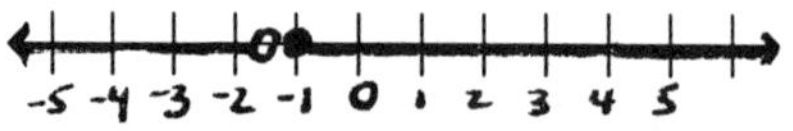 6. j 7. h 8. e

9. f 10. a 11. b 12. g

13.

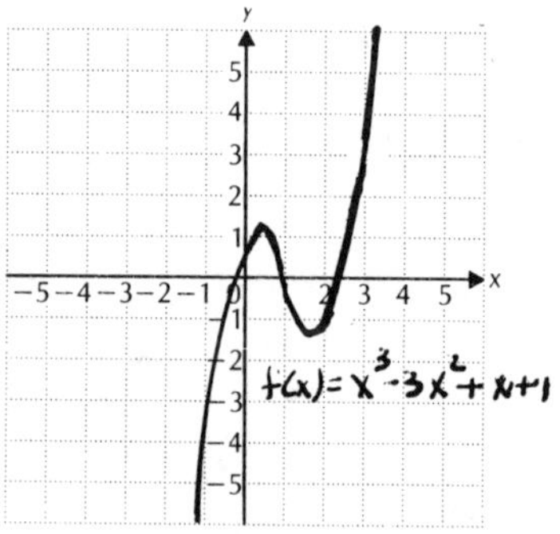

14.

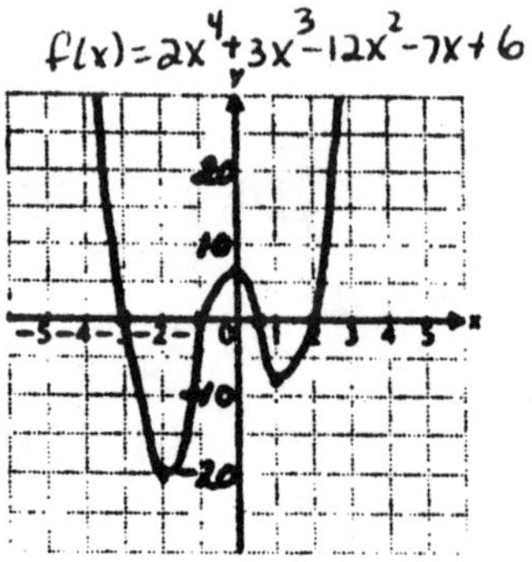

Chapter 4, Test Form E (continued)

15. 8 16. The quotient is $5x^4 - 10x^3 + 18x^2 - 35x + 68$. The remainder is -133. 17. 834 18. $(x - 5)(x - 1)(x + 1)(x + 3)(x + 5)$; 5, 1, -1, -3, -5

19. $x^4 - 2x^3 + 2x^2 - 8x - 8$ 20. $x(x + 2)(x - 5)^4(x + 7)^3$ 21. $-2 \pm 2i\sqrt{3}$

22. $2i, -\sqrt{7}, 4 + 5i$ 23. $\pm\left(\frac{1}{6}, \frac{1}{3}, \frac{1}{2}, \frac{2}{3}, 1, \frac{4}{3}, 2, 4\right)$ 24. 3 or 1 25. [-2,2]

26.

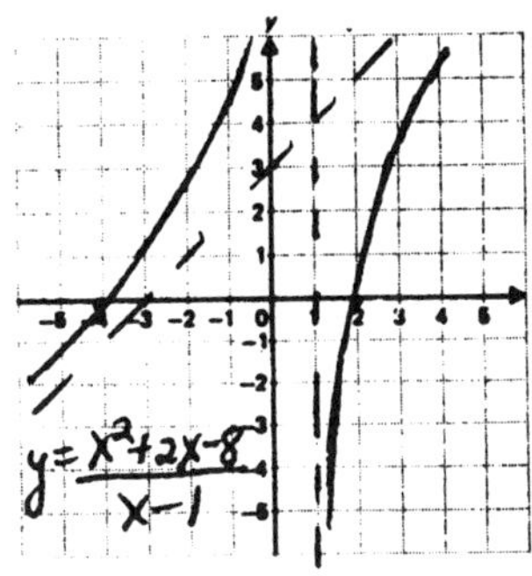

Chapter 4, Test Form F

1. d 2. b 3. a 4. a 5. b 6. a 7. c 8. d 9. c 10. b 11. a

12. d 13. c 14. b 15. c 16. a 17. d 18. b 19. c 20. a

Chapter 5, Test Form A

1. {(-1,2), (-3,5), (1,7), (2,-4), (4.5,3.2)} 2. $x = 3y - 3$ 3. b

4. $f^{-1}(x) = \frac{6 - x}{2}$ 5. 3 6.

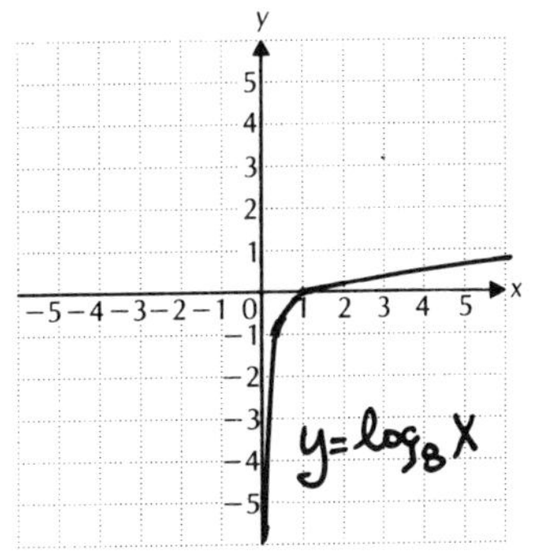

7.

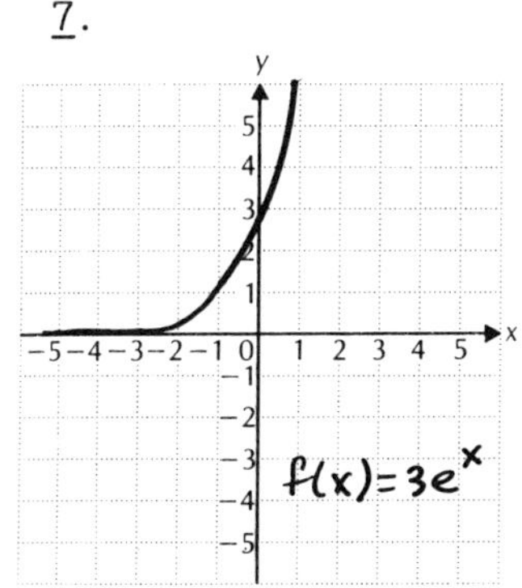

Chapter 5, Test Form A (continued)

8. 2.5   9. $\left(\sqrt{3}\right)^6 = 27$ 10. $\log_x 0.2401 = 4$ 11. $x + 2$ 12. $\log_a \dfrac{\sqrt[4]{x^3 z^5}}{\sqrt{y}}$

13. $2 \log a - \log b$ 14. 0.477   15. 1.447   16. 0.1945   17. $-\sqrt{2}, \sqrt{2}$

18. 64   19. $\frac{1}{2}$ 20. 1   21. 0.9163   22. Not defined   23. -11.2506

24. 0.7178   25. -1.3593   26. 4.6946   27. 7.3 yr   28. 6.34   29. 106,666

30. (a) 2.4 ft/sec; (b) 3,801,851   31. (a) $P(t) = 0.15e^{0.084t}$; (b) $1.22;

(c) 2005; (d) about 8 years   32. 3.5%   33. 10.2 yr   34. 4540 yr

35. 27 decibels   36. 4.6   37. $\pi^{4.2}$ 38. $\frac{11}{3}$

Chapter 5, Test Form B

1. {(-2,8), (-7,6), (3,8), (1,-5), (-1.2,5.6)}   2. $x = y^2 + 1$ 3. b

4. $f^{-1}(x) = \dfrac{3 - 2x}{x}$ 5. -2   6.   7.

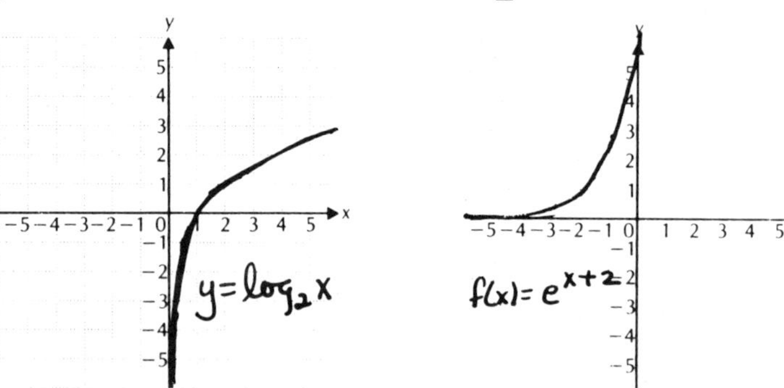

8. 1.7604   9. $\left(\sqrt{2}\right)^8 = 16$ 10. $\log_x 0.00243 = 5$ 11. 4x   12. $\log_a \dfrac{\sqrt[4]{xz}}{y^3}$

13. $2 \log x + \frac{1}{2} \log y$ 14. 0.602   15. 2.301   16. 0.1505   17. -3, 3   18. 36

19. 1   20. 2   21. 0.1823   22. 4.4048   23. 1.5933   24. -9.8837

25. -2.4895   26. Not defined   27. 11.9 yr   28. 38 decibels   29. 112,500

Chapter 5, Test Form B (continued)

30. (a) 3000; (b) 3180; (c) \$100,000,000   31. (a) $P(t) = 4.5e^{0.032t}$; (b) 7.5 million; (c) 1993   32. 2.0%   33. 7.4 yr   34. 7229 yr   35. 2.54   36. 7.9×10^{-7} 37. (a) 125.000000; (b) 156.590645; (c) 156.842871; (d) 156.969136   38. 4

Chapter 5, Test Form C

1. {(3,-1), (-7,4), (0,6), (5,2), (7.6,-1.3)}   2. $x = |y + 3|$ 3. a, b

4. $f^{-1}(x) = \dfrac{x + 2}{x - 1}$ 5. -2   6.

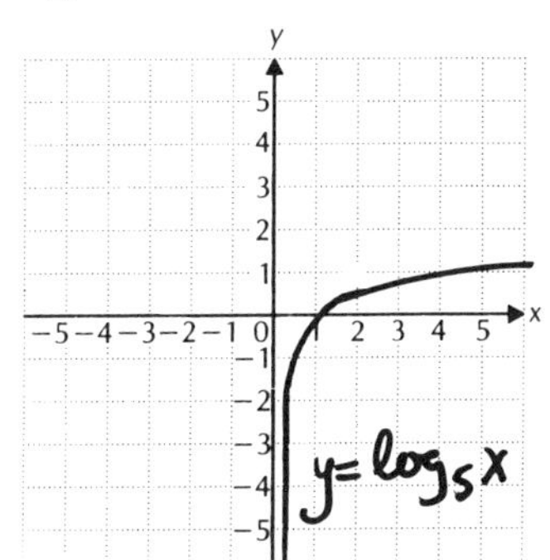

7.

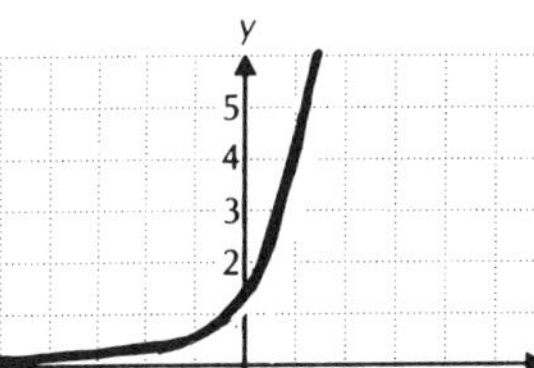

8. 3.3802   9. $\left(\sqrt{5}\right)^4 = 25$ 10. $\log_x 0.015625 = 6$ 11. $x - 1$ 12. $\log_a \dfrac{x^5 \sqrt[3]{y}}{z^2}$

13. log 5 + log c + 2 log d   14. 0.301   15. 1.681   16. 0.259   17. -2, 2

18. 125   19. $\frac{4}{5}$ 20. 9   21. -0.1054   22. -13.0736   23. 5.0052   24. -0.8625

25. Not defined   26. 1.6332   27. 9.0 yr   28. 5.17   29. 285,610

30. (a) 1.3 yr; (b) 5.8 yr   31. (a) $P(t) = 0.03e^{0.038t}$; (b) 58¢; (c) 2035 years   32. 3.8%   33. 6.9 yr   34. 3590 yr   35. 36 decibels

36. 1.3×10^{-5} 37. 16   38. -16, 16

Chapter 5, Test Form D

1. $\{(3,-4), (-5,2), (4,1), (4.3,6.2), (7,0)\}$ 2. $yx = 3$ 3. a, b

4. $f^{-1}(x) = x^2 - 3$ 5. -7 6.

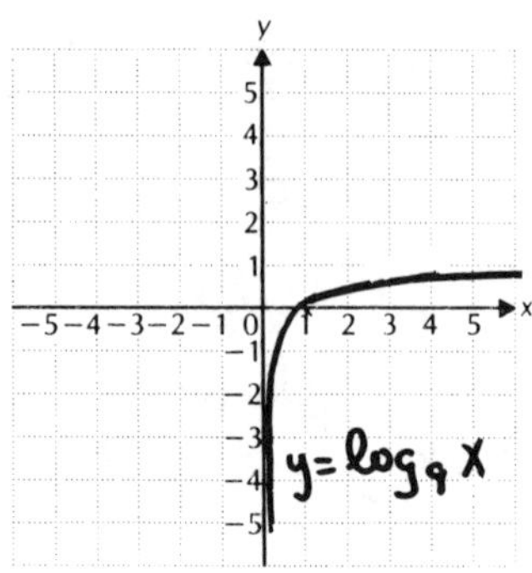

7. 

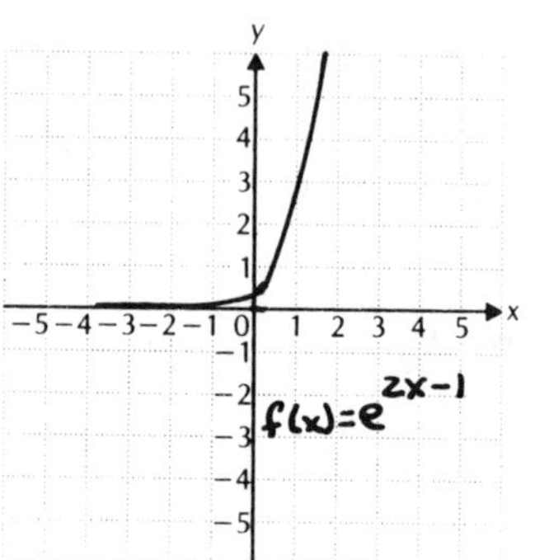

8. 4.6439 9. $\left(\sqrt{5}\right)^6 = 125$ 10. $\log_x 0.00032 = 5$ 11. $5x$ 12. $\log_a \dfrac{x^2y^4}{\sqrt[3]{z^2}}$

13. $\dfrac{2}{3} \log a + \dfrac{1}{3} \log b$ 14. 0.477 15. 2 16. 0.5395 17. -1, 1 18. 8

19. 3 20. 8 21. 0.3567 22. 0.4615 23. 1.8294 24. 4.2393 25. Not defined 26. -9.2926 27. 6.6 yr 28. 33 decibels 29. 104,509

30. (a) 75; (b) 63; (c) 60 months 31. (a) $P(t) = 100e^{0.182t}$; (b) 617; (c) 3.8 days 32. 1.9% 33. 8.0 yr 34. 2287 yr 35. 3.91 36. 5.7

37. $\left(\sqrt{7}\right)^{-3}$, or $7^{-3/2}$ 38. $\{x | x > e^{4/3}\}$

Chapter 5, Test Form E

1. a, b 2. $f\ (x)^{-1} = \sqrt{x - 3}$ 3. $f^{-1}(x) = (4x - 20)^2$ 4. 6

5.

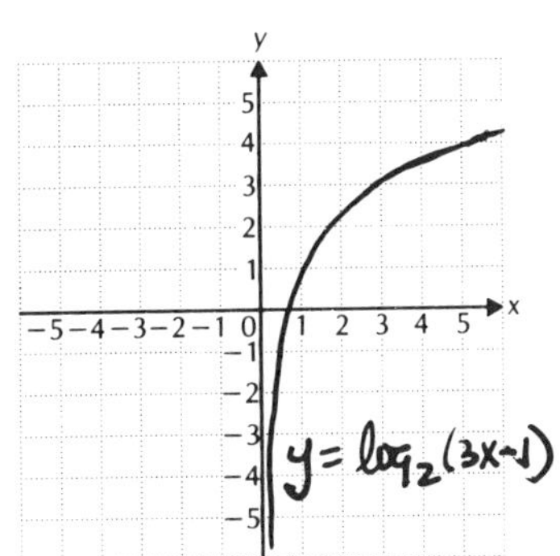

6. 

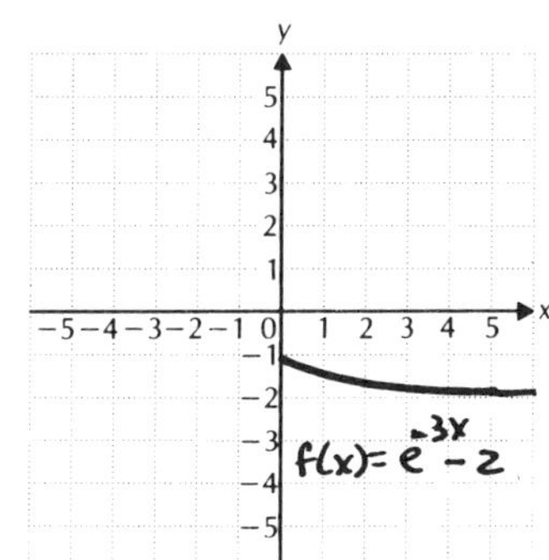

7. 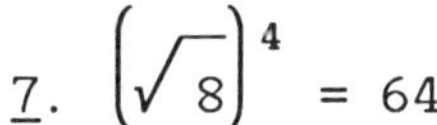 $\left(\sqrt{8}\right)^4 = 64$

8. $\log_a 0.00143 = -\frac{1}{3}$ 9. $5x + 2$ 10. x^3 11. 1.6340   12. -2, 4

13. $\log_a \frac{x^7 \sqrt[4]{z^3}}{\sqrt[3]{y}}$ 14. $\{x | x > 1\}$ 15. $\frac{1}{3}$ 16. $\frac{6}{13}$ 17. 6   18. $8\sqrt{10}$

19. 1.6371   20. -8.8116   21. Not defined   22. 4.8769   23. 0.544   24. 0.690

25. 1.623   26. (a) 12.2%; (b) 11.6%   27. (a) 182,211; (b) in 2021   28. About 3463 years   29. (a) 2.5 ft/sec; (b) 86,441

Chapter 5, Test Form F

1. b   2. d   3. c   4. b   5. a   6. c   7. d   8. b   9. c   10. c   11. a

12. d   13. b   14. c   15. d   16. b   17. b   18. a   19. c   20. b   21. d

22. a   23. c   24. b

Chapter 6, Test Form A

1. Origin: $\left(\frac{2}{3}, -\frac{\sqrt{5}}{3}\right)$; u-axis: $\left(-\frac{2}{3}, -\frac{\sqrt{5}}{3}\right)$; v-axis: $\left(\frac{2}{3}, \frac{\sqrt{5}}{3}\right)$

2.

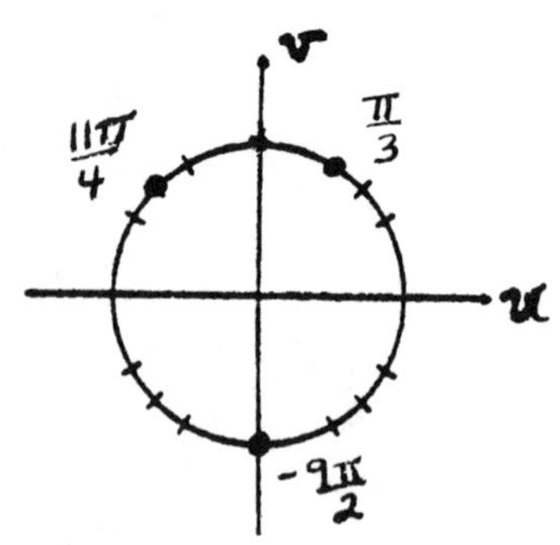

3.

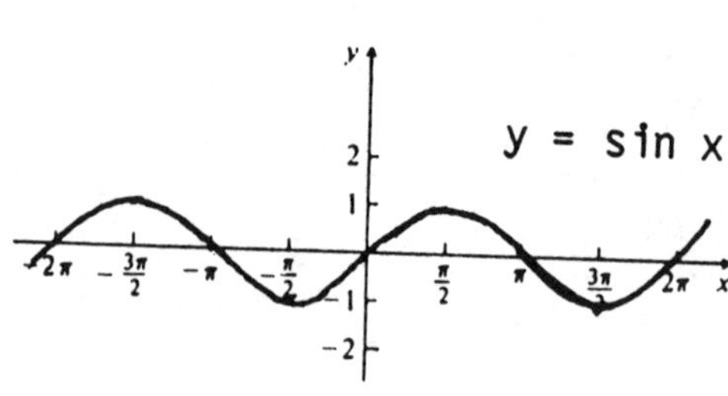

4. All reals from -1 to 1 inclusive 5. 2π 6. All reals 7. 1

8.

x	$\frac{\pi}{3}$	$\frac{7\pi}{4}$	$-\frac{3\pi}{2}$
sin x	$\frac{\sqrt{3}}{2}$	$-\frac{\sqrt{2}}{2}$	1
cos x	$\frac{1}{2}$	$\frac{\sqrt{2}}{2}$	0

9.

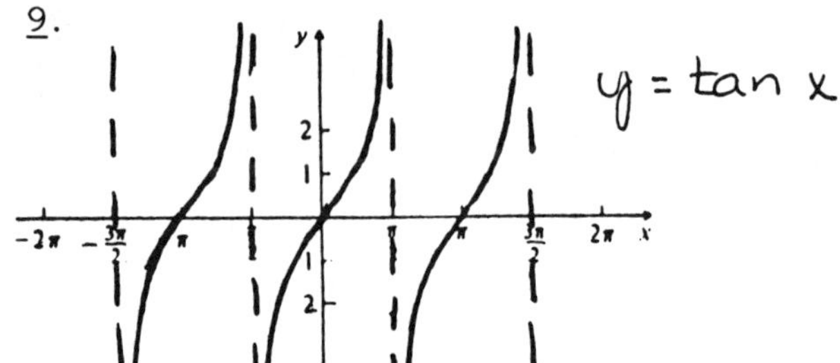

10. π 11. All reals except $\frac{\pi}{2} + k\pi$ 12. I, IV 13. II, $-\frac{7\pi}{6}$, -3.67

14. I, $\frac{3\pi}{10}$, 0.94 15. III, $\frac{5\pi}{4}$, 3.93 16. 405° 17. 171.89° 18. 20.94 cm

19. $\frac{1}{4}$, 14° 20. 440 21. 8 22. $\sin\theta = -\frac{5}{\sqrt{61}}$, or $-\frac{5\sqrt{61}}{61}$;

$\cos\theta = -\frac{6}{\sqrt{61}}$, or $-\frac{6\sqrt{61}}{61}$; $\tan\theta = \frac{5}{6}$; $\cot\theta = \frac{6}{5}$; $\sec\theta = -\frac{\sqrt{61}}{6}$;

$\csc\theta = -\frac{\sqrt{61}}{5}$ 23. $-\sqrt{3}$ 24. $-\frac{\sqrt{3}}{2}$ 25. $-\frac{1}{2}$ 26. $\cos\theta = -\frac{2\sqrt{2}}{3}$;

$\tan\theta = -\frac{\sqrt{3}}{4}$; $\cot\theta = -2\sqrt{2}$; $\sec\theta = -\frac{3\sqrt{2}}{4}$; $\csc\theta = 3$ 27. $28^\circ 21'$

28. 89.6° 29. -0.450025 30. 0.440177 31. 1.73205 32. 1.05922

33. -2.75963 34. -0.932515 35. 57.83°, 1.0094 36. 27.66°, 0.4828

37. 143.62° 38. -1.3850 39.

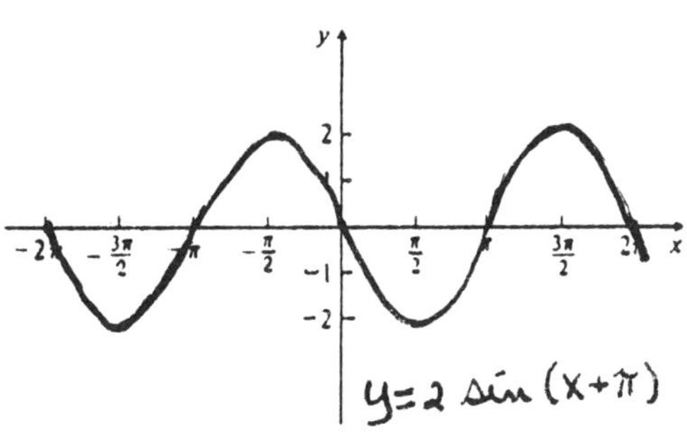

40. 2π 41. $-\pi$ 42.

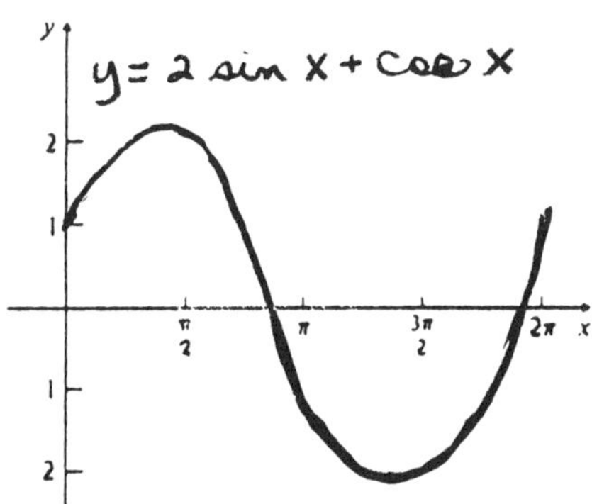

43. $\pm \frac{\sqrt{15}}{4}$

44. No, $\cos x = \frac{10}{8}$, but cosines are never greater than 1.

Chapter 6, Test Form B

1. Origin: $\left(\frac{1}{2}, \frac{\sqrt{3}}{4}\right)$; u-axis: $\left(-\frac{1}{2}, \frac{\sqrt{3}}{4}\right)$; v-axis: $\left(\frac{1}{2}, -\frac{\sqrt{3}}{4}\right)$

2.

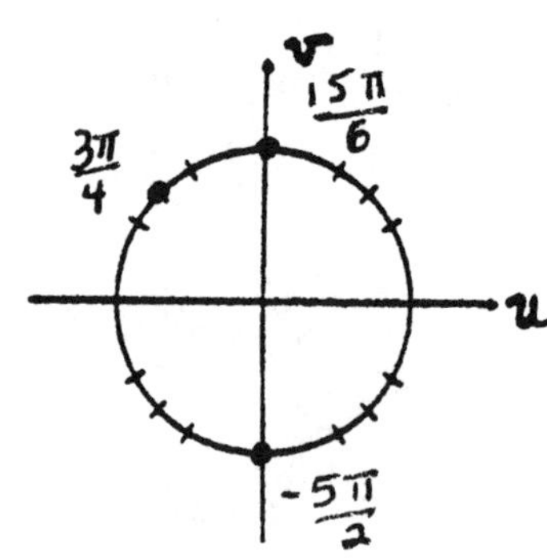

3.

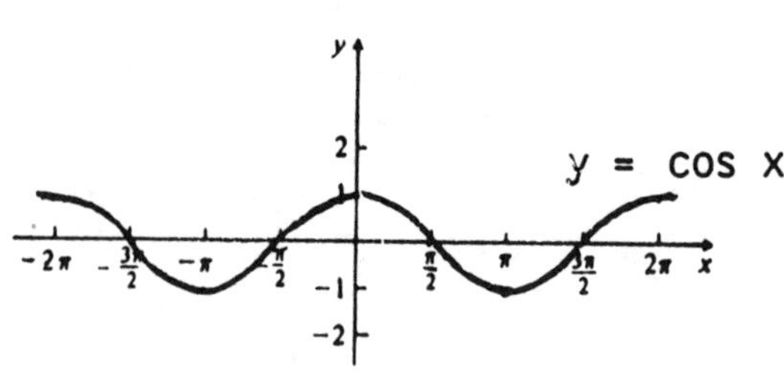

4. All reals from −1 to 1 inclusive 5. 2π 6. All reals 7. 1

8.

x	$\frac{3\pi}{4}$	$\frac{5\pi}{6}$	$-\frac{\pi}{2}$
sin x	$\frac{\sqrt{2}}{2}$	$\frac{1}{2}$	−1
cos x	$-\frac{\sqrt{2}}{2}$	$-\frac{\sqrt{3}}{2}$	0

9.

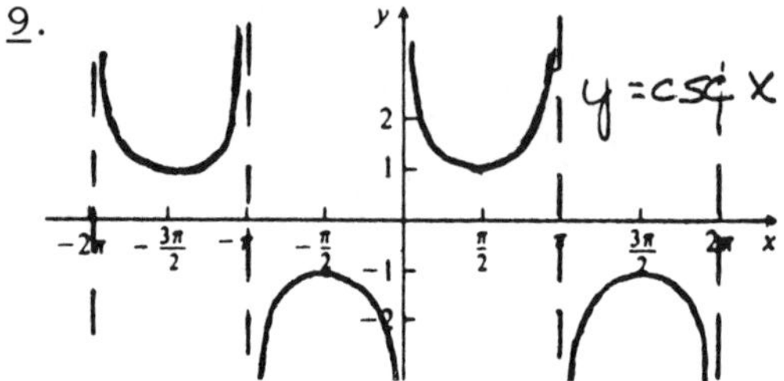

10. 2π 11. All reals except $k\pi$ 12. I, II 13. II, $\frac{2\pi}{3}$, 2.09

14. II, $\frac{41\pi}{60}$, 2.15 15. IV, $\frac{7\pi}{4}$, 5.50 16. 210° 17. 229.18° 18. 23.56 cm

19. $\frac{2}{5}$, 23° 20. 942 21. 1333 22. $\sin\theta = -\frac{5}{\sqrt{34}}$, or $-\frac{5\sqrt{34}}{34}$;

$\cos\theta = \frac{3}{\sqrt{34}}$, or $\frac{3\sqrt{34}}{34}$; $\tan\theta = -\frac{5}{3}$; $\cot\theta = -\frac{3}{5}$; $\sec\theta = \frac{\sqrt{34}}{3}$;

$\csc\theta = -\frac{\sqrt{34}}{5}$ 23. $-\frac{\sqrt{3}}{3}$ 24. $\frac{\sqrt{2}}{2}$ 25. $\frac{\sqrt{3}}{2}$ 26. $\sin\theta = -\frac{\sqrt{3}}{2}$;

$\tan\theta = -\sqrt{3}$; $\cot\theta = -\frac{\sqrt{3}}{3}$; $\sec\theta = 2$; $\csc\theta = -\frac{2\sqrt{3}}{3}$ 27. $57^\circ 27'$

28. 17.25° 29. 0.013889 30. 8.44993 31. -0.587785 32. 1.65399

33. -3.91480 34. 2.36620 35. 60.26°, 1.0517 36. 78.46°, 1.3694

37. 346.78° 38. 2.5068 39.

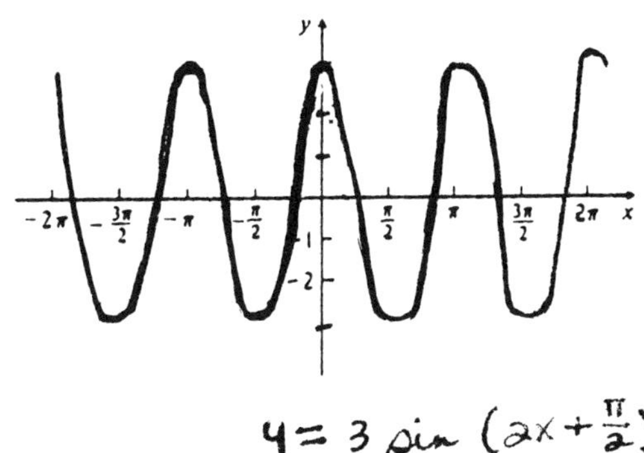

40. π 41. $-\frac{\pi}{4}$ 42.

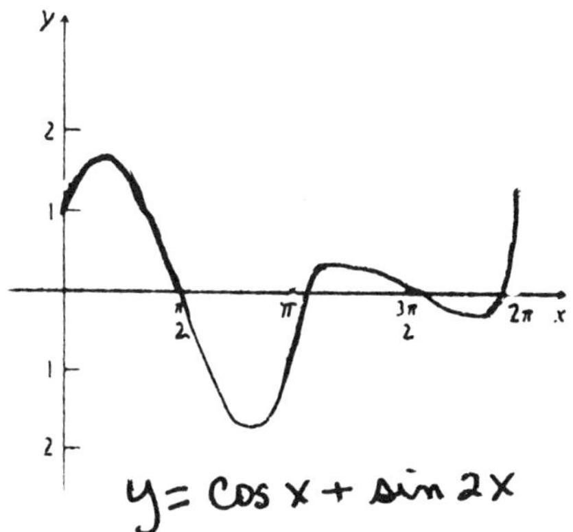

43. 83.33, 150

44. sin x = -0.9569, cos x = -0.2904

Chapter 6, Test Form C

1. Origin: $\left(-\frac{4}{7}, \frac{\sqrt{33}}{7}\right)$; u-axis: $\left(\frac{4}{7}, \frac{\sqrt{33}}{7}\right)$; v-axis: $\left(-\frac{4}{7}, -\frac{\sqrt{33}}{7}\right)$

2.

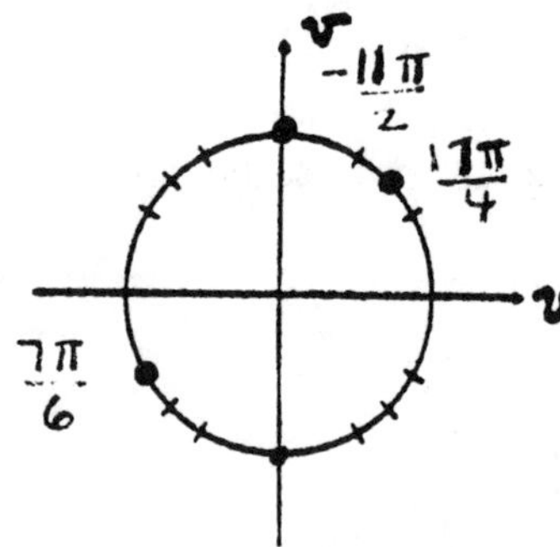

3. y = sin x

4. All reals from -1 to 1 inclusive 5. 2π 6. All reals 7. 1

8.

x	$\frac{\pi}{4}$	$\frac{7\pi}{6}$	$-\frac{5\pi}{2}$
sin x	$\frac{\sqrt{2}}{2}$	$-\frac{1}{2}$	-1
cos x	$\frac{\sqrt{2}}{2}$	$-\frac{\sqrt{3}}{2}$	0

9.

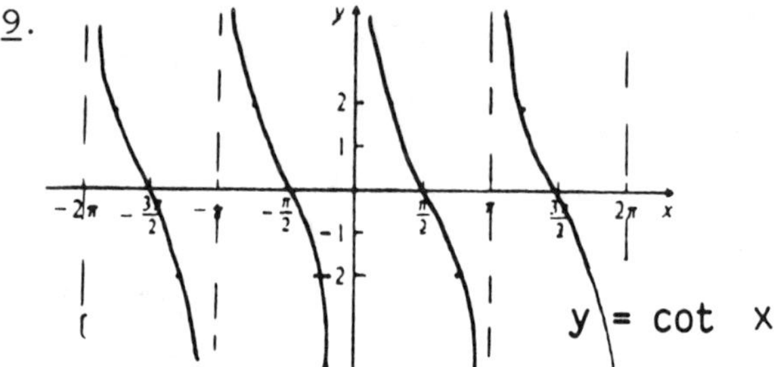

10. π 11. All reals except $k\pi$ 12. I, III 13. I, $-\frac{5\pi}{3}$, -5.24

14. III, $\frac{16\pi}{15}$, 3.35 15. I, $\frac{\pi}{4}$, 0.79 16. 300° 17. 286.48° 18. 16.76 cm

19. $\frac{4}{5}$, 46° 20. 2262 21. 61 22. $\sin\theta = \frac{3}{\sqrt{13}}$, or $\frac{3\sqrt{13}}{13}$;

$\cos\theta = -\frac{2}{\sqrt{13}}$, or $-\frac{2\sqrt{13}}{13}$; $\tan\theta = -\frac{3}{2}$; $\cot\theta = -\frac{2}{3}$; $\sec\theta = -\frac{\sqrt{13}}{2}$;

$\csc\theta = \frac{\sqrt{13}}{3}$ 23. 1 24. $\frac{1}{2}$ 25. $-\frac{\sqrt{2}}{2}$ 26. $\sin\theta = -\frac{3}{5}$; $\cos\theta = -\frac{4}{5}$;

$\cot\theta = \frac{4}{3}$; $\sec\theta = -\frac{5}{4}$; $\csc\theta = -\frac{5}{3}$ 27. $101^\circ 45'$ 28. 124.9°

29. 4.91796 30. -0.258244 31. 0.707107 32. 0.080751 33. -1.00238

34. -5.75877 35. 19.81°, 0.3457 36. 83.64°, 1.4598 37. 254.36°

38. 3.3788 39.

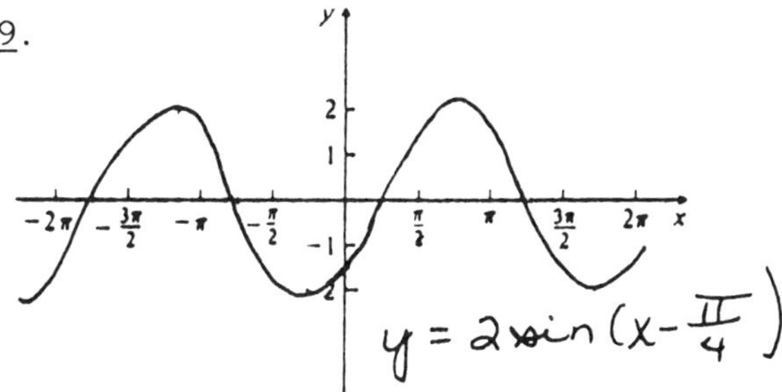

40. 2π 41. $\frac{\pi}{4}$ 42.

43. $\pm \frac{2\sqrt{6}}{5}$

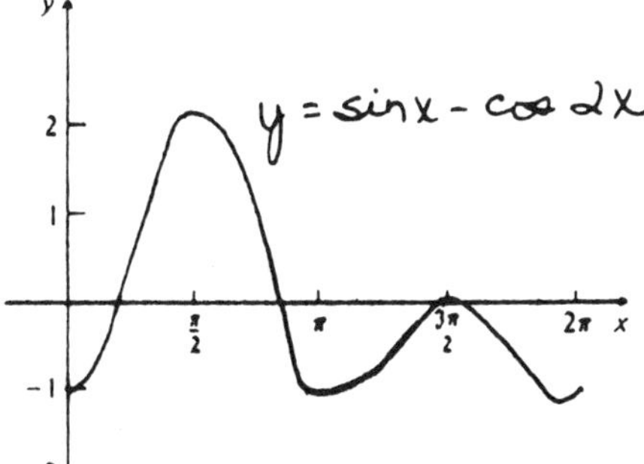

44. a) 0.98079; b) 0.98079

Chapter 6, Test Form D

1. Origin: $\left(-\frac{\sqrt{5}}{3}, -\frac{2}{3}\right)$; u-axis: $\left(\frac{\sqrt{5}}{3}, -\frac{2}{3}\right)$; v-axis: $\left(-\frac{\sqrt{5}}{3}, \frac{2}{3}\right)$

2.

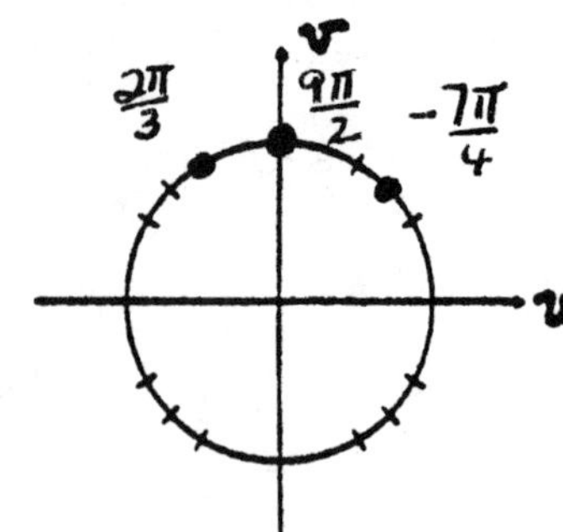

3.

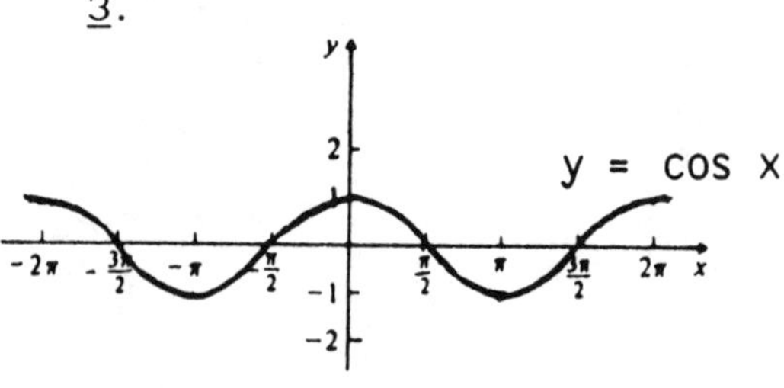

4. All reals from -1 to 1 inclusive 5. 2π 6. All reals 7. 1

8.

x	$\frac{2\pi}{3}$	$\frac{7\pi}{2}$	$-\frac{5\pi}{4}$
sin x	$\frac{\sqrt{3}}{2}$	-1	$\frac{\sqrt{2}}{2}$
cos x	$-\frac{1}{2}$	0	$-\frac{\sqrt{2}}{2}$

9.

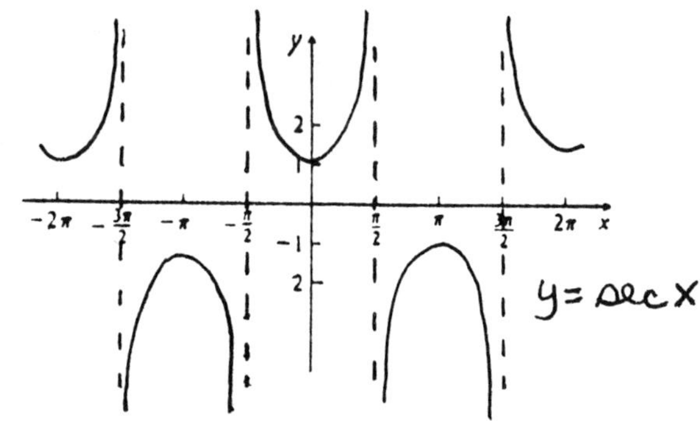

10. 2π 11. All reals except $\frac{\pi}{2} + k\pi$ 12. III, IV 13. III, $\frac{4\pi}{3}$, 4.19

14. III, $\frac{67\pi}{60}$, 3.51 15. I, $\frac{29\pi}{60}$, 1.52 16. 270° 17. 343.77° 18. 70.69 cm

19. $\frac{3}{7}$, 25° 20. 503 21. 267 22. $\sin\theta = -\frac{6}{\sqrt{61}}$, or $\frac{-6\sqrt{61}}{61}$;

$\cos\theta = -\frac{5}{\sqrt{61}}$, or $\frac{-5\sqrt{61}}{61}$; $\tan\theta = \frac{6}{5}$; $\cot\theta = \frac{5}{6}$; $\sec\theta = -\frac{\sqrt{61}}{5}$;

$\csc\theta = -\frac{\sqrt{61}}{6}$ 23. 0 24. $-\frac{\sqrt{2}}{2}$ 25. $\frac{\sqrt{3}}{2}$ 26. $\sin\theta = -\frac{\sqrt{5}}{3}$;

$\tan\theta = -\frac{\sqrt{5}}{2}$; $\cot\theta = -\frac{2\sqrt{5}}{5}$; $\sec\theta = \frac{3}{2}$; $\csc\theta = -\frac{3\sqrt{5}}{5}$ 27. $89^\circ 3'$

28. 153.4° 29. 0.826104 30. 14.5127 31. -0.923880 32. -5.59277

33. 1.33387 34. -2.45859 35. 77.68°, 1.3558 36. 77.32°, 1.3494

37. 22.59° 38. 4.6038 39.

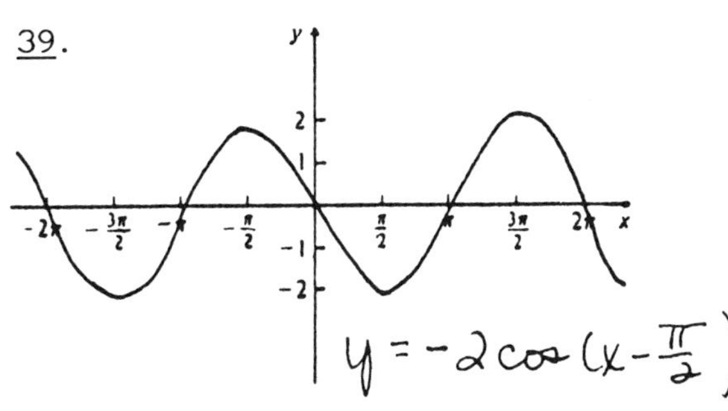

40. 2π 41. $\frac{\pi}{2}$ 42.

43. 133.33, 33.33

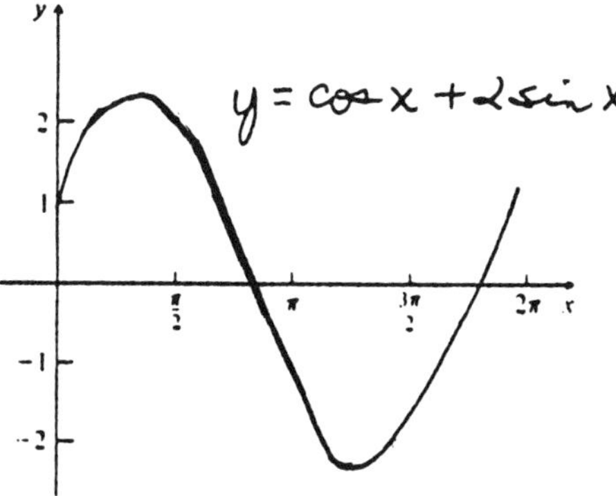

44. $\cos x = -0.3488$, $\tan x = -2.6870$

Chapter 6, Test Form E

1. $-\frac{4\pi}{3}, \frac{2\pi}{3}$ 2. $\left(\frac{5}{8}, \frac{\sqrt{39}}{8}\right)$

3.

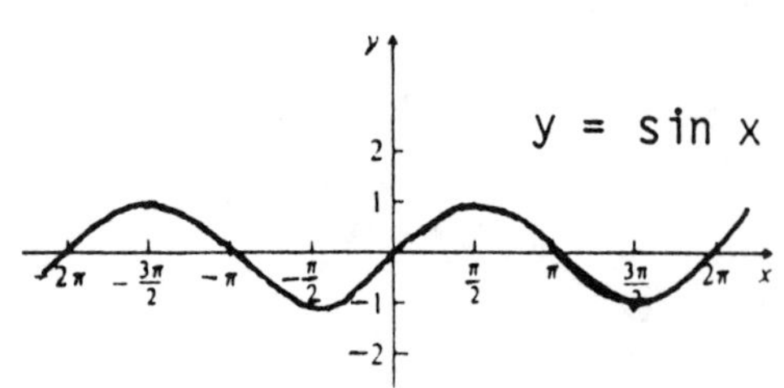

4.

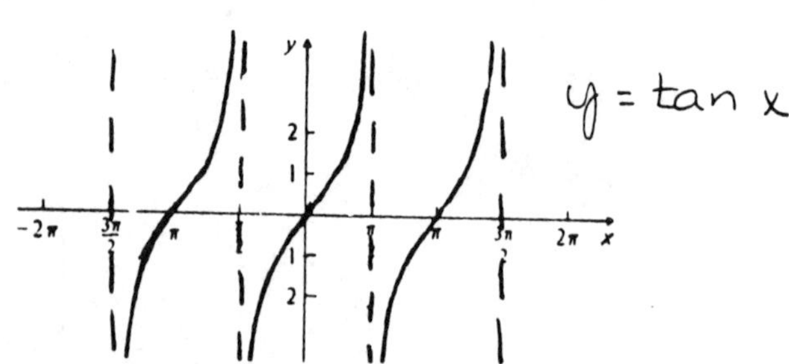

5. All reals   6. All reals from −1 to 1 inclusive   7. 2π 8. 1

9. All real numbers except $k\pi$ 10. π 11. All real numbers 1 and greater, and all real numbers −1 and less   12. All real numbers except $\frac{\pi}{2} + k\pi$

13.

x	$\frac{2\pi}{3}$	$\frac{-9\pi}{4}$	$\frac{5\pi}{6}$
sin x	$\frac{\sqrt{3}}{2}$	$-\frac{\sqrt{2}}{2}$	$\frac{1}{2}$
tan x	$-\sqrt{3}$	−1	$-\frac{\sqrt{3}}{3}$
sec x	−2	$\sqrt{2}$	$\frac{-2\sqrt{3}}{3}$
csc x	$\frac{2\sqrt{3}}{3}$	$-\sqrt{2}$	2

14. II, $-\frac{19\pi}{6}$ 15. $390°$ 16. $458°$

17. 3200   18. $\frac{3\pi}{4}$, or 2.36

19. $\sin\theta = -\frac{1}{\sqrt{26}}$, or $-\frac{\sqrt{26}}{26}$;

$\cos\theta = -\frac{5}{\sqrt{26}}$, or $-\frac{5\sqrt{26}}{26}$; $\tan\theta = \frac{1}{5}$;

$\cot\theta = 5$; $\sec\theta = -\frac{\sqrt{26}}{5}$; $\csc\theta = -\sqrt{26}$

20. $\frac{\sqrt{3}}{2}$ 21. −1   22. $\frac{\sqrt{3}}{2}$ 23. 1   24. a) $\frac{\sqrt{10}}{3}$, b) $-\frac{\sqrt{10}}{10}$, c) $-\frac{1}{3}$

25. $54°15'$ 26. $93.15°$ 27. a) 0.3249, b) 0.4266, c) 1.2725, d) 0.2483, e) −1.1076, f) 0.2349   28. $41.20°$, 0.72   29. $308.61°$

30. 

31. $y = 1 + \cos(2x - \pi)$

32. $\frac{\pi}{2}$

33.

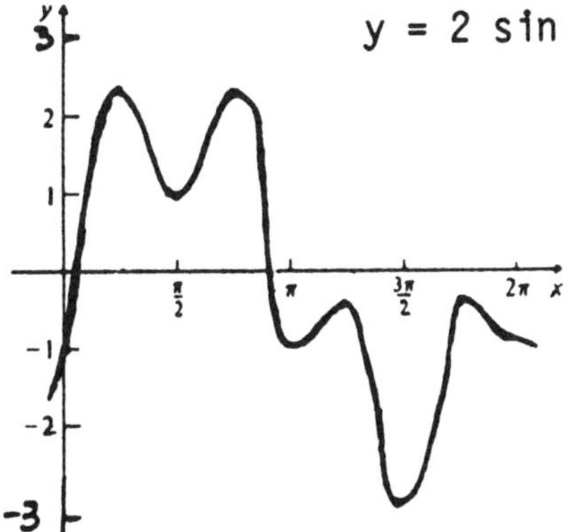

Chapter 6, Test Form F

1. b 2. b 3. d 4. b 5. b 6. a 7. b 8. b 9. d 10. c 11. c

12. d 13. b 14. d 15. c 16. d 17. a 18. a 19. c 20. b 21. a

22. b 23. a 24. b 25. d 26. a 27. d 28. a 29. c 30. d 31. b

32. c 33. b 34. a 35. d 36. c 37. b

Chapter 7, Test Form A

1. $\cos \pi \cos x - \sin \pi \sin x$ 2. $\dfrac{\tan 19^\circ + \tan 35^\circ}{1 - \tan 19^\circ \tan 35^\circ}$ 3. $\sin 70^\circ$

4. $\dfrac{\sqrt{6} - \sqrt{2}}{4}$ 5. $-2 + \sqrt{3}$ 6. $-\csc x$ 7. 1 8. $\sec^2 x$ 9. $-\cos x$

10. $\sec x$ 11. $\tan x$ 12. $\tan x = \pm\sqrt{\sec^2 x - 1}$ 13. $\csc x$ 14. -1

15. $\dfrac{\sqrt{\tan x \cos x}}{\cos x}$ 16. $\sin 2\theta = -\dfrac{24}{25}$; $\cos 2\theta = \dfrac{7}{25}$; $\tan 2\theta = -\dfrac{24}{7}$;

quadrant IV 17. $\dfrac{\sqrt{2 + \sqrt{3}}}{2}$ 18. $\csc 2\theta$ 19.

$\sec^2 \theta + \csc^2 \theta$	$\sec^2 \theta \csc^2 \theta$
$\dfrac{1}{\cos^2 \theta} + \dfrac{1}{\sin^2 \theta}$	$\dfrac{1}{\cos^2 \theta} \dfrac{1}{\sin^{-2} \theta}$
$\dfrac{\sin^2 \theta + \cos^2 \theta}{\cos^2 \theta \sin^2 \theta}$	$\dfrac{1}{\cos^2 \theta \sin^{-2} \theta}$
$\dfrac{1}{\cos^2 \theta \sin^2 \theta}$	

20. $\dfrac{\pi}{3}$ 21. 37.86° 22. $\dfrac{\pi}{3}$ 23. $\dfrac{\pi}{3}$ 24. $\dfrac{1}{2}$

25. -1 26. $\dfrac{2\pi}{3}, \dfrac{4\pi}{3}$

27. $0, \dfrac{\pi}{6}, \dfrac{5\pi}{6}, \pi, \dfrac{7\pi}{6}, \dfrac{11\pi}{6}$

28. 2.1588, 2.3562, 5.3004, 5.4978

29. $\sin(90 - \theta) = 0.6691$, $\cos(90 - \theta) = 0.7431$, $\tan(90 - \theta) = 0.9004$,

$\cot(90 - \theta) = 1.111$, $\sec(90 - \theta) = 1.346$, $\csc(90 - \theta) = 1.495$

30. $\sin\left(x + \dfrac{5\pi}{2}\right) = \sin x \cos \dfrac{5\pi}{2} + \cos x \sin \dfrac{5\pi}{2} = \sin x\ (0) + \cos x\ (1) = \cos x$

Chapter 7, Test Form B

1. $\sin \pi \cos x - \cos \pi \sin x$ 2. $\dfrac{\tan 27^\circ - \tan 34^\circ}{1 + \tan 27^\circ \tan 34^\circ}$ 3. $\cos 90^\circ$, or 0

4. $\dfrac{\sqrt{6} + \sqrt{2}}{4}$ 5. $2 - \sqrt{3}$ 6. $\csc x$ 7. 1 8. $\csc^2 x$ 9. $\cos x$ 10. $\tan x$

11. $-\sin x$ 12. $\cot x = \pm\sqrt{\csc^2 x - 1}$ 13. $\csc x$ 14. $\csc^2 x$

15. $\dfrac{\sqrt{\sin x \tan x}}{\tan x}$ 16. $\sin 2\theta = -\dfrac{24}{25}$; $\cos 2\theta = \dfrac{7}{25}$; $\tan 2\theta = -\dfrac{24}{7}$; quadrant IV

17. $\dfrac{\sqrt{2 - \sqrt{3}}}{2}$ 18. $\cos 2\theta$ 19.

$(\tan \theta)(1 + \cot^2 \theta)$	$(\cot \theta)(1 + \tan^2 \theta$
$\tan \theta \csc^2 \theta$	$\cot \theta \sec^2 \theta$
$\dfrac{\tan \theta}{\sin^2 \theta}$	$\dfrac{\cot \theta}{\cos^2 \theta}$
$\dfrac{1}{\sin \theta \cos \theta}$	$\dfrac{1}{\sin \theta \cos \theta}$

20. $-\dfrac{\pi}{6}$, or $\dfrac{5\pi}{6}$ 21. 82.89°

22. $\dfrac{\pi}{4}$ 23. $-\dfrac{\pi}{2}$ 24. $-\dfrac{\sqrt{3}}{3}$

25. $\dfrac{\pi}{6}$ 26. $\dfrac{\pi}{2}, \dfrac{7\pi}{6}, \dfrac{11\pi}{6}$

27. $\dfrac{\pi}{3}, \dfrac{\pi}{2}, \dfrac{2\pi}{3}, \dfrac{4\pi}{3}, \dfrac{3\pi}{2}, \dfrac{5\pi}{3}$

28. 1.7992, 2.5327, 4.9407, 5.6743 29. $\sin (90 - \theta) = 0.5592$, $\cos (90 - \theta) = 0.8290$, $\tan (90 - \theta) = 0.6745$, $\cot (90 - \theta) = 1.483$, $\sec (90 - \theta) = 1.206$, $\csc (90 - \theta) = 1.788$ 30. $\sin^2 x$

Chapter 7, Test Form C

1. $\sin \pi \cos x + \cos \pi \sin x$ 2. $\dfrac{\tan 37^\circ + \tan 45^\circ}{1 - \tan 37^\circ \tan 45^\circ}$ 3. $\sin 55^\circ$

4. $\dfrac{\sqrt{6} + \sqrt{2}}{4}$ 5. $2 - \sqrt{3}$ 6. $\sec x$ 7. 1 8. $\sec^2 x$ 9. $\cot x$ 10. $\cos x$

11. $\csc x$ 12. $\sec x = \pm\sqrt{1 + \tan^2 x}$ 13. $\sin x$ 14. -1 15. $\dfrac{\sqrt{\sec x \cot x}}{\cot x}$

16. $\sin 2\theta = -\dfrac{120}{169}$; $\cos 2\theta = \dfrac{119}{169}$; $\tan 2\theta = -\dfrac{120}{119}$; quadrant IV 17. $\sqrt{2} + 1$

18. $\sin 4y$ 19.

$\sin 2x$	$\dfrac{2 \cot x}{\csc^2 x}$
$2 \sin x \cos x$	$2 \cot x \sin^2 x$
	$2 \sin x \cos x$

20. $-\dfrac{\pi}{2}$ 21. 36.21° 22. $\dfrac{\pi}{3}$

23. $\dfrac{\pi}{3}$ 24. $\dfrac{\sqrt{2}}{2}$ 25. $-\dfrac{\sqrt{3}}{3}$

26. $\dfrac{\pi}{3}, \dfrac{5\pi}{3}$

Chapter 7, Test Form C (continued)

27. $0, \frac{\pi}{4}, \frac{3\pi}{4}, \pi, \frac{5\pi}{4}, \frac{7\pi}{4}$ 28. 0.3649, 1.2059, 3.5065, 4.3475

29. $\sin(90 - \theta) = 0.2756$, $\cos(90 - \theta) = 0.9613$, $\tan(90 - \theta) = 0.2867$,

$\cot(90 - \theta) = 3.487$, $\sec(90 - \theta) = 1.040$, $\csc(90 - \theta) = 3.628$

30. $\sin\theta = \sqrt{\frac{1}{2} + \frac{\sqrt{5}}{6}}$, $\cos\theta = \sqrt{\frac{1}{2} - \frac{\sqrt{5}}{6}}$, $\tan\theta = \sqrt{\frac{3 + \sqrt{5}}{3 - \sqrt{5}}}$

Chapter 7, Test Form D

1. $\cos\frac{\pi}{2}\cos x + \sin\frac{\pi}{2}\sin x$ 2. $\frac{\tan 92^\circ - \tan 18^\circ}{1 + \tan 92^\circ \tan 18^\circ}$ 3. $\cos 77^\circ$

4. $\frac{\sqrt{6} - \sqrt{2}}{4}$ 5. $\frac{\sqrt{3}}{3}$ 6. $\sec x$ 7. 1 8. $\csc^2 x$ 9. $\csc x$ 10. $\cot x$

11. $\sin x$ 12. $\cos x = \pm\sqrt{1 - \sin^2 x}$ 13. $\cot x$ 14. $\sec^2 x$

15. $\frac{\sqrt{\csc x \tan x}}{\tan x}$ 16. $\sin 2\theta = -\frac{240}{289}$, $\cos 2\theta = \frac{161}{289}$, $\tan 2\theta = -\frac{240}{161}$,

quadrant IV 17. $-\frac{\sqrt{2 + \sqrt{2}}}{2}$ 18. $\frac{1}{2}\sin 4x$ 19.

$\sin x$	$\frac{\tan x}{\sqrt{1 + \tan^2 x}}$
	$\frac{\tan x}{\sec x}$
	$\tan x \cos x$
	$\sin x$

20. $\frac{\pi}{6}$ 21. 69.78° 22. $\frac{\pi}{6}$ 23. $\frac{\pi}{3}$ 24. $\frac{\sqrt{3}}{2}$

25. $\frac{2\pi}{3}$ 26. $\frac{\pi}{6}, \frac{5\pi}{6}, \frac{3\pi}{2}$ 27. $0, \frac{\pi}{4}, \frac{3\pi}{4}, \pi, \frac{5\pi}{4}, \frac{7\pi}{4}$

28. 0.5299, 1.2859, 3.6715, 4.4275

29. $\sin(90 - \theta) = 0.7771$, $\cos(90 - \theta) = 0.6293$, $\tan(90 - \theta) = 1.235$,

$\cot(90 - \theta) = 0.8098$, $\sec(90 - \theta) = 1.589$, $\csc(90 - \theta) = 1.287$

30. $30^\circ, 150^\circ, 210^\circ, 330^\circ$

Chapter 7, Test Form E

1. $\dfrac{\tan 75^\circ - \tan 13^\circ}{1 - \tan 75^\circ \tan 13^\circ}$ 2. $\cos(x - \pi)$ 3. $-\dfrac{\sqrt{2} + \sqrt{6}}{4}$ 4. $\sqrt{7 + 4\sqrt{3}}$

5. $\dfrac{-\sqrt{10} - 2\sqrt{2}}{6}$ 6. $\sec \theta$ 7. $\csc^2 \theta$ 8. $\cos \theta$ 9. $\sec^2 \theta$ 10. $\sin \theta$

11. 1   12. $\cos \theta$ 13. 1   14. $\cot x$ 15. 0   16. $\dfrac{\sqrt{\csc x \tan x}}{\tan x}$

17. $\sin 2\theta = -\dfrac{6\sqrt{14}}{23}$, $\cos 2\theta = -\dfrac{5}{23}$, $\tan 2\theta = \dfrac{6\sqrt{14}}{5}$, quadrant III   18. $\cos \theta$

19. $2 \tan \theta$ 20.

$\dfrac{2\cos^2\theta - 1}{\sin\theta\cos\theta}$	$\cot\theta - \tan\theta$
$\dfrac{\cos 2\theta}{\sin\theta\cos\theta}$	
$\dfrac{\cos^2\theta - \sin^2\theta}{\sin\theta\cos\theta}$	
$\dfrac{\cos\theta}{\sin\theta} - \dfrac{\sin\theta}{\cos\theta}$	
$\cot\theta - \tan\theta$	

21. 52.98° 22. 16.35°

23. $-\dfrac{\pi}{3}$ 24. $\dfrac{\pi}{4}$ 25. 1

26. $\dfrac{\pi}{4}$ 27. $\dfrac{\pi}{6}, \dfrac{\pi}{3}, \dfrac{2\pi}{3}, \dfrac{5\pi}{6}, \dfrac{7\pi}{6}, \dfrac{4\pi}{3}, \dfrac{5\pi}{3}, \dfrac{11\pi}{6}$

28. $0, \dfrac{\pi}{6}, \dfrac{5\pi}{6}, \pi, \dfrac{7\pi}{6}, \dfrac{11\pi}{6}$

29. 20.2°, 200.2°

Chapter 7, Test Form F

1. d   2. b   3. b   4. a   5. d   6. c   7. b   8. a   9. c   10. b   11. d

12. a   13. c   14. a   15. d   16. b   17. d   18. a   19. d   20. b   21. d

22. b   23.

$\csc 2\theta + \cot 2\theta$	$\cot\theta$
$\dfrac{1}{\sin 2\theta} + \dfrac{\cos 2\theta}{\sin 2\theta}$	
$\dfrac{1 + \cos 2\theta}{\sin 2\theta}$	
$\dfrac{1 + 2\cos^2\theta - 1}{\sin 2\theta}$	
$\dfrac{2\cos^2\theta}{2\sin\theta\cos\theta}$	
$\dfrac{\cos\theta}{\sin\theta}$	
$\cot\theta$	

Chapter 8, Test Form A

1. $B = 50.4^\circ$, $a = 17.8$, $c = 27.9$ 2. $A = 49.6^\circ$, $B = 40.4^\circ$, $b = 14$ 3. 88 m

4. $B = 101^\circ$, $b = 6.13$, $c = 2.34$ 5. 10   6. 185 cm   7. 18.3 m^2

8. 16.1 newtons, 60° 9. $\left(-\dfrac{15\sqrt{3}}{2}, -\dfrac{15}{2}\right)$ 10. $(10, 53^\circ)$ 11. (a) $\langle -1, -10 \rangle$;

(b) 15.3   12. (a) N. 5, E. 4; (b) 6.4, N 39° E   13. $\left(\dfrac{7\sqrt{3}}{2}, -\dfrac{7}{2}\right)$

14. $x^2 + y^2 = 81$ 15. $\sqrt{3}$ cis 45°, $\sqrt{3}$ cis 225°; or $\dfrac{\sqrt{6}}{2} + \dfrac{\sqrt{6}}{2}i$, $-\dfrac{\sqrt{6}}{2} - \dfrac{\sqrt{6}}{2}i$

16.

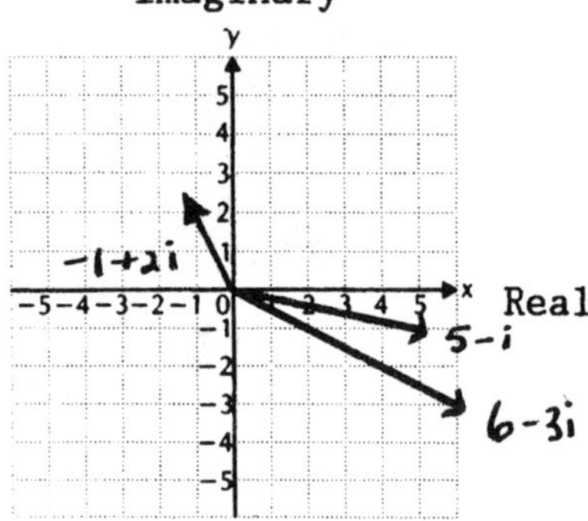

17. $\sqrt{2} + \sqrt{2}i$ 18. $\sqrt{2}$ cis 135°

19. 10 cis 70° 20. $\dfrac{4}{3}$ cis 65°

21. $4\sqrt{2}$ cis 135° 22. 1 cis 90°, 1 cis 210°, 1 cis 330°; or i, $-\dfrac{\sqrt{3}}{2} - \dfrac{1}{2}i$, $\dfrac{\sqrt{3}}{2} - \dfrac{1}{2}i$

23.

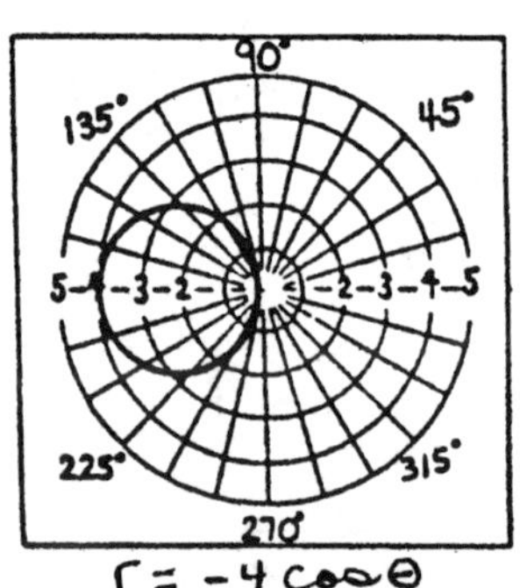

24.

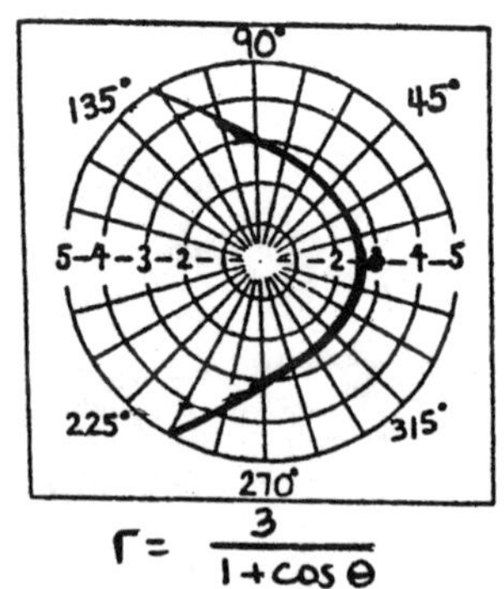

25. 4.5

1. $A = 70.5^\circ$, $b = 11.3$, $c = 33.9$ 2. $A = 56.3^\circ$, $B = 33.7^\circ$, $c = 14$ 3. 37°

4. $C = 37^\circ$, $b = 8.1$, $c = 6.3$ 5. 12 6. 3 and 5 7. 14.9 cm^2

8. 18 newtons, 34° 9. $\left(-9\sqrt{2}, 9\sqrt{2}\right)$ 10. $(17.5, 301^\circ)$ 11. (a) $<-2,-1>$;

(b) 18.4 12. (a) N. 1, E. 12; (b) 12.0, N 85° E 13. $\left(-\sqrt{2}, -\sqrt{2}\right)$

14. $x^2 + y^2 = 121$ 15. $2\sqrt{3}$ cis 45°, $2\sqrt{3}$ cis 225°; or $\sqrt{6} + \sqrt{6}i$,

$-\sqrt{6} - \sqrt{6}i$

16. Imaginary

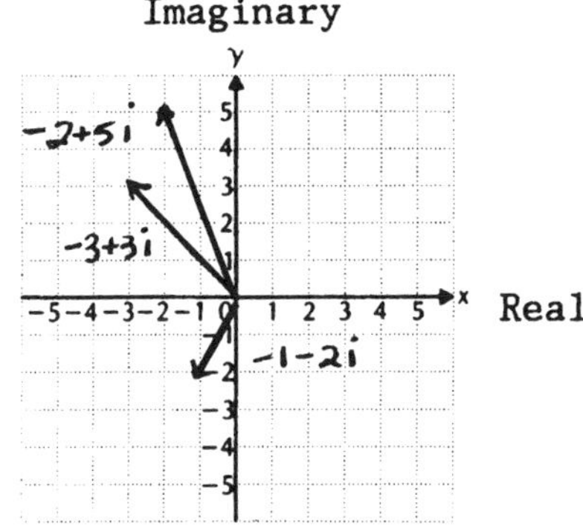

17. $\frac{5}{2} - \frac{5\sqrt{3}}{2}i$ 18. 2 cis 330°

19. 32 cis 35° 20. 3 cis 75°

21. 4096 cis 0° 22. 1 cis 0°,

1 cis 120°, 1 cis 240°; or 1, $-\frac{1}{2} + \frac{\sqrt{3}}{2}i$,

$-\frac{1}{2} - \frac{\sqrt{3}}{2}i$

23.

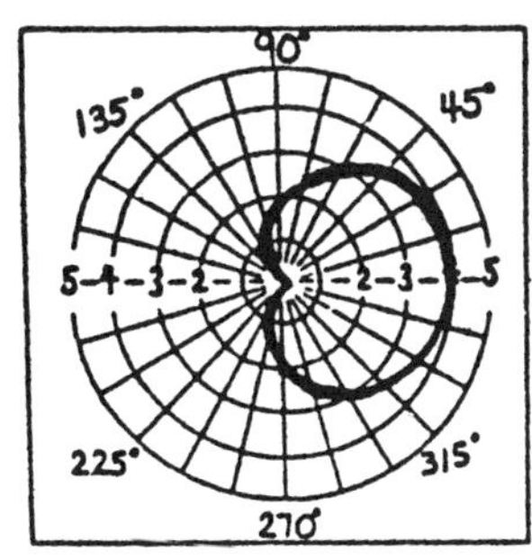

$r = 2 + 2\cos\theta$

24.

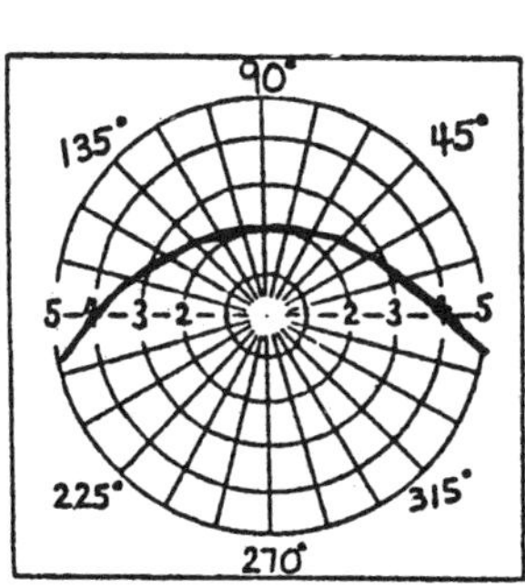

$r = \frac{4}{1 + \sin\theta}$

25. 29.23°, 150.77°

Chapter 8, Test Form C

1. $B = 28.6^\circ$, $a = 10.1$, $b = 5.5$ 2. $A = 61.7^\circ$, $B = 28.3^\circ$, $a = 17$ 3. 22°

4. $C = 61^\circ$, $a = 12.5$, $c = 12.1$ 5. 13 6. 67 cm 7. 17.4 in.^{2}

8. 17 newtons, 62° 9. $\left(-5, 5\sqrt{3}\right)$ 10. $(8.1, 120^\circ)$ 11. (a) $\langle -9, 8 \rangle$; (b) 7.6

12. (a) S. 4, W. 9; (b) 9.8, S 66° W 13. $\left(-2, -2\sqrt{3}\right)$ 14. $x^2 + y^2 = 144$

15. 3 cis 45°, 3 cis 225°; or $\frac{3\sqrt{2}}{2} + \frac{3\sqrt{2}}{2}i$, $-\frac{3\sqrt{2}}{2} - \frac{3\sqrt{2}}{2}i$

16.

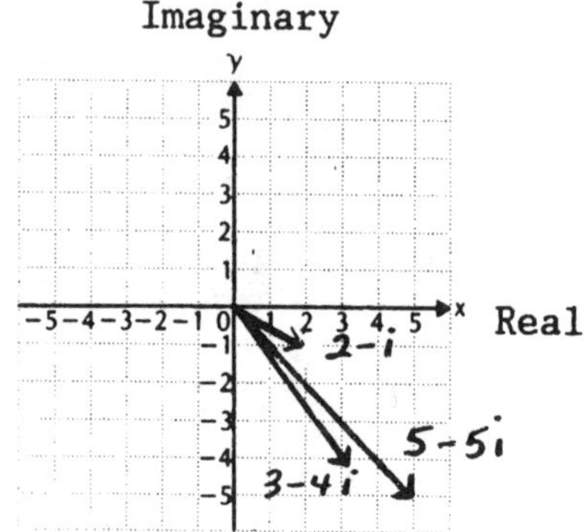

17. $-3\sqrt{3} - 3i$ 18. $\sqrt{2}$ cis 225°

19. 21 cis 47° 20. 2 cis 110°

21. 32 cis 210° 22. 3 cis 60°, 3 cis 180°, 3 cis 300°; or $\frac{3}{2} + \frac{3\sqrt{3}}{2}i$, -3, $\frac{3}{2} - \frac{3\sqrt{3}}{2}i$

23.

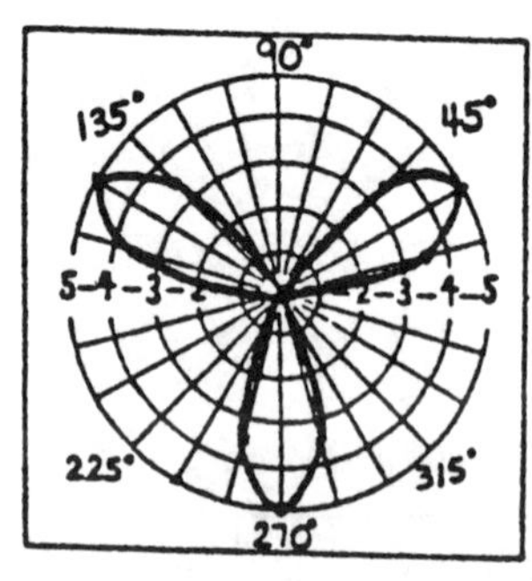

$r = 5 \sin 3\theta$

24.

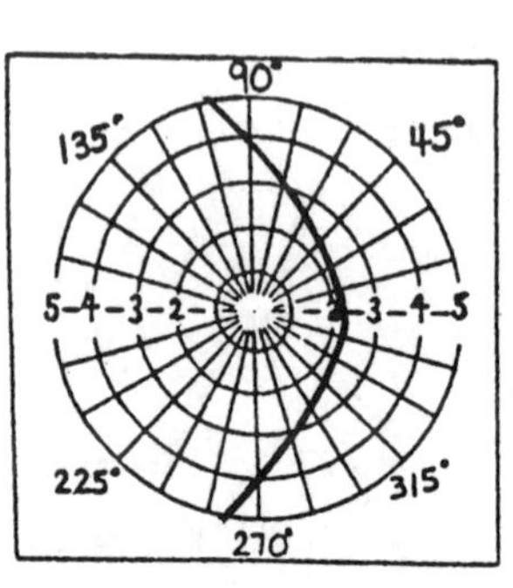

$r = \frac{4}{1 + \cos\theta}$

25. $\langle \frac{8}{5}, \frac{6}{5} \rangle$

Chapter 8, Test Form D

1. $A = 31.8°$, $a = 11.3$, $b = 18.2$ 2. $A = 20.6°$, $B = 69.4°$, $c = 9$ 3. 308 ft

4. $A = 52°$, $a = 29.3$, $b = 37.1$ 5. 21 6. 6.8 ft, 7.8 ft 7. 150.3 ft^2

8. 24.4 newtons, 35° 9. $\left(\frac{11\sqrt{2}}{2}, -\frac{11\sqrt{2}}{2}\right)$ 10. (12.5, 241°) 11. (a) <2,-8>;

(b) 10 12. (a) N. 9, E. 3; (b) 9.5, N 18° E 13. $\left(\frac{5}{2}, \frac{5\sqrt{3}}{2}\right)$

14. $x^2 + y^2 = 169$ 15. $\sqrt{15}$ cis 45°, $\sqrt{15}$ cis 225°;

or $\frac{\sqrt{30}}{2} + \frac{\sqrt{30}}{2}i$, $-\frac{\sqrt{30}}{2} - \frac{\sqrt{30}}{2}i$

16.

Imaginary

-4+5i

-1+3i

-3+2i

Real

17. $2\sqrt{2} - 2\sqrt{2}i$ 18. 16 cis 30°

19. 20 cis 33° 20. $\frac{5}{3}$ cis 65°

21. 32 cis 150° 22. 2 cis 90°,

2 cis 210°, 2 cis 330°; or 2i,

$-\sqrt{3} - i$, $\sqrt{3} - i$

23.

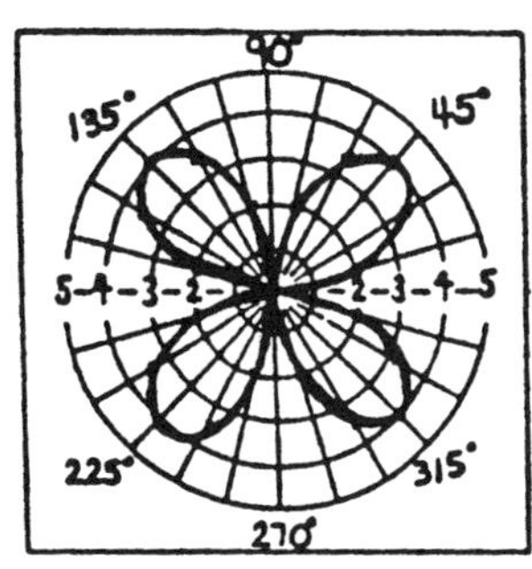

$r = 4 \sin 2\theta$

24.

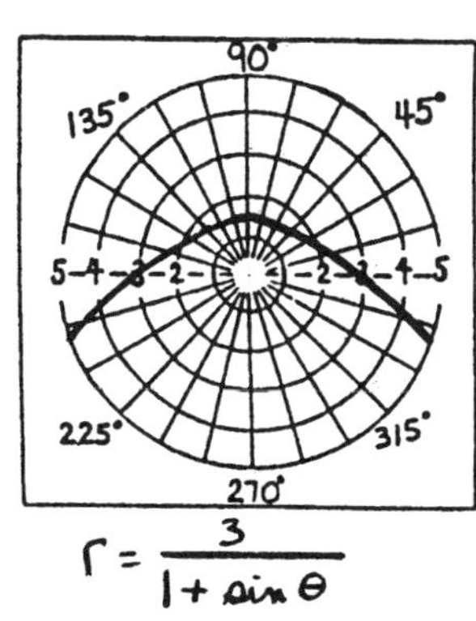

$r = \frac{3}{1 + \sin\theta}$

25. 46.06 m^2

1. $A = 74.8^\circ$, $b = 9.3$, $c = 35.4$ 2. $A = 62.5^\circ$, $B = 27.5^\circ$, $c = 8.5$

3. 102.5 ft   4. $A = 51.8^\circ$, $a = 0.461$, $b = 0.528$ 5. 29° 6. 13.2 m

7. 69.2 newtons, 36° 8. $(15.3, 101.3^\circ)$ 9. $\langle -7.4, 22.8 \rangle$

10. (a) $\sqrt{5} + 3\sqrt{34}$; (b) $\langle -1, 1 \rangle$ 11. 16.52 m^2 12. $\left(7\sqrt{2}, 315^\circ\right)$

13. $r^2 = 400$, $r = \pm 20$ 14. $x^2 + 10y = 25$ 15. $2 \text{ cis } 30^\circ$, $2 \text{ cis } 210^\circ$; or $\sqrt{3} + i$, $-\sqrt{3} - i$

16.

Imaginary

1+4i   5+6i   -4-2i

Real

17. -41

18. $48 \text{ cis } 220^\circ$

19. $\sqrt{3} \text{ cis } 300^\circ$

20. $32 \text{ cis } 210^\circ$

21. 2 cis 0, 2 cis 60, 2 cis 120, 2 cis 180, 2 cis 240, 2 cis 300; or 2, $1 + \sqrt{3}i$, $-1 + \sqrt{3}i$, -2, $-1 - \sqrt{3}i$, $1 - \sqrt{3}i$

22.

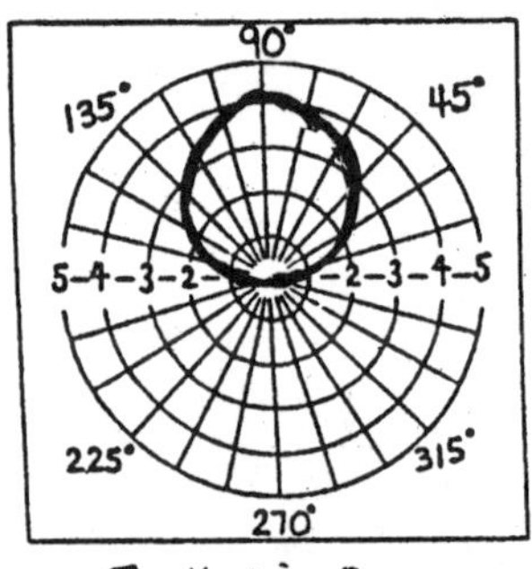

$r = 4 \sin \theta$

23.

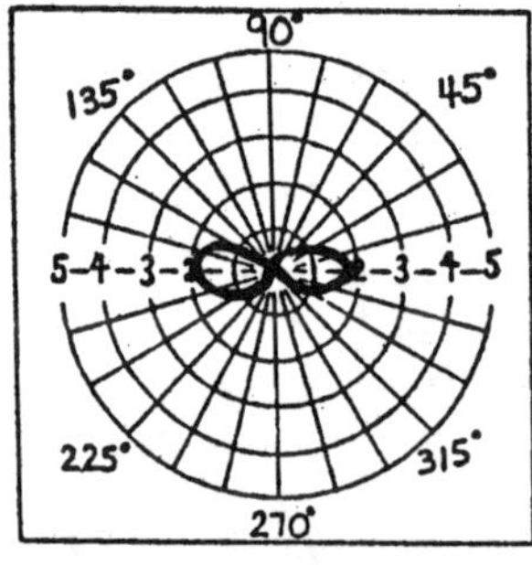

$r^2 = 4 \cos 2\theta$

Chapter 8, Test Form F

1. c 2. d 3. b 4. a 5. d 6. b 7. c 8. a 9. b 10. d 11. b

12. c 13. b 14. b 15. d 16. c 17. a 18. d 19. b 20. a 21. c

22. d 23. d 24. b

Chapter 9, Test Form A

1. (4,1) 2. Boat: 21 km/h; stream: 4 km/h 3. 21 black, 14 red 4. 5, 9, 15

5. $\left(\frac{1}{2}, -\frac{1}{4}\right)$ 6. $\left(\frac{2 + 5y}{4}, y, \frac{3y - 6}{4}\right)$, (3,2,0), (-2,-2,-3), (8,6,3), etc. Answers may vary. 7. (0,1,3) 8. Consistent, independent 9. Consistent, dependent

10. $f(x) = -x^2 + 2x + 3$ 11. 11 12. 27 13. (3,-2) 14. (1,-2,3)

15. Not possible 16. A or $\begin{bmatrix} -1 & 4 \\ 3 & 2 \end{bmatrix}$ 17. $\begin{bmatrix} 0 & 4 \\ 3 & 3 \end{bmatrix}$ 18. $\begin{bmatrix} 4 & 1 & -1 \\ 4 & 2 & 9 \\ 5 & 4 & 7 \end{bmatrix}$

19. $\begin{bmatrix} 0 & -1 \\ 1 & 5 \end{bmatrix}$ 20. $\begin{bmatrix} 5 & 8 \\ 23 & 18 \end{bmatrix}$ 21. Not possible 22. $\begin{bmatrix} -1 & 0 & 4 \end{bmatrix}$

23. $\begin{bmatrix} -1 & 5 & -8 \\ 8 & 1 & 6 \\ -5 & 2 & -4 \end{bmatrix}$ 24. $\begin{bmatrix} -1 & 5 \\ 2 & -3 \end{bmatrix}$ 25. $\begin{bmatrix} \frac{1}{5} & -\frac{2}{5} \\ \frac{1}{5} & \frac{3}{5} \end{bmatrix}$ 26. Does not exist

27. $\begin{bmatrix} -\frac{5}{3} & 1 & \frac{1}{3} \\ -\frac{13}{3} & 2 & \frac{2}{3} \\ -\frac{11}{3} & 2 & \frac{1}{3} \end{bmatrix}$ 28. $a_{12} = 2$; $M_{12} = \begin{vmatrix} 3 & 2 \\ -1 & 0 \end{vmatrix} = 2$; $A_{12} = -2$ 29. 10 30. 0

31. -8 32. 28 33. $(a - b)(c - b)(c - a)$ 34. $\begin{bmatrix} 3 & -2 \\ 1 & 3 \end{bmatrix}\begin{bmatrix} x \\ y \end{bmatrix} = \begin{bmatrix} -11 \\ 11 \end{bmatrix}$; (-1,4)

Chapter 9, Test Form A (continued)

35.

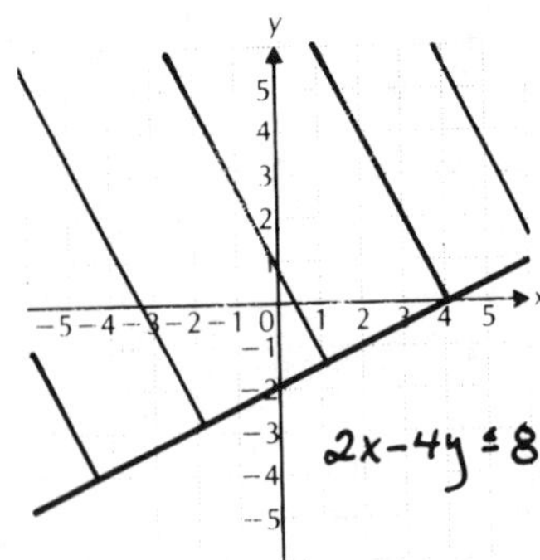

36. Minimum 20 at (2,0); maximum $58\frac{1}{3}$ at $\left(4, \frac{11}{3}\right)$ 37. Bread: 300; cakes: 400; maximum profit: $1800

38. $\frac{5}{3x - 1} - \frac{2}{x - 4}$ 39. $\left(-1, -\frac{1}{4}\right)$

Chapter 9, Test Form B

1. (1,-4) 2. Airplane: 775 km/h; wind: 125 km/h 3. 24 L of 10%; 36 L of 35%

4. $A = 70^\circ$, $B = 62^\circ$, $C = 48^\circ$ 5. $\left(\frac{2}{3}, -\frac{1}{2}\right)$ 6. (1,2,1)

7. (14 + 2z,18 + 3z,z); (14,18,0), (16,21,1), (10,12,-2), etc. Answers may vary. 8. Inconsistent, independent 9. Consistent, independent

10. $f(x) = -2x^2 - x + 1$ 11. -9 12. 28 13. (-5,4) 14. (0,-3,2)

15. Not possible 16. $\begin{bmatrix} 3 & -2 \\ 4 & 1 \end{bmatrix}$ 17. $\begin{bmatrix} 4 & -2 \\ 4 & 2 \end{bmatrix}$ 18. $\begin{bmatrix} -1 & 7 & -3 \\ 6 & -1 & 0 \\ -1 & 9 & 2 \end{bmatrix}$

19. $\begin{bmatrix} 5 & -6 \\ 5 & -2 \end{bmatrix}$ 20. $\begin{bmatrix} -2 & 11 \\ 5 & 8 \end{bmatrix}$ 21. Not possible 22. $\begin{bmatrix} -2 & -1 & 3 \end{bmatrix}$

23. $\begin{bmatrix} 4 & 2 & -3 \\ 3 & -2 & 6 \\ 1 & 6 & 1 \end{bmatrix}$ 24. $\begin{bmatrix} -2 & 4 \\ -1 & 3 \end{bmatrix}$ 25. $\begin{bmatrix} \frac{2}{7} & \frac{1}{7} \\ -\frac{3}{7} & \frac{2}{7} \end{bmatrix}$ 26. $\begin{bmatrix} \frac{1}{25} & -\frac{4}{25} & \frac{12}{25} \\ \frac{6}{25} & \frac{1}{25} & -\frac{3}{25} \\ \frac{4}{25} & \frac{9}{25} & -\frac{2}{25} \end{bmatrix}$

27. Does not exist 28. $a_{23} = -1$; $M_{23} = \begin{vmatrix} 3 & -2 \\ 1 & -2 \end{vmatrix} = -4$; $A_{23} = 4$

29. 25 30. 12 31. 50 32. 1

33. $(x - y)(z - y)(x - z)$ 34. $\begin{bmatrix} 3 & -1 \\ 1 & 2 \end{bmatrix}\begin{bmatrix} x \\ y \end{bmatrix} = \begin{bmatrix} 9 \\ -4 \end{bmatrix}$; $(2,-3)$

35.

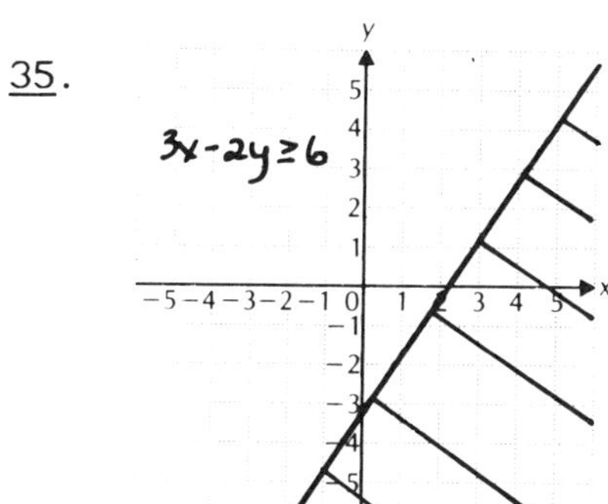

36. Minimum -3 at (-1,0); maximum 25 at $\left(-1,\frac{7}{2}\right)$ 37. Type A: 7, Type B: 3; maximum score: 85 38. $\frac{3}{4x - 3} - \frac{2}{x + 5}$

39. $m = \frac{1}{2}$; $b = -3$

Chapter 9, Test Form C

1. (2,-1) 2. 5 3. Bill: 5; Ann: 10 4. \$580 at 5%; \$1000 at 8%; \$340 at 10% 5. $\left(-\frac{1}{3},\frac{1}{4}\right)$ 6. $\left(\frac{19 - 7y}{13}, y, \frac{11 + 11y}{13}\right)$; $\left(\frac{19}{13}, 0, \frac{11}{13}\right)$, $\left(\frac{5}{13}, 2, \frac{33}{13}\right)$, (2,-1,0), etc. Answers may vary. 7. (1,2,1) 8. Consistent, dependent

9. Consistent, independent 10. $f(x) = x^2 - 4x + 1$ 11. 8 12. -24

13. (-1,-3) 14. (-2,1,1) 15. Not possible 16. A or $\begin{bmatrix} -3 & 1 \\ 0 & 2 \end{bmatrix}$

17. $\begin{bmatrix} -2 & 1 \\ 0 & 3 \end{bmatrix}$ 18. $\begin{bmatrix} 7 & -2 & -1 \\ 8 & -4 & 10 \\ 2 & -1 & 5 \end{bmatrix}$ 19. $\begin{bmatrix} 0 & 2 \\ 2 & 6 \end{bmatrix}$ 20. $\begin{bmatrix} -4 & 4 \\ 13 & -14 \end{bmatrix}$

21. Not possible 22. $\begin{bmatrix} -5 & 2 & -7 \end{bmatrix}$ 23. $\begin{bmatrix} -4 & 2 & -5 \\ 4 & 1 & 5 \\ 4 & -8 & 7 \end{bmatrix}$

24. $\begin{bmatrix} -3 & -1 \\ -2 & -4 \end{bmatrix}$ 25. $\begin{bmatrix} -\frac{2}{5} & \frac{3}{10} \\ \frac{1}{5} & \frac{1}{10} \end{bmatrix}$

26. Does not exist 27. $\begin{bmatrix} \frac{9}{17} & -\frac{3}{17} & \frac{5}{17} \\ \frac{2}{17} & \frac{5}{17} & \frac{3}{17} \\ \frac{12}{17} & -\frac{4}{17} & \frac{1}{17} \end{bmatrix}$ 28. $a_{31} = 2$;

$M_{31} = \begin{vmatrix} -3 & 2 \\ 5 & -2 \end{vmatrix} = -4$; $A_{31} = -4$ 29. -14 30. 0 31. -30 32. 42

33. $(x - y)(z - y)(x - z)$ 34. $\begin{bmatrix} 2 & -4 \\ 3 & 2 \end{bmatrix} \begin{bmatrix} x \\ y \end{bmatrix} = \begin{bmatrix} 6 \\ 1 \end{bmatrix}$; (1,-1)

35.

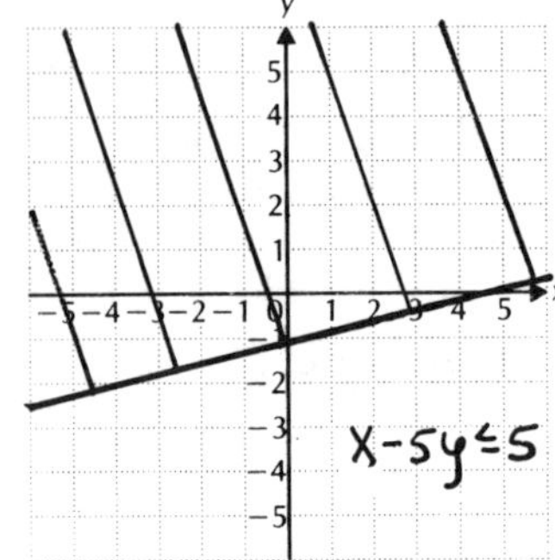

36. Minimum -12 at (0,-1); maximum 64 at (2,4) 37. 40 slacks, 160 skirts; maximum profit: $2880

38. $\frac{5}{2x + 1} - \frac{3}{x - 4}$ 39. (1,-2,0,3)

Chapter 9, Test Form D

1. (-3,1) 2. 1080 km 3. 11 dimes, 16 nickels 4. A: 350; B: 270; C: 420

5. $\left(-\frac{3}{4}, \frac{1}{2}\right)$ 6. (1,-1,3) 7. $\left(\frac{-4 - 3y}{2}, y, \frac{12 + 7y}{2}\right)$, (-2,0,6), $\left(-\frac{7}{2}, 1, \frac{19}{2}\right)$, $\left(-\frac{1}{2}, -1, \frac{5}{2}\right)$, etc. Answers may vary. 8. Consistent, independent 9. Consistent, dependent 10. $f(x) = 2x^2 - x - 3$ 11. 12 12. 20 13. (7,-2)

14. (5,0,-1) 15. Not possible 16. A or $\begin{bmatrix} -2 & -3 \\ 1 & 4 \end{bmatrix}$ 17. $\begin{bmatrix} -1 & -3 \\ 1 & 5 \end{bmatrix}$

18. $\begin{bmatrix} -2 & 4 & 4 \\ 3 & 1 & 0 \\ 10 & -1 & 7 \end{bmatrix}$ 19. $\begin{bmatrix} 0 & 0 \\ -3 & -2 \end{bmatrix}$ 20. $\begin{bmatrix} 14 & 2 \\ 2 & 26 \end{bmatrix}$ 21. Not possible

22. $\begin{bmatrix} 2 & -3 & -5 \end{bmatrix}$ 23. $\begin{bmatrix} 2 & -1 & 5 \\ -9 & 2 & 12 \\ 11 & 1 & -1 \end{bmatrix}$ 24. $\begin{bmatrix} -2 & -3 \\ 4 & 6 \end{bmatrix}$ 25. $\begin{bmatrix} -\frac{3}{14} & \frac{1}{14} \\ \frac{1}{7} & \frac{2}{7} \end{bmatrix}$

26. $\begin{bmatrix} -1 & 0 & -1 \\ 1 & \frac{1}{5} & \frac{2}{5} \\ -1 & \frac{1}{5} & -\frac{3}{5} \end{bmatrix}$ 27. Does not exist 28. $a_{23} = 5$;

$M_{23} = \begin{vmatrix} 4 & 0 \\ 2 & -3 \end{vmatrix} = -12$; $A_{23} = 12$ 29. -77 30. 0 31. -3 32. 14

33. $(a - b)(c - b)(c - a)$ 34. $\begin{bmatrix} 5 & -1 \\ 2 & 4 \end{bmatrix} \begin{bmatrix} x \\ y \end{bmatrix} = \begin{bmatrix} 11 \\ 22 \end{bmatrix}$; (3,4)

35. 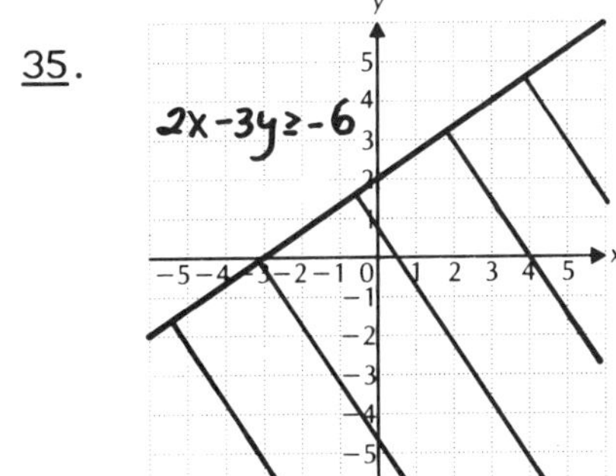

36. Minimum -36 at (-2,-3); maximum 114 at (8,-3) 37. 60 hamburgers; 40 hot dogs; maximum profit: $65

38. $\frac{7}{5x - 3} - \frac{2}{x + 4}$ 39. -2, 2

Chapter 9, Test Form E

1. $\left(\frac{1}{2}, -\frac{1}{2}\right)$ 2. (5,-6) 3. $8500 at 6%, $6500 at 9% 4. 350 5. (2y,y); (0,0), (2,1), (-4,-2), etc. Answers may vary. 6. (1,-2,1)

Chapter 9, Test Form E (continued)

7. Consistent, independent   8. $f(x) = -2x^2 + 5x - 4$ 9. -7   10. (3,-1,4)

11. 312   12. -108   13. 0   14. 270   15. $(x - y)(z - y)(z - x)$

16. $\begin{bmatrix} -3 & 4 \\ 2 & -5 \end{bmatrix} \begin{bmatrix} x \\ y \end{bmatrix} = \begin{bmatrix} 10 \\ -9 \end{bmatrix}$; (-2,1)   17. f   18. ℓ   19. a   20. j   21. g

22. d   23. k   24.

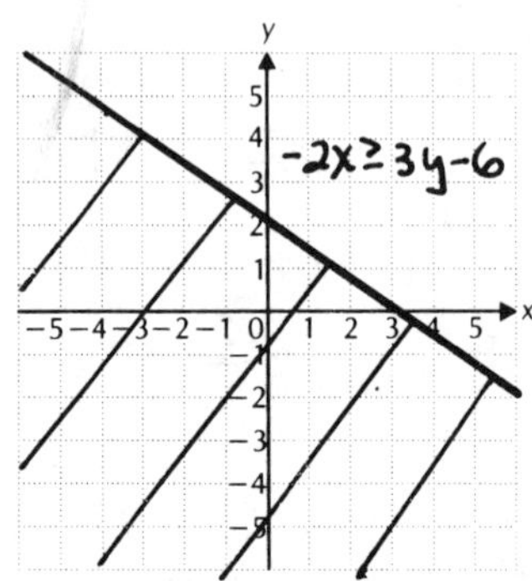

25. Minimum 15 at (0,1); maximum $\frac{125}{3}$ at $\left(\frac{4}{3}, 1\right)$

26. 3 dresses, 2 suits; maximum profit: $145

Chapter 9, Test Form F

1. a   2. b   3. b   4. d   5. d   6. d   7. b   8. c   9. d   10. c   11. a

12. d   13. b   14. c   15. d   16. a   17. d   18. d   19. a   20. d   21. a

22. b   23. b   24. b   25. d   26. c   27. b   28. d   29. a   30. a   31. b

32. a   33. c

Chapter 10, Test Form A

1. 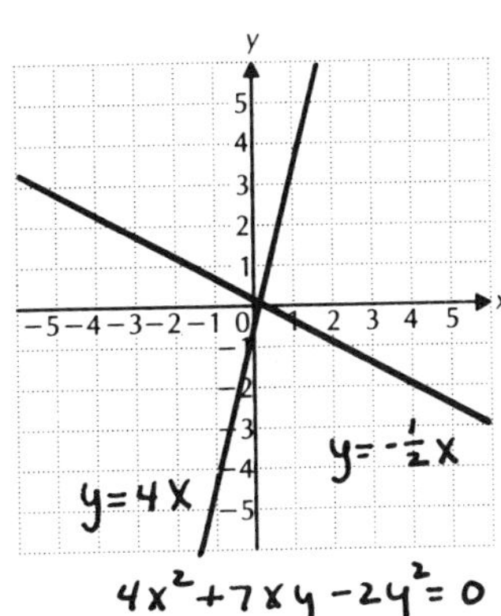

2. C: (2,3); V: (3,3), (1,3), (2,6), (2,0);

F: $\left(2, 3 + 2\sqrt{2}\right)$, $\left(2, 3 - 2\sqrt{2}\right)$

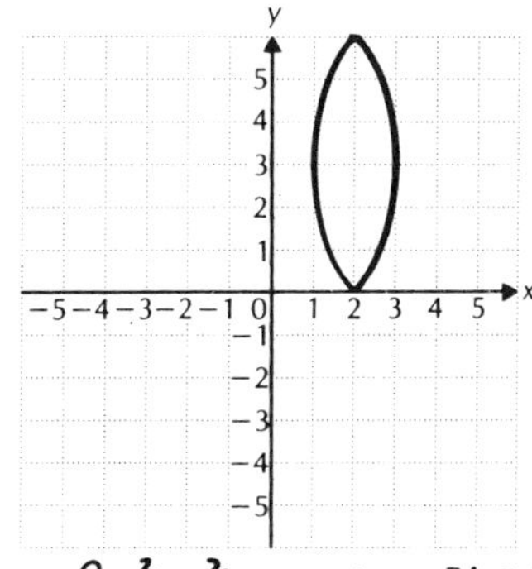

3. $\frac{x^2}{4} + \frac{y^2}{16} = 1$ 4. C: (0,0); V: (0,4), (0,-4); F: $\left(0, 2\sqrt{5}\right)$, $\left(0, -2\sqrt{5}\right)$;

A: $y = 2x$, $y = -2x$ 5.

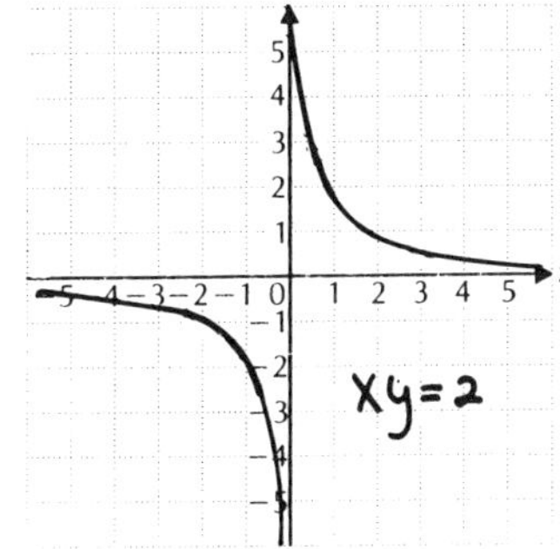

6. $y^2 = 24x$

7. V: (-1,3);

F: (-1,4); D: $y = 2$

8. (-8,6), (8,-6) 9. (1,1), (-1,-1) 10. 7 and 10 11. 9 ft by 12 ft

12. Numerator 11, denominator 10; or numerator 10, denominator 11

13. 2 ft by 12 ft 14. 4 and 18, or -4 and -18 15. 3 ft, 5 ft

16. Parabola 17. Hyperbola 18. Circle 19. Ellipse 20. Parabola

21. Hyperbola 22. $\frac{(x - 3)^2}{4} + \frac{(y + 1)^2}{9} = 1$ 23. $(x - 2)^2 = 4(y + 1)$

24. $\left(\frac{1}{3},\frac{1}{6}\right)$, $\left(-\frac{1}{3},\frac{1}{6}\right)$, $\left(\frac{1}{3},-\frac{1}{6}\right)$, $\left(-\frac{1}{3},-\frac{1}{6}\right)$

Chapter 10, Test Form B

1.

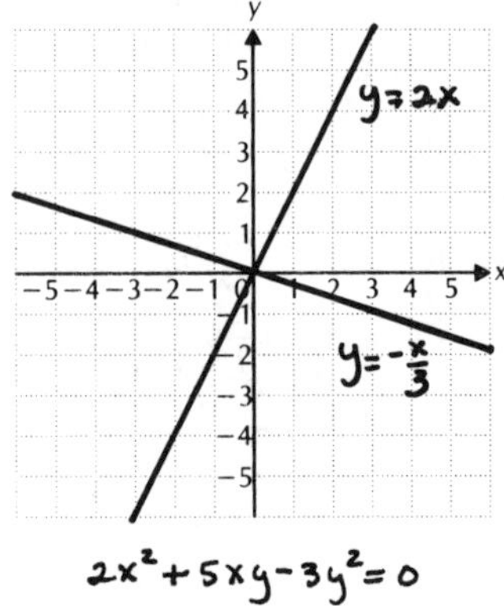

2. C: (1,-1); V: (5,-1), (-3,-1), (1,-3), (1,1);

F: $\left(1 + 2\sqrt{3},-1\right)$, $\left(1 - 2\sqrt{3},-1\right)$

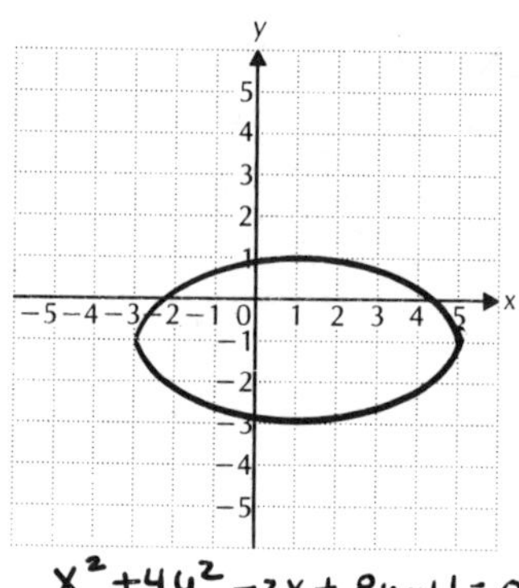

3. $\frac{x^2}{25} + y^2 = 1$ 4. C: (0,0); V: (3,0), (-3,0); F: (5,0), (-5,0);

A: $y = \frac{4}{3}x$, $y = -\frac{4}{3}x$ 5.

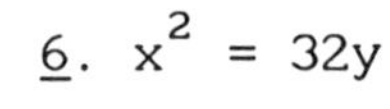

6. $x^2 = 32y$

7. V: (4,-7); F: (3,-7); D: x = 5 8. (0,4) 9. (5,0), (-4,3), (-4,-3)

10. 10 and 15 11. 5 cm by 7cm 12. Numerator 9, denominator 5; or numerator 5, denominator 9 13. 5 by 9 14. 6 and 11, or -6 and -11

15. 7 ft, 8 ft 16. Circle 17. Ellipse 18. Parabola 19. Hyperbola

20. Parabola 21. Hyperbola 22. $\frac{x^2}{16} - \frac{y^2}{9} = 1$ 23. 2 and 5

24. $(x - 1)^2 + (y + 2)^2 = 1$

Chapter 10, Test Form C

1.

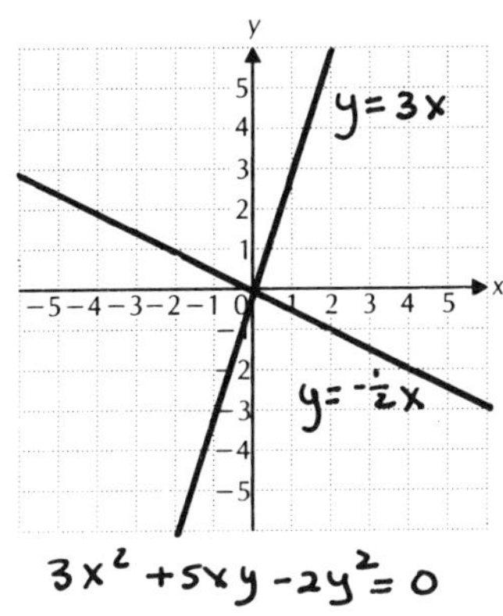

2. C: (2,-3); V: (5,-3), (-1,-3), (2,-1), (2,-5);

F: $\left(2 + \sqrt{5},-3\right)$, $\left(2 - \sqrt{5},-3\right)$

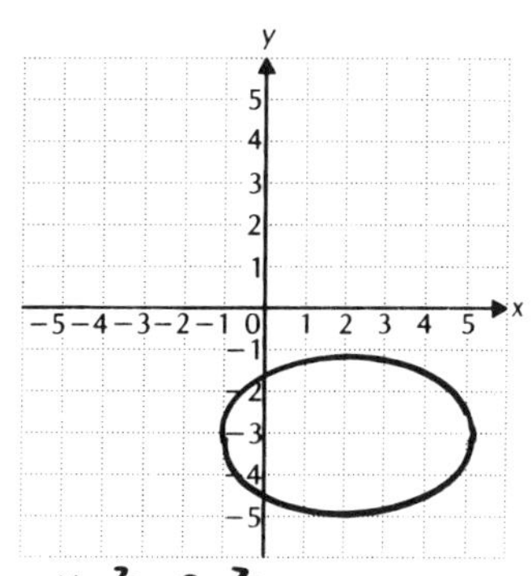

3. $\frac{x^2}{36} + \frac{y^2}{25} = 1$ 4. C: (0,0); V: (2,0), (-2,0); F: $\left(\sqrt{13},0\right)$, $\left(-\sqrt{13},0\right)$;

A: $y = \frac{3}{2}x$, $y = -\frac{3}{2}x$ 5.

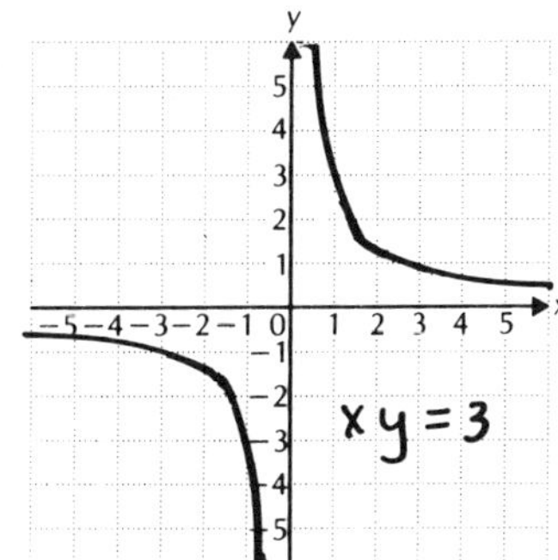

6. $y^2 = -12x$

Chapter 10, Test Form C (continued)

7. V: (1,-5); F: (1,-6); D: $y = -4$ 8. (-1,2), (-4,1) 9. (2,2), (2,-2), (-2,2), (-2,-2) 10. 3 and 12 11. 4 in. by 8 in. 12. Numerator 8, denominator 13; or numerator 13, denominator 8 13. 5 yd by 12 yd 14. 5 and 15; or -5 and -15 15. 2 cm, 7 cm 16. Circle 17. Parabola 18. Hyperbola 19. Ellipse 20. Hyperbola 21. Parabola

22. $\frac{(x+2)^2}{25} + \frac{(y-3)^2}{16} = 1$ 23. 3 and 4 24. $(x+3)^2 + (y-2)^2 = 1$

Chapter 10, Test Form D

1.

$y = -x$

$5x^2 + 3xy - 2y^2 = 0$

2. C: (2,1); V: (6,1), (-2,1), (2,4), (2,-2);
F: $\left(2 + \sqrt{7}, 1\right)$, $\left(2 - \sqrt{7}, 1\right)$

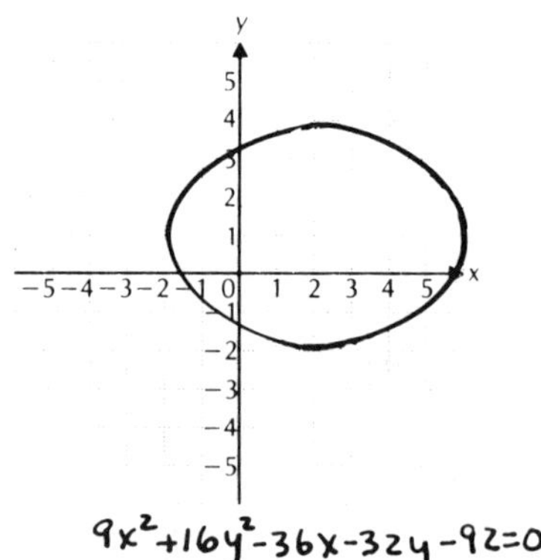

3. $\frac{x^2}{16} + \frac{y^2}{9} = 1$ 4. C: (0,0); V: (0,3), (0,-3); F: $\left(0, \sqrt{34}\right)$, $\left(0, -\sqrt{34}\right)$;
A: $y = \frac{3}{5}x$, $y = -\frac{3}{5}x$

5. 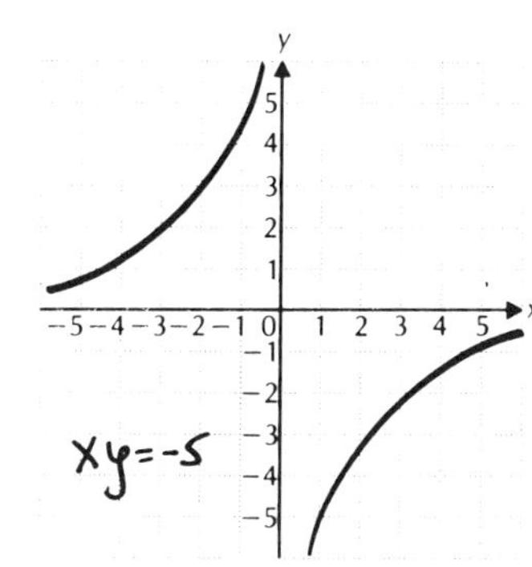

6. $x^2 = -24y$ 7. V: (2,3); F: (3,3); D: x = 1

8. (8,2), (5,1) 9. (3,3), (-3,-3), (9,1), (-9,-1)

10. 4 and 9 11. 40 yd by 115 yd 12. Numerator 7, denominator 16; or numerator 16, denominator 7

13. 4 by 15 14. 6 and 16; or -6 and -16

15. 4 ft, 11 ft 16. Hyperbola 17. Parabola

18. Ellipse 19. Hyperbola 20. Parabola 21. Circle 22. $\frac{y^2}{4} - \frac{x^2}{9} = 1$

23. $(y - 2)^2 = 4(x + 3)$ 24. $\frac{5}{7}$

Chapter 10, Test Form E

1. ℓ 2. d 3. n 4. b 5. j 6. i 7. Parabola 8. Hyperbola

9. Circle 10. Ellipse 11. 12.

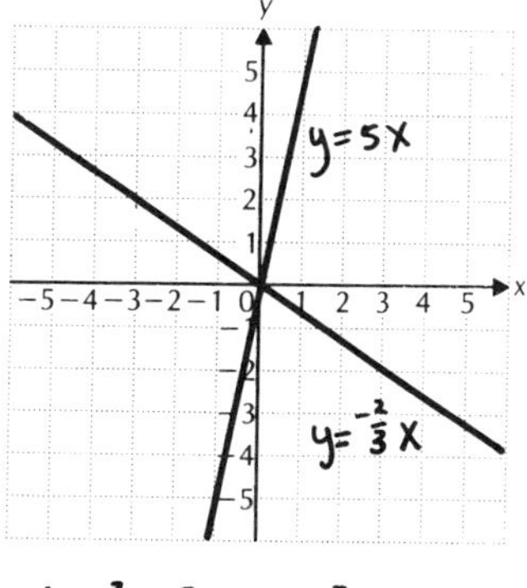

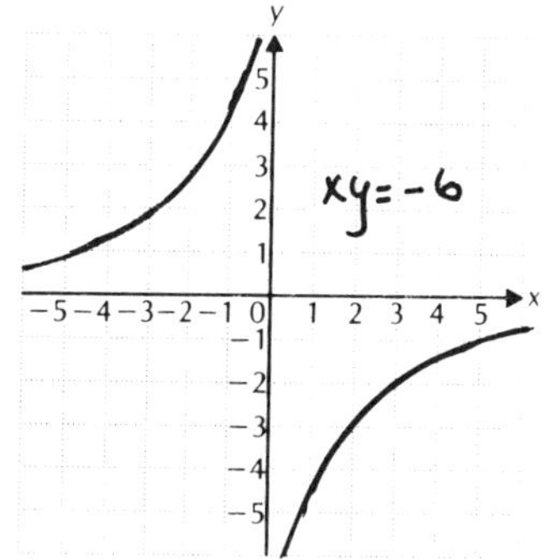

Chapter 10, Test Form E (continued)

13.

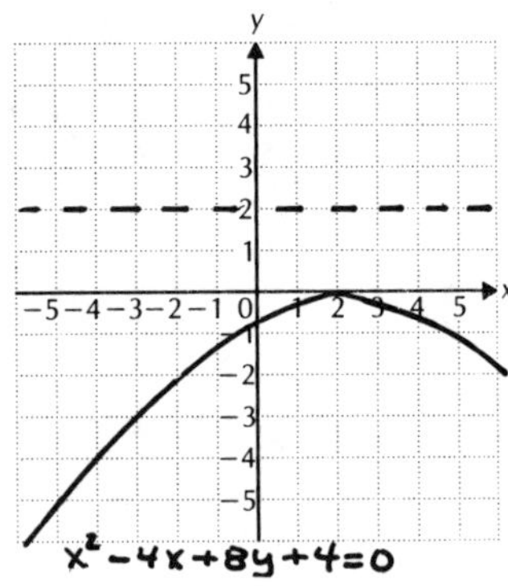

14.

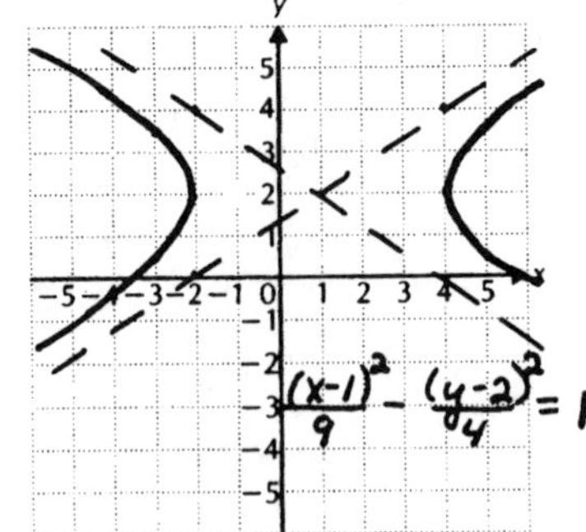

15. (4,0), (0,2)

16. (2,2), (2,-2), (-2,2), (-2,-2)   17. 6 and -12, or -6 and 12

18. $(y + 4)^2 = -5(x - 3)$

Chapter 10, Test Form F

1. c   2. a   3. b   4. b   5. d   6. b   7. b   8. d   9. a   10. c   11. a

12. b   13. b   14. a   15 c

Chapter 11, Test Form A

1. 7   2. 4   3. 22nd   4. $\frac{27}{4}, \frac{17}{2}, \frac{41}{4}$ 5. $-\frac{1}{1024}$ 6. $\frac{63}{8}$ 7. b   8. $\frac{216}{7}$

9. \$305, \$4800   10. $\frac{1}{81}$ ft   11. $\frac{8}{45}$

12. S_n: $1\cdot 2 + 2\cdot 4 + 3\cdot 8 + \ldots + n\cdot 2^n = (n - 1)2^{n+1} + 2$

S_1: $2 = (1 - 1)2^{1+1} + 2$

S_k: $1\cdot 2 + 2\cdot 4 + 3\cdot 8 + \ldots + k\cdot 2^k = (k - 1)2^{k+1} + 2$

S_{k+1}: $1\cdot 2 + 2\cdot 4 + 3\cdot 8 + \ldots + k\cdot 2^k + (k + 1)2^{k+1} = (k + 1 - 1)2^{k+1+1} + 2$
$= k\cdot 2^{k+2} + 2$

1. Basis step: $(1 - 1)2^{1+1} + 2 = 2$, so S_1 is true.

2. Induction step: Assume S_k. Then add $(k + 1)\cdot 2^{k+1}$ on both sides.

$$\begin{aligned}1\cdot 2 + 2\cdot 4 + 3\cdot 8 + \ldots + k\cdot 2^k + (k + 1)2^{k+1} &= (k - 1)2^{k+1} + 2 + (k + 1)2^{k+1}\\ &= 2^{k+1}(k - 1 + k + 1) + 2\\ &= 2^{k+1}(2k) + 2\\ &= 2^{k+2}\cdot k + 2\\ &= k\cdot 2^{k+2} + 2\end{aligned}$$

Chapter 11, Test Form A (continued)

13. 3, 10, 101, 10,202 14. (a) 7^4, or 2401; (b) 840; (c) 210 15. 462

16. 4704 17. 2520 18. 2^7, or 128 19. $-540a^3b^3$

20. $x^4 + 4\sqrt{3}x^3 + 18x^2 + 12\sqrt{3}x + 9$ 21. $\frac{1}{9}$ 22. $\frac{2}{13}$ 23. $\frac{420}{4199}$ 24. 3, 6.25, 12.703704, 25.628906, 51.53632 25. 45

Chapter 11, Test Form B

1. 14 2. 7 3. 10th 4. $\frac{39}{4}, \frac{27}{2}, \frac{69}{4}$ 5. 9.6 6. $\frac{211}{81}$ 7. b, c

8. $\frac{3}{2}$ 9. 5184 ft 10. $1464.10 11. $\frac{8}{33}$

12. $S_n = \frac{1}{1\cdot2\cdot3} + \frac{1}{2\cdot3\cdot4} + \ldots + \frac{1}{n(n+1)(n+2)} = \frac{n(n+3)}{4(n+1)(n+2)}$

$S_1 = \frac{1}{1\cdot2\cdot3} = \frac{1(4)}{4(2)(3)}$

$S_k = \frac{1}{1\cdot2\cdot3} + \frac{1}{2\cdot3\cdot4} + \ldots + \frac{1}{k(k+1)(k+2)} = \frac{k(k+3)}{4(k+1)(k+2)}$

$S_{k+1} = \frac{1}{1\cdot2\cdot3} + \frac{1}{2\cdot3\cdot4} + \ldots + \frac{1}{k(k+1)(k+2)} + \frac{1}{(k+1)(k+2)(k+3)} = \frac{(k+1)(k+4)}{4(k+2)(k+3)}$

1. Basis step: $\frac{1}{6} = \frac{4}{24}$, so S_1 is true

2. Induction step: Assume S_k. Then add $\frac{1}{(k+1)(k+2)(k+3)}$ on both sides.

$$\frac{1}{1\cdot2\cdot3} + \frac{1}{2\cdot3\cdot4} + \ldots + \frac{1}{k(k+1)(k+2)} + \frac{1}{(k+1)(k+2)(k+3)}$$
$$= \frac{k(k+3)}{4(k+1)(k+2)} + \frac{1}{(k+1)(k+2)(k+3)}$$
$$= \frac{k(k+3)(k+3)+4}{4(k+1)(k+2)(k+3)}$$
$$= \frac{k^3+6k^2+9k+4}{4(k+1)(k+2)(k+3)}$$
$$= \frac{(k+1)(k+1)(k+4)}{4(k+1)(k+2)(k+3)}$$
$$= \frac{(k+1)(k+4)}{4(k+2)(k+3)}$$

13. -4, 17, -25, 59   14. (a) 8^3, or 512; (b) 336; (c) 6   15. 495   16. 1120

17. 5040   18. 2^9, or 512   19. $21{,}504a^5b^2$ 20. $25\sqrt{5} - 125a + 50\sqrt{5}a^2 - 50a^3 + 5\sqrt{5}a^4 - a^5$ 21. $\frac{1}{9}$ 22. $\frac{1}{4}$ 23. $\frac{105}{286}$ 24. \$2600, \$1690, \$1098.50, \$714.03, \$464.12   25. 9

Chapter 11, Test Form C

1. 4   2. 8   3. 13th   4. $\frac{27}{4}, \frac{19}{2}, \frac{49}{4}$ 5. $\frac{3}{32}$ 6. -47   7. a   8. $\frac{70}{9}$

9. 127.4   10. About 29,387   11. $\frac{77}{9}$

12. S_n: $\frac{1}{3} + \frac{1}{15} + \frac{1}{35} + \ldots + \frac{1}{4n^2 - 1} = \frac{n}{2n + 1}$

S_1: $\frac{1}{3} = \frac{1}{2(1) + 1}$

S_k: $\frac{1}{3} + \frac{1}{15} + \frac{1}{35} + \ldots + \frac{1}{4k^2 - 1} = \frac{k}{2k + 1}$

S_{k+1}: $\frac{1}{3} + \frac{1}{15} + \frac{1}{35} + \ldots + \frac{1}{4k^2 - 1} + \frac{1}{4(k + 1)^2 - 1} = \frac{k + 1}{2(k + 1) + 1}$

1. Basis step: $\frac{1}{3} = \frac{1}{3}$, so S_1 is true.

2. Induction step: Assume S_k. Then add $\frac{1}{4(k + 1)^2 - 1}$ on both sides.

$$\begin{aligned}
\frac{1}{3} + \frac{1}{15} + \frac{1}{35} + \ldots + \frac{1}{4k^2 - 1} + \frac{1}{4(k + 1)^2 - 1} &= \frac{k}{2k + 1} + \frac{1}{4(k + 1)^2 - 1} \\
&= \frac{k}{2k + 1} + \frac{1}{4(k^2 + 2k + 1) - 1} \\
&= \frac{k}{2k + 1} + \frac{1}{4k^2 + 8k + 3} \\
&= \frac{k}{2k + 1} + \frac{1}{(2k + 1)(2k + 3)} \\
&= \frac{k(2k + 3) + 1}{(2k + 1)(2k + 3)} \\
&= \frac{2k^2 + 3k + 1}{(2k + 1)(2k + 3)} \\
&= \frac{(2k + 1)(k + 1)}{(2k + 1)(2k + 3)} \\
&= \frac{k + 1}{2k + 3} \\
&= \frac{k + 1}{2(k + 1) + 1}
\end{aligned}$$

Chapter 11, Test Form C (continued)

13. 4, 5, 7, 11   14. (a) 6^5, or 7776; (b) 720; (c) 6   15. 4845   16. 199,584

17. 1260   18. 2^5, or 32   19. $-1458ab^5$ 20. $343 + 294\sqrt{7}a + 735a^2 + 140\sqrt{7}a^3 + 105a^4 + 6\sqrt{7}a^5 + a^6$ 21. $\frac{1}{6}$ 22. $\frac{2}{13}$ 23. $\frac{175}{2431}$ 24. 2, 5, 8, 11

25. 10

Chapter 11, Test Form D

1. 12   2. 5   3. 27th   4. $\frac{27}{4}, \frac{21}{2}, \frac{57}{4}$ 5. $\frac{4}{729}$ 6. $\frac{312}{125}$ 7. a, c   8. $-\frac{32}{5}$

9. 3550   10. \$16,383   11. $\frac{60}{11}$

12. S_n: $1^3 + 3^3 + 5^3 + \ldots + (2n - 1)^3 = n^2(2n^2 - 1)$

S_1: $1^3 = 1^2(2(1)^2 - 1)$

S_k: $1^3 + 3^3 + 5^3 + \ldots + (2k - 1)^3 = k^2(2k^2 - 1)$

S_{k+1}: $1^3 + 3^3 + 5^3 + \ldots + (2k - 1)^3 + [2(k + 1) - 1]^3 = (k + 1)^2[2(k + 1)^2 - 1]$

1. Basis step: $1^3 = 1^2(2(1)^2 - 1)$, so S_1 is true.

2. Induction step: Assume S_k. Then add $[2(k + 1) - 1]^3$ on both sides.

$$1^3 + 3^3 + 5^3 + \ldots + (2k - 1)^3 + [2(k + 1) - 1]^3 =$$
$$= k^2(2k^2 - 1) + [2(k + 1) - 1]^3$$
$$= 2k^4 - k^2 + 8k^3 + 12k^2 + 6k + 1$$
$$= 2k^4 + 8k^3 + 11k^2 + 6k + 1$$
$$= (k^2 + 2k + 1)(2k^2 + 4k + 1)$$
$$= (k + 1)^2[2(k + 1)^2 - 1]$$

13. 4, 18, 88, 438   14. (a) 9^4, or 6561; (b) 3024; (c) 42   15. 792   16. 160

17. 840   18. 2^6, or 64   19. $720x^3y^2$ 20. $x^5 - 5\sqrt{6}x^4 + 60x^3 - 60\sqrt{6}x^2 + 180x - 36\sqrt{6}$ 21. $\frac{1}{12}$ 22. $\frac{1}{2}$ 23. $\frac{90}{221}$ 24. 6, 4, 2   25. 5

Chapter 11, Test Form E

1. 285 2. 57 3. 50th 4. 10q + 15p 5. 145.8 6. $\frac{133}{1944}$ 7. $\frac{31}{4}$

8. $\frac{26}{3}, \frac{34}{3}, 14, \frac{50}{3}, \frac{58}{3}$ 9. \$2912 10. 650 m 11. 386,428 12. $\frac{181}{45}$

13. S_n: $1 + 3 + 6 + \ldots + \frac{1}{2}n(n + 1) = \frac{n(n + 1)(n + 2)}{6}$

S_1: $1 = \frac{1(1 + 1)(1 + 2)}{6}$

S_k: $1 + 3 + 6 + \ldots + \frac{1}{2}k(k + 1) = \frac{k(k + 1)(k + 2)}{6}$

S_{k+1}: $1 + 3 + 6 + \ldots + \frac{1}{2}k(k + 1) + \frac{1}{2}(k + 1)(k + 2) = \frac{(k + 1)(k + 2)(k + 3)}{6}$

1. Basis step: $\frac{1(1 + 1)(1 + 2)}{6} = 1$, so S_1 is true.

2. Induction step: Assume S_k. Then add $\frac{1}{2}(k + 1)(k + 2)$ on both sides.

$$1 + 3 + 6 + \ldots + \frac{1}{2}k(k + 1) + \frac{1}{2}(k + 1)(k + 2) =$$

$$= \frac{k(k + 1)(k + 2)}{6} + \frac{1}{2}(k + 1)(k + 2)$$
$$= \frac{k(k + 1)(k + 2)}{6} + \frac{3(k + 1)(k + 2)}{6}$$
$$= \frac{k(k + 1)(k + 2) + 3(k + 1)(k + 2)}{6}$$
$$= \frac{(k + 1)(k + 2)(k + 3)}{6}$$

14. 480 15. 7!, or 5040; 6!, or 720 16. 1260 17. (a) 5^4, or 625; (b) 120; (c) 72 18. 56 19. $17{,}500x^3y^4$ 20. $32x^5 + 80x^4y^2 + 80x^3y^4 + 40x^2y^6 + 10xy^8 + y^{10}$ 21. $\frac{1}{18}$ 22. $\frac{1}{221}$ 23. $\frac{77}{207}$

Chapter 11, Test Form F

1. a 2. c 3. d 4. b 5. a 6. b 7. c 8. d 9. c 10. a 11. b

12. S_n: $2 + 6 + 18 + \ldots + 2(3^{n-1}) = 3^n - 1$

S_1: $2 = 3^1 - 1$

S_k: $2 + 6 + 18 + \ldots + 2(3^{k-1}) = 3^k - 1$

S_{k+1}: $2 + 6 + 18 + \ldots + 2(3^{k-1}) + 2(3^k) = 3^{k+1} - 1$

1. Basis step: $3^1 - 1 = 2$, so S_1 is true.

2. Induction step: Assume S_k. Then add $2(3^k)$ on both sides.

$$\begin{aligned} 2 + 6 + 18 + \ldots + 2(3^{k-1}) + 2(3^k) &= 3^k - 1 + 2(3^k) \\ &= 3(3^k) - 1 \\ &= 3^{k+1} - 1 \end{aligned}$$

13. c   14. d   15. c   16. b   17. d   18. a   19. c   20. b   21. d   22. b

23. d   24. c   25. d

Final Exam, Test Form A

1. 15, 0   2. $-\sqrt[5]{8}$, $\sqrt{40}$ 3. 15   4. 15, 0, -18   5. 1.333×10^{-5}

6. 3.2×10^{-5} 7. $\dfrac{x^2 + 6x + 12}{(x - 2)(x + 2)(x + 3)}$ 8. $\{x \mid x \le -\frac{13}{5}\}$ 9. $\left\{-\frac{1}{2}, \frac{3}{7}\right\}$

10. $x^2 + \sqrt{3}x - 6 = 0$ 11. $y = \dfrac{2400xz}{w^2}$ 12. $\{x \mid x \le 9\}$ 13. $y = -4x - 5$

14. $\sqrt{13}$, or 3.6056   15. $(f + g)(x) = x^2 + 2x - 1$, $(f - g)(x) = -x^2 + 2x - 7$,

$fg(x) = 2x^3 - 4x^2 + 6x - 12$, $(f/g)(x) = \dfrac{2x - 4}{x^2 + 3}$, $f \circ g(x) = 2x^2 + 2$,

$g \circ f(x) = 4x^2 - 16x + 19$ 16. All reals, all reals, all reals, all reals, all reals, all reals, all reals, all reals   17. c   18. a and b

19. $\{x \mid -1 \le x \le 6\}$ 20. $\{x \mid -\frac{7}{2} < x < \frac{1}{3}\}$ 21. b = 15 cm, h = 15 cm

22. $(x + 3)(x - 1 + \sqrt{2})(x - 1 - \sqrt{2})$; -3, $1 - \sqrt{2}$, $1 + \sqrt{2}$

23. $(x - 2)(x + 3)(x + 1)^2(x - 4)^3$

24. 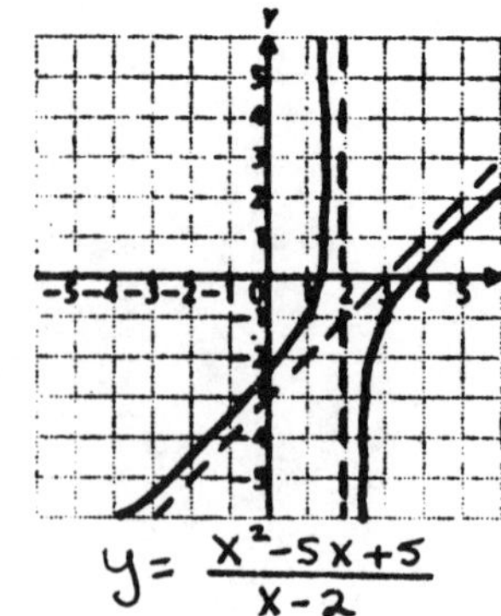

25. $x = 2y - 5$ 26. 1.8502 27. 64 28. About 9 years 29. 4.1973 30. 0.8544

31. Origin: $\left(-\frac{2}{3}, \frac{\sqrt{5}}{3}\right)$; u-axis: $\left(\frac{2}{3}, \frac{\sqrt{5}}{3}\right)$; v-axis: $\left(-\frac{2}{3}, -\frac{\sqrt{5}}{3}\right)$ 32.

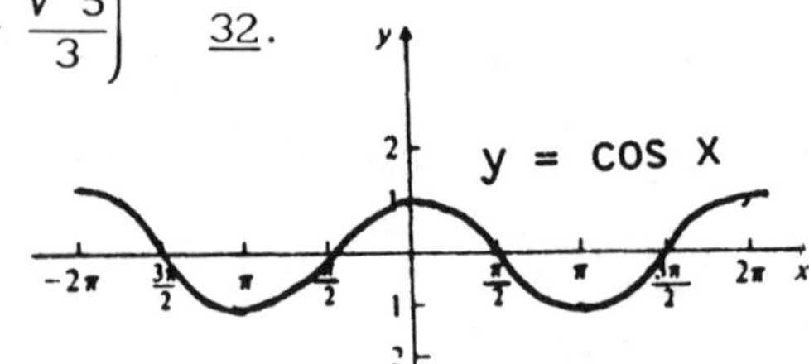

33. All reals from -1 to 1 inclusive

34. 2π 35. I, IV 36. 480°

37. 114.59° 38. 0.2622

39. 3.0789 40. 377 ft/min 41. 1 42. csc x 43. $\tan^2 x$ 44. 18.8396°

45. $\frac{1}{2}$ 46. $\sin(90^\circ - \theta) = 0.6018$, $\cos(90^\circ - \theta) = 0.7986$, $\tan(90^\circ - \theta) = 0.7536$, $\cot(90^\circ - \theta) = 1.3270$, $\sec(90^\circ - \theta) = 1.2522$, $\csc(90^\circ - \theta) = 1.6617$ 47. 168 m 48. 13 49. (a) $\langle -3, -11 \rangle$; (b) 9.2

50. $\left(-3\sqrt{2}, -3\sqrt{2}\right)$ 51. $\frac{8}{5}$ cis 24° 52. $3\sqrt{2}$ cis 135° 53. (4,3) 54. (2,1,1)

55. $f(x) = -2x^2 + 3x - 4$ 56. 60 57. $\begin{bmatrix} 7 & -1 \\ 4 & 12 \end{bmatrix}$ 58. $\begin{bmatrix} \frac{2}{5} & -\frac{1}{5} \\ \frac{1}{15} & \frac{2}{15} \end{bmatrix}$

59. $\frac{2}{x+1} + \frac{1}{x+2}$ 60. $\frac{x^2}{4} + \frac{y^2}{9} = 1$

61. 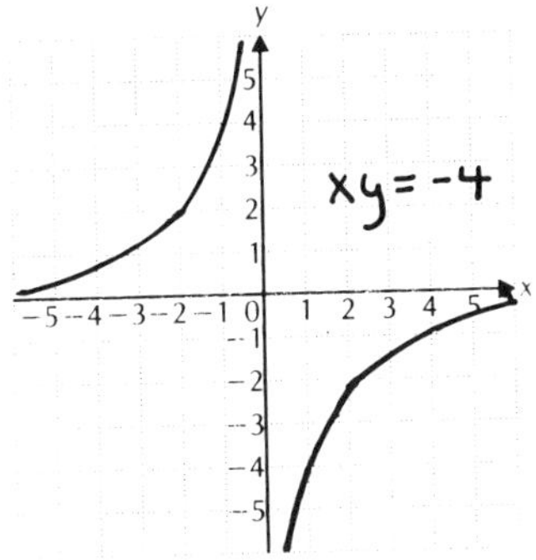

62. 3 and 11   63. 4 m x 15 m   64. Circle

65. 12,075   66. 150   67. (a) 9^5, or 59,049;

(b) 15,120; (c) 6561   68. 3, 7, 23, 87, 343

69. $x^5 + 5\sqrt{2}x^4 + 20x^3 + 20\sqrt{2}x^2 + 20x + 4\sqrt{2}$

70. $\frac{20}{221}$

Final Exam, Test Form B

1. -9.4   2. 70   3. 68   4. -28   5. $(2x - 5)(3x + 4)$

6. $(a + 7)(a^2 - 7a + 49)$ 7. $(x + 2y)(6 - x)$ 8. $\frac{2}{13} - \frac{3}{13}i$ 9. $\left\{-\frac{9}{4}, \frac{7}{3}\right\}$

10. 15   11. 6 in. x 18 in.   12. Two real   13. No   14. {3, 4, 8}

15. {-2, 1, 7}   16. $(x - 5)^2 + (y + 1)^2 = 9$

17. 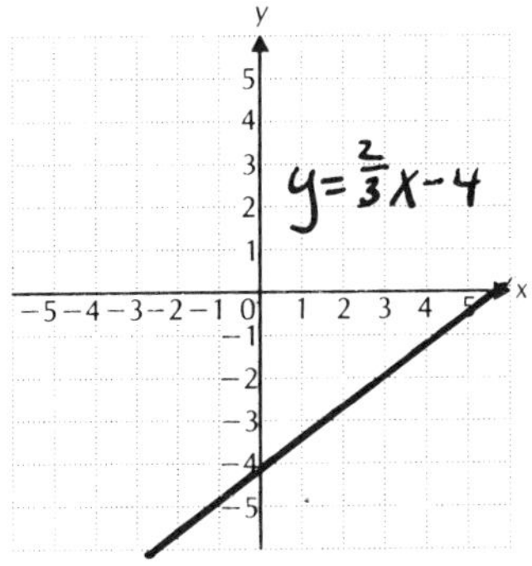

18. $(-\infty, -5]$

19. 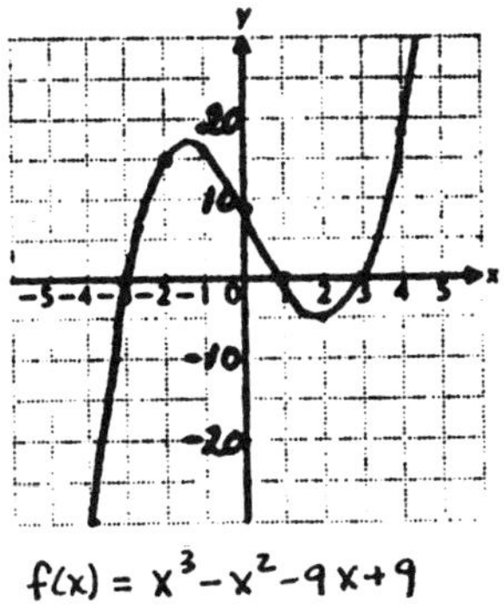

20. $\left(\frac{7}{2}, 0\right)$, (-3,0)   21. 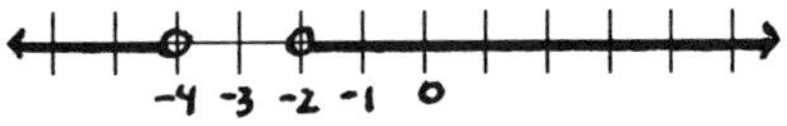

22. Yes

23. {1,2,3,5,7,9,13,15}   24. {-7,4}   25. a) $P(t) = 200{,}000e^{0.034t}$;

b) 280,989; c) 2010

26.

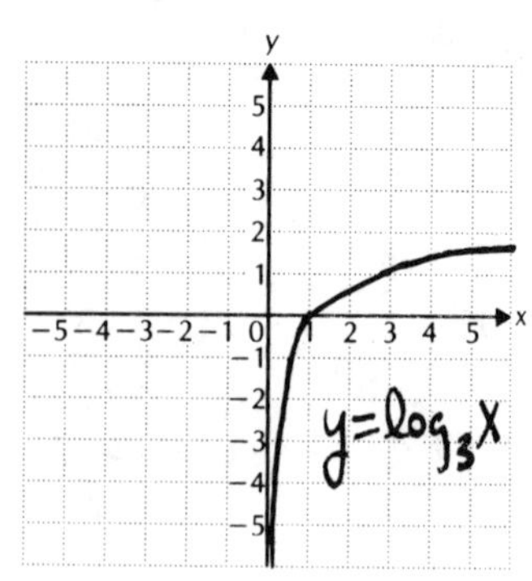

27. $\log_a \dfrac{x^4}{\sqrt{yz^2}}$ 28. -8.2171 29. Not defined

30. 3.98×10^{-6}

31.

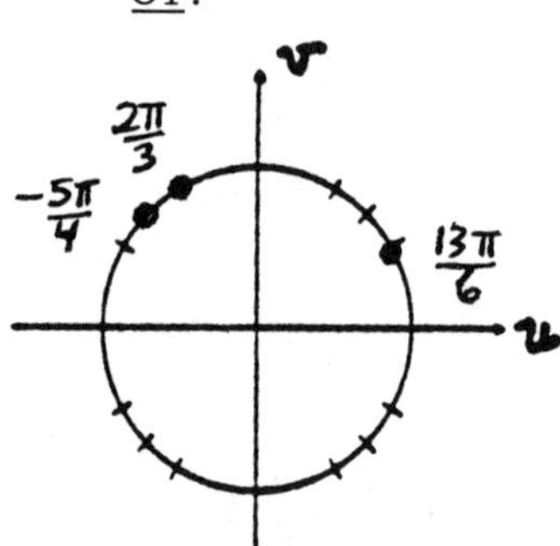

32. I, $\frac{5\pi}{12}$, 1.3090 33. II, $\frac{67\pi}{90}$, 2.3387

34. 8π cm, or 25.1 cm 35. $\sin\theta = \dfrac{3}{\sqrt{34}}$, or $\dfrac{\sqrt{34}}{8}$;

$\cos\theta = -\dfrac{5}{\sqrt{34}}$, or $\dfrac{-5\sqrt{34}}{34}$, $\tan\theta = -\dfrac{3}{5}$; $\cot\theta = -\dfrac{5}{3}$; $\sec\theta = -\dfrac{\sqrt{34}}{5}$;

$\csc\theta = \dfrac{\sqrt{34}}{3}$ 36. 0 37. $\dfrac{\sqrt{2}}{2}$ 38. 13.2° 39. 1 40. 2.0466

41. $\dfrac{\tan 23^\circ + \tan 42^\circ}{1 - \tan 23^\circ \tan 42^\circ}$ 42. 0 43. $\dfrac{\sqrt{\sin x \tan x}}{\tan x}$ 44. 2.3562

45. $\frac{\pi}{3}, \frac{\pi}{2}, \frac{2\pi}{3}, \frac{4\pi}{3}, \frac{3\pi}{2}, \frac{5\pi}{3}$ 46. $\csc x = \pm\sqrt{1 + \cot^2 x}$ 47. $B = 47.5^\circ$,

$a = 13.9$, $c = 20.6$ 48. 18.2 cm^2 49. $\langle -4\sqrt{3}, 4\rangle$ 50. $x^2 + y^2 = 64$

51. $-\dfrac{3\sqrt{2}}{2} - \dfrac{3\sqrt{2}}{2}i$ 52. 2 cis 90°, 2 cis 210°, 2 cis 330°; or $2i$, $-\sqrt{3} - i$,

$\sqrt{3} - i$ 53. $\frac{5}{3}$ km/h 54. $\left(\frac{8}{7}, \frac{11}{7}\right)$

55.

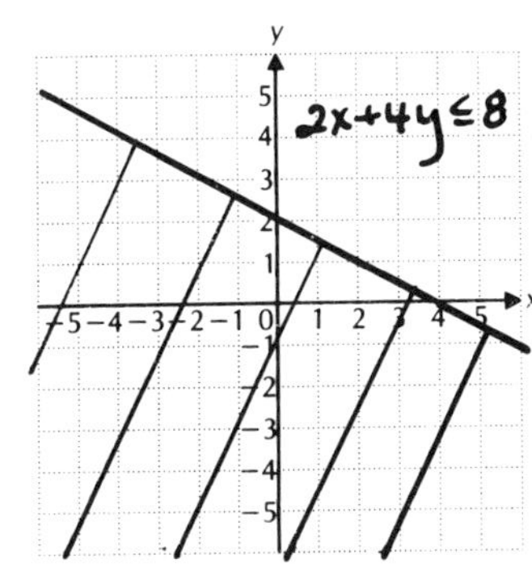

56. $a_{21} = 1$; $M_{21} = \begin{vmatrix} 0 & 4 \\ 3 & 7 \end{vmatrix} = -12$; $A_{21} = 12$

57. -21 58. $\begin{bmatrix} 33 & 5 \\ 9 & 6 \end{bmatrix}$ 59. Inconsistent, independent

60.

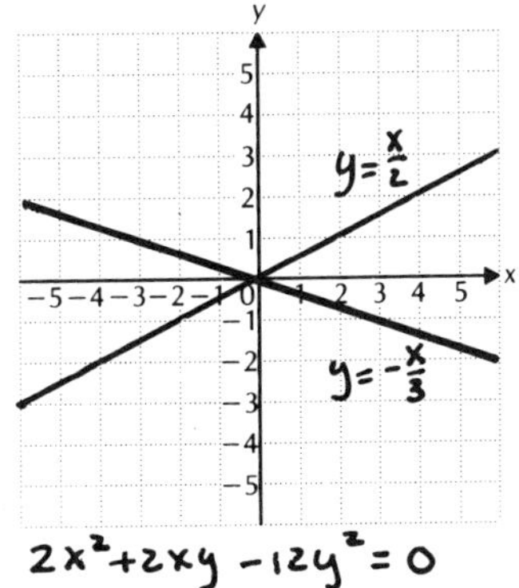

61. C: (0,0); V: (3,0), (-3,0); F: (5,0), (-5,0); A: $y = \frac{4}{3}x$, $y = -\frac{4}{3}x$

62. (1,2), (6,3) 63. 11 and 12, -11 and -12 64. Hyperbola 65. 13th

66. a, c 67. $\frac{1}{6}$ 68. 2^8, or 256 69. $84{,}375x^2y^4$ 70. $\frac{2}{13}$

Final Exam, Test Form C

1. 0.00006129 2. 41,000 3. 2150 4. 0.0129 5. $\frac{(x - 5)(x + 3)}{(x + 1)(x + 2)}$

6. $\frac{49 + 14\sqrt{x} + x}{49 - x}$ 7. 11 8. $-i$ 9. $3 + 11i$ 10. $-\frac{5}{17} + \frac{14}{17}i$ 11. 50

12. 6 mph 13. b 14. d 15. a, b, c, e, f 16. 18 17. $2a^2 + 3a - 2$

18. $2h - 5 + 4a$

19. 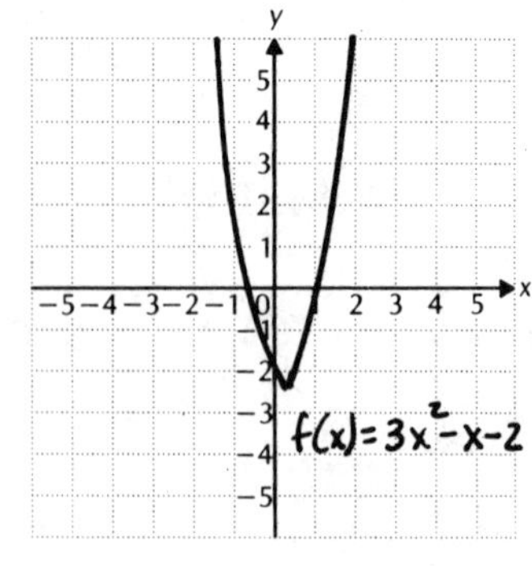

20. $\{x|x < 4 \text{ or } x > 12\}$ 21. $\{x|x < -7 \text{ or } x > 3\}$

22. $\{x|x < -3 \text{ or } x > -2\}$ 23. 93

24. $(x + 3)(x - 1 + \sqrt{2})(x - 1 - \sqrt{2})$; -3, $1 - \sqrt{2}$, $1 + \sqrt{2}$ 25. a and b 26. 3x 27. $\frac{5}{4}$ 28. 3

29. 3590 yrs 30. a) 80; b) 70; c) 45 months

31.

x	$\frac{2\pi}{3}$	$\frac{7\pi}{6}$	$-\frac{\pi}{4}$
sin x	$\frac{\sqrt{3}}{2}$	$-\frac{1}{2}$	$-\frac{\sqrt{2}}{2}$
cos x	$-\frac{1}{2}$	$-\frac{\sqrt{3}}{2}$	$\frac{\sqrt{2}}{2}$

32. 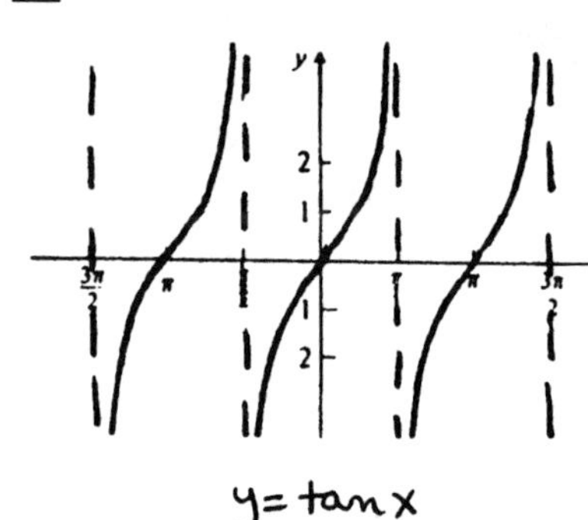

33. π 34. All reals except $\frac{\pi}{2} + k\pi$

35. 0.75 radian, 43° 36. $83^\circ 15'$ 37. 235°

38. 1.9198 39. 2.7217 40. 0.8813 41. $2 - \sqrt{3}$

42. sin x 43. $\frac{\sqrt{2} + \sqrt{6}}{4}$ 44. 0.45 45. $\frac{\pi}{2}$

46. csc x 47. b = 18.7; A = 38.7°; B = 51.3° 48. 11.7 newtons, 59°

49. 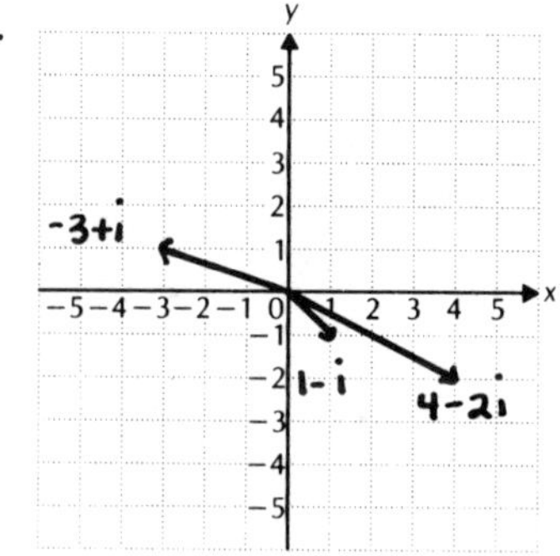

50. $(10.8, 68^\circ)$ 51. 12 cis 95°

52. $25\sqrt{5}$ cis 317° 53. 19 54. $(1 + z, -3z, z)$: (1,0,0); (3,-6,2); (0,3,-1), etc. Answers may vary. 55. $\frac{49}{4}$ 56. $\begin{bmatrix} 5 & -1 \\ 1 & 3 \end{bmatrix} \begin{bmatrix} x \\ y \end{bmatrix} = \begin{bmatrix} 11 \\ -1 \end{bmatrix}$; (2,-1) 57. Minimum -36 at (-3,7); maximum 0 at (0,0) 58. $\begin{bmatrix} 4 & -5 \\ 2 & 9 \end{bmatrix}$ 59. $\begin{bmatrix} -\frac{3}{2} & 2 \\ -\frac{1}{2} & 1 \end{bmatrix}$

60. 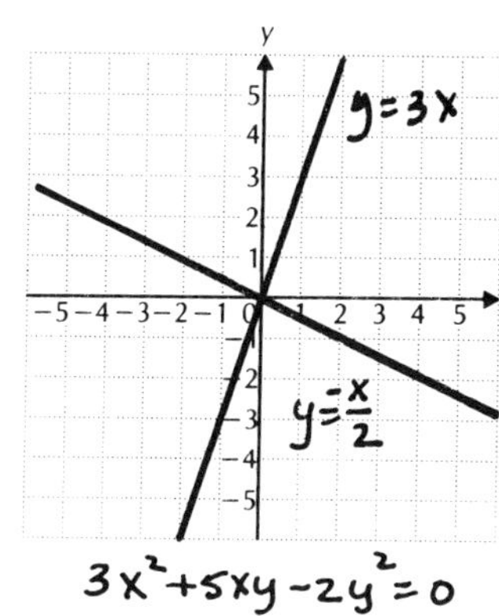

61. V: $(-\frac{1}{2},5)$; F: $(2,5)$; D: $x = -3$ 62. $(3,-2)$, $(-3,2)$, $\left(2\sqrt{3},-\sqrt{3}\right)$, $\left(-2\sqrt{3},\sqrt{3}\right)$ 63. 5 cm, 9 cm

64. Ellipse 65. 90 66. $\frac{29}{4}$, $\frac{19}{2}$, $\frac{47}{4}$ 67. $-\frac{20}{3}$

68. $\frac{269}{33}$ 69. 1120 70. $\frac{75}{364}$

Final Exam, Test Form D

1. 3.4×10^{-6} 2. 8.9417×10^{2} 3. $-12x^3y^{-2}$ 4. -4 5. 1

6. $25a^6 - 20a^3b^2 + 4b^4$ 7. 800 g/cm 8. $\frac{3}{2}$, 1, -5 9. $\{y|y > -2\}$

10. $\frac{11}{3}$ 11. 4 12. $V_2 = \frac{P_1V_1T_2}{P_2T_1}$ 13. 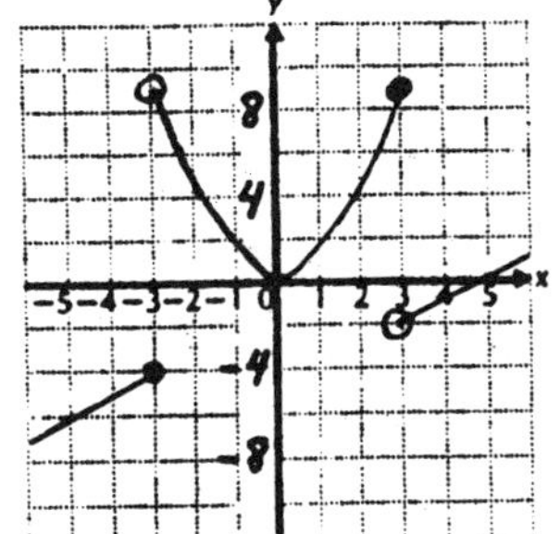

14. $P(x) = -0.5x^2 + 12x - 360$

15. $m = -\frac{1}{2}$, y-intercept is $\frac{15}{4}$

16. $y = -\frac{1}{4}x + \frac{11}{4}$

17. Center: $(-3,5)$; radius: 4

18. a, c 19. a) $f(x) = \left(x - \frac{5}{2}\right)^2 + \frac{1}{4}$; b) $\left(\frac{5}{2},\frac{1}{4}\right)$, c) maximum: $\frac{1}{4}$

20. a) $f(x) = 2(x - 1)^2 - 5$; b) $(1,-5)$; c) minimum: -5 21. $\{2,12\}$

22. The quotient is $2x^3 - 4x^2 - 4x - 5$. The remainder is 2.

23. $x^3 - 3x^2 + 4x - 12$ 24. Positive: 3 or 1; negative: 2 or 0

25. $\{(-3,2), (-2,4), (5,1), (-7,-2), (-1.5,2.1)\}$ 26. -5

27. $2 \log a + \log b - 5 \log c$ 28. 39 decibels 29. 0.699 30. 0.954

31.

32. All reals 33. 1 34. 1 rad/sec

35. $\cos\theta = -\frac{\sqrt{3}}{2}$, $\tan\theta = -\frac{\sqrt{3}}{3}$, $\cot\theta = -\frac{3}{\sqrt{3}}$, or $-\sqrt{3}$, $\sec\theta = -\frac{2}{\sqrt{3}}$, or $-\frac{2\sqrt{3}}{3}$, $\csc\theta = 2$

36. 25.6°, 0.4468 37. 25°, 0.4363 38. 6

39. $\frac{2\pi}{3}$ 40. $-\frac{\pi}{9}$ 41. $\sin(-4^\circ)$ 42. $-\csc x$

43. 1 44. $\cot x$ 45. $\sin 2\theta = \frac{24}{25}$; $\cos 2\theta = -\frac{7}{25}$; $\tan 2\theta = -\frac{24}{7}$; II 46. 30°

47.

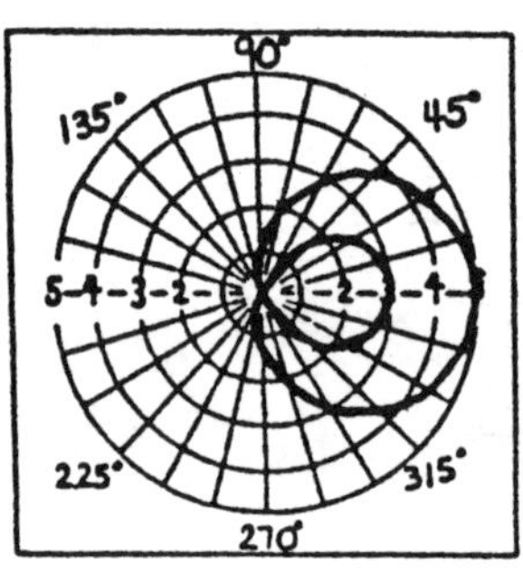

48. $b = 18.3$; $A = 42.8^\circ$; $B = 47.2^\circ$ 49. 47.4 cm

50. $\langle 4\sqrt{2}, -4\sqrt{2} \rangle$ 51. $\sqrt{10}$ cis 45°, $\sqrt{10}$ cis 225°, or $\sqrt{5} + \sqrt{5}i$, $-\sqrt{5} - \sqrt{5}i$ 52. $x^2 + y^2 = 144$

53. 30° 54. $(1,0,-3)$ 55. A, or $\begin{bmatrix} 5 & -3 \\ 2 & 0 \end{bmatrix}$

56. A, or $\begin{bmatrix} 5 & -3 \\ 2 & 0 \end{bmatrix}$ 57. $\begin{bmatrix} -3 & -2 & 4 \end{bmatrix}$ 58. $\begin{bmatrix} 13 \end{bmatrix}$

59. 20 pies and 20 cakes 60. C: $(2,-3)$; V: $(2,-1)$, $(2,-5)$, $(5,-3)$, $(-1,-3)$; F: $\left(2 - \sqrt{5}, -3\right)$, $\left(2 + \sqrt{5}, -3\right)$ 61. $y^2 = 16x$ 62. 8 ft x 12 ft 63. $(2,1)$, $(2,-1)$, $(-2,1)$, $(-2,-1)$ 64. Parabola 65. 209 66. 19 67. 6144

68. \$1023 69. 3024 70. $x^4 - 4\sqrt{3}x^3 + 18x^2 - 12\sqrt{3}x + 9$

Final Exam, Test Form E

1. $\sqrt[60]{p^{20}q^{45}r^{24}}$ 2. $\frac{2x + 7\sqrt{xy} + 3y}{x - 9y}$ 3. $\frac{x^2 - 3x + 9}{x}$ 4. $5x^2 - 17x - 12 = 0$

5. 7   6. $y = \frac{3.75\ xz^2}{w}$ 7. $\frac{23}{2}$ 8. No   9. $\{x|x \geq 2,\ x \neq 3\}$

10. a) $f(x) = 2\left(x - \frac{3}{2}\right)^2 - \frac{7}{2}$; b) $\left(\frac{3}{2}, -\frac{7}{2}\right)$; c) minimum at $-\frac{7}{2}$

11. $\{x|x < \frac{3}{2}$ or $x > 6\}$ 12. $x|x < 2$ or $x > 3\}$ 13. a) 1.146; b) 0.954

14. -3.2651   15. 4   16. 84.9 ft   17. $\frac{\pi}{4}$ 18. a) $\sec^2 x$; b) $\cos x$ 19. $\csc^2 x$

20. -3π 21. 6.67   22. -1.0192   23. $\frac{\sqrt{3}}{2}$

24. $\cot 2\theta = \frac{\cot^2 \theta - 1}{2 \cot \theta}$

$\frac{\cos 2\theta}{\sin 2\theta}$	$\frac{\frac{\cos^2 \theta}{\sin^2 \theta} - 1}{2\frac{\cos \theta}{\sin \theta}}$
	$\frac{\cos^2 \theta - \sin^2 \theta}{\sin^2 \theta} \cdot \frac{\sin \theta}{2 \cos \theta}$
	$\frac{\cos 2\theta}{2 \sin \theta \cos \theta}$
	$\frac{\cos 2\theta}{\sin 2\theta}$

25. $\frac{\pi}{4}, \frac{\pi}{2}, \frac{3\pi}{4}, \frac{5\pi}{4}, \frac{3\pi}{2}, \frac{7\pi}{4}$

26. 11.5   27. $r = 11$

28. $4\sqrt{2}$ cis $315°$ 29. $\frac{7}{41} + \frac{22}{41}i$

30. 2 cis $60°$, 2 cis $180°$, 2 cis $300°$; or $1 + \sqrt{3}i$, -2, $1 - \sqrt{3}i$ 31. (2,1,-3)

32. Maximum 125 at (0,5)   33. $f(x) = -x^2 + 3x - 4$

34. -3   35. $\begin{bmatrix} \frac{1}{4} & -\frac{1}{2} & \frac{1}{2} \\ \frac{1}{8} & \frac{1}{4} & -\frac{1}{4} \\ \frac{1}{4} & \frac{1}{2} & \frac{1}{2} \end{bmatrix}$ 36. $\begin{bmatrix} -9 & 22 \\ 14 & -4 \end{bmatrix}$ 37. (0,1), (0,-1)

38. $x^2 + y^2 = 25$ 39. (0,5), (0,-5)   40. one   41. $2i$, $-\sqrt{5}$, $3 + 2i$

42. -1, 2   43. 6.1   44. $\frac{128}{5}$

45. S_n: $1 + 6 + 11 + \ldots + (5n - 4) = \frac{n(5n - 3)}{2}$

S_1: $1 = \frac{1(5\cdot 1 - 3)}{2}$

S_k: $1 + 6 + 11 + \ldots + (5k - 4) = \frac{k(5k - 3)}{2}$

S_{k+1}: $1 + 6 + 11 + \ldots + (5k - 4) + [5(k + 1) - 4] = \frac{(k + 1)[5(k + 1) - 3]}{2}$

$$1 + 6 + 11 + \ldots + (5k - 4) + (5k + 1) = \frac{(k + 1)(5k + 2)}{2}$$

1. Basis step: $\frac{1(5 - 3)}{2} = 1$, so S_1 is true.

2. Induction step. Assume S_k. Then add $(5k + 1)$ on both sides.

$$\begin{aligned} 1 + 6 + 11 + \ldots + (5k - 4) + (5k + 1) &= \frac{k(5k - 3)}{2} + (5k + 1) \\ &= \frac{k(5k - 3)}{2} + \frac{2(5k + 1)}{2} \\ &= \frac{k(5k - 3) + 2(5k + 1)}{2} \\ &= \frac{5k^2 - 3k + 10k + 2}{2} \\ &= \frac{5k^2 + 7k + 2}{2} \\ &= \frac{(k + 1)(5k + 2)}{2} \end{aligned}$$

46. 120 47. $16x^4 - 32\sqrt{3}x^3y + 72x^2y^2 - 24\sqrt{3}xy^3 + 9y^4$ 48. $\frac{96}{323}$

49.

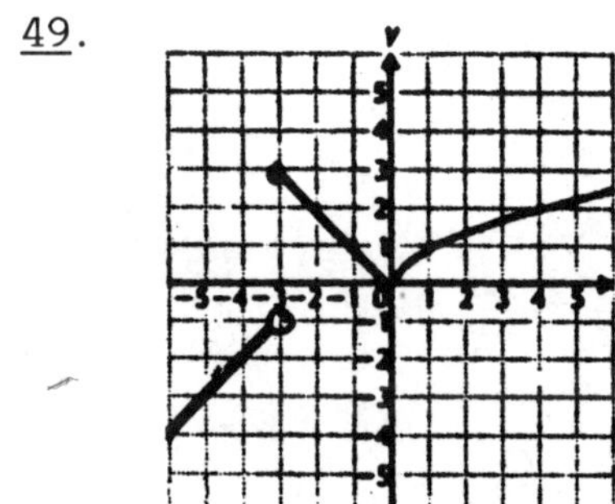

50.

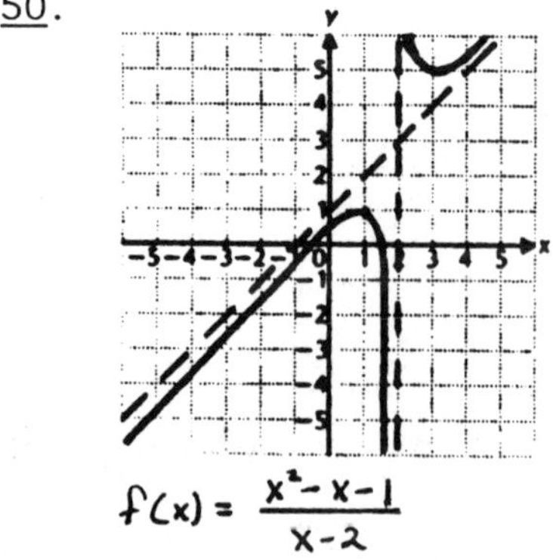

51.

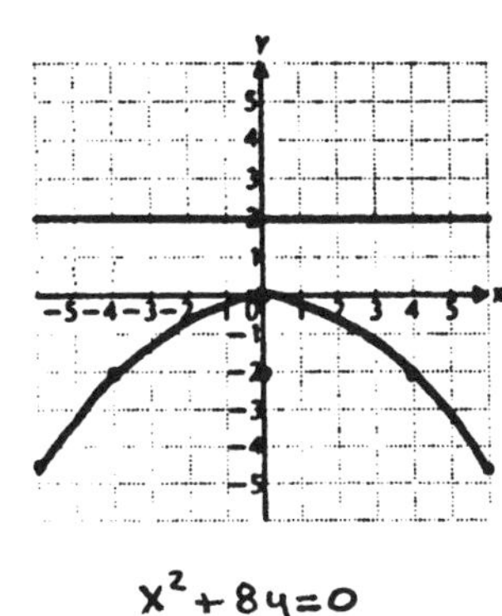

52.

$y = \sin x - \cos x$

53. 11 yr   54. (5.4, 338°)   55. 78   56. 22 ft/sec   57. 10 ft by 24 ft

58. -84   59. $\frac{4}{3x + 1} + \frac{1}{x + 1}$ 60. 29.2567°

Final Exam, Test Form F

1. b   2. a   3. a   4. b   5. d   6. b   7. c   8. c   9. b   10. d   11. a

12. c   13. a   14. a   15. d   16. c   17. d   18. c   19. b   20. b   21. c

22. b   23. d   24. c   25. d   26. b   27. c   28. a   29. d   30. b   31. c

32. a   33. d   34. b   35. c   36. a   37. a   38. d   39. a   40. d   41. a

42. d   43. c   44. b

45. $S_n: 1 + 3 + 3^2 + \ldots + 3^{n-1} = \dfrac{3^n - 1}{2}$

$S_1: 1 = \dfrac{3^1 - 1}{2}$

$S_k: 1 + 3 + 3^2 + \ldots + 3^{k-1} = \dfrac{3^k - 1}{2}$

$S_{k+1}: 1 + 3 + 3^2 + \ldots + 3^{k-1} + 3^k = \dfrac{3^{k+1} - 1}{2}$

1. Basis step: $\dfrac{3^1 - 1}{2} = 1$, so S_1 is true.

2. Induction step: Assume S_k. Then add 3^k on both sides.

$$
\begin{aligned}
1 + 3 + 3^2 + \ldots + 3^{k-1} + 3^k &= \frac{3^k - 1}{2} + 3^k \\
&= \frac{3^k - 1}{2} + \frac{2 \cdot 3^k}{2} \\
&= \frac{3 \cdot 3^k - 1}{2} \\
&= \frac{3^{k+1} - 1}{2}
\end{aligned}
$$

46. a 47. a 48. d

ALGEBRA AND TRIGONOMETRY VIDEOTAPE SERIES

Tape	Section	Title
TAPE 1	11.7	The Binomial Theorem
39.39		
TAPE 2	2.4	The Complex Numbers
39.30		
TAPE 3	1.9	Handling Dimension Symbols
18.46		
TAPE 4	3.1	Graphs of Equations
35.00		
TAPE 5	3.3	Functions
35.00		
TAPE 6	3.6	Symmetry
35.00	3.8	Transformations
TAPE 7	3.5	Some Special Classes of Functions
32.00		
TAPE 8	3.4	Lines and Linear Functions
45.01	3.2, 3.4	Parallel and Perpendicular Lines; The Distance Formula
TAPE 9	4.1	Quadratic Functions
	4.1	Mathematical Models
TAPE 10	4.2	Sets, Sentences, and Inequalities
35.02	4.3	Equations and Inequalities with Absolute Value
TAPE 11	4.4	Quadratic and Rational Inequalities
21.49		
TAPE 12	4.5, 4.6	Polynomials and Polynomial Functions
40.44	4.6	The Remainder and Factor Theorems
TAPE 13	4.7	Theorems about Roots
46.26	4.7	Rational Roots
TAPE 32	8.4	Vectors
33.08	8.4	Vectors and Coordinates
TAPE 33	8.5	Polar Coordinates
30.12		
TAPE 34	8.6	Complex Numbers: Graphical Representation and Polar Notation
32.54		
	8.7	DeMoivre's Theorem
TAPE 35	9.1	Systems of Equations in Two Variables
21.11		

TAPE 36	9.2	Systems of Equations in Three or More Variables
32.36	9.3	Special Cases
TAPE 37	9.4	Matrices and Systems of Equations
29.55		
TAPE 38	9.4	The Algebra of Matrices
30.09		
TAPE 39	9.5	Determinants and Cramer's Rule
27.00		
TAPE 40	9.5	Determinants of Higher Order
36.02	9.5	Properties of Determinants
TAPE 41	9.6	Inverses of Matrices
30.45		
TAPE 42	9.7	Systems of Inequalities
30.42	9.7	Linear Programming
TAPE 43	3.2, 10.1	Conic Sections
45.58	10.1	Ellipses
TAPE 44	10.2	Hyperbolas
37.31	10.3	Parabolas
TAPE 45	10.4	Systems of First-Degree and Second-Degree Equations
30.34	10.5	Systems of Second-Degree Equations
TAPE 46	11.1	Sequences and Sums
36.71	11.2	Arithmetic Sequences and Series
TAPE 47	11.3	Geometric Sequences and Series
33.62	11.3	Infinite Geometric Sequences and Series
TAPE 48	11.4	Mathematical Induction
24.25		
TAPE 49	11.5	Permutations
28.05	11.6	Combinations
TAPE 50	11.8	Probability
25.24		